AUXIN MOLECULAR BIOLOGY

Edited by

CATHERINE PERROT-RECHENMANN

and

GRETCHEN HAGEN

Reprinted from *Plant Molecular Biology*, Volume 49 Nos. 3–4, 2002

Springer Science+Business Media, LLC

A C.I.P. Catalogue record for this book is available from the Library of Congress

ISBN 978-1-4020-0646-3   ISBN 978-94-010-0377-3 (eBook)
DOI 10.1007/978-94-010-0377-3

Printed on acid-free paper

CONTENTS

Cover illustration
Whole mount immunolocalization with anti-AtPIN1 antibody in *Arabidopsis* root tip. FITC coupled secondary antibody was used for signal visualization (depicted in yellow). AtPIN1 is positioned at the basal side of stele and pericycle cells.
From Friml and Palme, this issue, pp. 273–284.

Plant Molecular Biology **49:** v, 2002.
Perrot-Rechenmann and Hagen (Eds.), Auxin Molecular Biology.

Preface

The plant hormone auxin plays a fundamental role in the growth and development of plants. The molecular mechanisms by which auxin controls such diverse processes as cell division, cell elongation and differentiation have yet to be clearly established, but they are under intensive investigation in laboratories around the world. In May 2000, representatives from the major labs engaged in research on the mechanisms of auxin action gathered at an EMBO workshop on the island of Corsica. It was a small and unique meeting, devoted to auxin biology. The goals of the meeting were to communicate results, generate scientific dialogue, advance collective knowledge and stimulate ideas for future research.

One overwhelming conclusion of the meeting was that the auxin field is experiencing a 'revolution'. In the past, many individual laboratories have used a limited number of experimental approaches (e.g. molecular biology, genetics, biochemistry, physiology) to study auxin action. These parallel approaches have been valuable for generating basic information, but few unifying principles have emerged. Recently, however, advances such as the completion of the *Arabidopsis* genome sequence and the introduction of new technologies (e.g. more rapid and efficient mutant gene-mapping methods) have greatly accelerated the rate of progress in the field. This has led to the discovery of multiple points of convergence of data that previously appeared to be unlinked. For example, many of the auxin response mutants have now been shown to contain mutations in genes that encode transcription factors involved in auxin-regulated gene expression. In addition, new concepts have been introduced, such as the importance of protein degradation in auxin action. As a consequence, complex models of the auxin response pathway are being proposed and continually updated; new challenges and questions have emerged; the potential benefits of integrating a variety of experimental approaches have been underscored.

The major objective of this book is to communicate some of the exciting advances in auxin biology to a wide audience of scientists. Articles included in this text represent a cross-section of topics that were highlighted at the Corsica meeting. These topics cover aspects of auxin biosynthesis, metabolism and transport; auxin perception and signal transduction; auxin-regulated gene expression; protein degradation in auxin signaling; auxin effects on physiological responses, growth and development; cross-talk between auxin and other plant-signaling pathways. The articles are written as 'reviews' of a particular topic area, and updated to include recently published and unpublished results. As such, they represent our current understanding of some of the molecular mechanisms that impact auxin action. Certainly, this understanding is still rudimentary and rapidly evolving, and aspects of the current models will undoubtedly be challenged and altered.

We wish to thank the authors for their contribution to this volume. Their time and effort is gratefully acknowledged. We also thank the scientists who agreed to review the manuscripts. Their comments and suggestions were appreciated and helpful.

Plant Molecular Biology **49**: 249–272, 2002.
Perrot-Rechenmann and Hagen (Eds.), Auxin Molecular Biology.
© 2002 *Kluwer Academic Publishers.*

Biosynthesis, conjugation, catabolism and homeostasis of indole-3-acetic acid in *Arabidopsis thaliana*

Karin Ljung[1,+], Anna K. Hull[2,+], Mariusz Kowalczyk[1,+], Alan Marchant[3,+], John Celenza[2],
Jerry D. Cohen[4] and Göran Sandberg[1,*]
[1]*Department of Forest Genetics and Plant Physiology, Swedish University of Agricultural Sciences, 901 83 Umeå,
Sweden (*author for correspondence; e-mail goran.sandberg@genfys.slu.se); [2]Department of Biology, 2 Cum-
mington Street, Boston, MA 02215, USA; [3]School of Biosciences, Plant Science Division, Sutton Bonnington
Campus, University of Nottingham, Loughborough LE12 5RD, UK; [4]Department of Horticultural Science, University of Minnesota, 1970 Folwell Ave., 305 Alderman Hall, Saint Paul, MN 55108, USA; [+]these authors contributed
equally*

Received 29 August 2001; accepted 1 October 2001

Key words: auxin metabolism, biosynthesis, conjugation, catabolism, homeostasis

Introduction

It was once proposed that there are only two kinds
of biology: elegant genetics and sloppy biochemistry (E.C. Pauling, unpublished). For those who
study auxin metabolism in *Arabidopsis*, this geneti-
cist's view of the different approaches to biological
research has particular resonance. *Arabidopsis* has
the advantage of providing a model molecular genetic system in a plant that uses the indole ring to
produce diverse compounds, such as the glucosinolate glucobrassicin, the phytoalexin camalexin and the
phytohormone indole-3-acetic acid (IAA). This model
plant genetic system offers unique opportunities to apply new approaches to answer long-standing questions
regarding auxin. However, studies in *Arabidopsis* can
often present us with confounding problems when it
comes to careful dissection of the network of indolic
pathways in either normal or mutant plants. In this
review, we focus our attention on IAA metabolism
in *Arabidopsis*. However, by necessity we have been
obliged to draw complementary information from the
literature on other species to delineate as completely as
possible the most current views on processes responsible for IAA production and its regulation.

IAA biosynthesis

Although the physiological role of IAA is well documented, the precise routes of its biosynthesis remain

elusive. Only within the past twenty years have precise details of the IAA biosynthetic pathways begun
to emerge. This information comes largely from (1)
stable isotope labelling of biosynthetic intermediates
and (2) cloning of genes encoding IAA biosynthetic
enzymes. While isotope labelling clearly supports two
separate paths for IAA biosynthesis; the tryptophan
(Trp)-dependent and Trp-independent pathways; only
genes encoding Trp-dependent biosynthetic enzymes
have been cloned to date. Here, the evidence for both
pathways will be discussed.

Tryptophan-dependent IAA biosynthesis

IAA is synthesized in *Agrobacterium* by a well-
defined pathway beginning with Trp and involving
indole-3-acetamide (IAM) as an intermediate. However, IAM is not considered to be a normal intermediate in plants and instead pathways that use
indole-3-pyruvic acid (IPA) or indole-3-acetaldoxime
(IAOx) as intermediates have been proposed, based
on various biochemical analyses (reviewed in Bartel,
1997; Normanly and Bartel, 1999; and illustrated in
Figure 1). Although IPA has been detected as an intermediate in IAA biosynthesis, little is known about
the enzymatic activities used in this pathway. With the
recent development of a method for accurately measuring IPA levels in *Arabidopsis* (Tam and Normanly,
1998), this may soon change. This new methodol-
ogy, combined with the amenability of *Arabidopsis* to

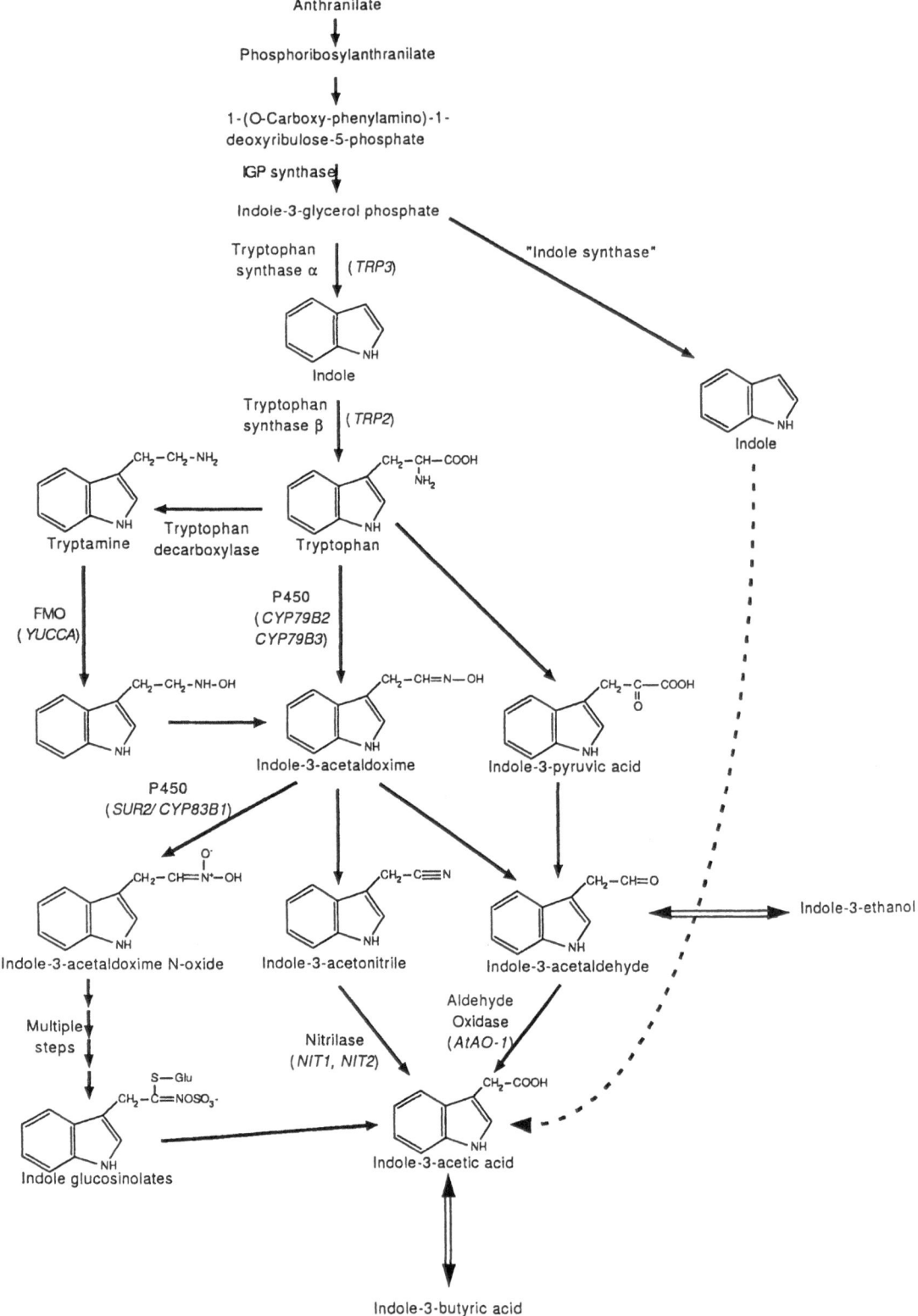

Figure 1. Pathways for IAA biosynthesis. Routes involved in IAA biosynthesis in *Arabidopsis*. Enzyme activities and *Arabidopsis* genes (italics) shown are described in the text or have been reviewed previously (Bartel, 1997; Normanly and Bartel, 1999). The dashed line indicates the Trp-independent pathway leading from indole produced by indole synthase, proposed in Ouyang *et al.* (2000). Double-headed arrows represent pathways that may also contribute to IAA biosynthesis, but are described elsewhere. Key: IGP synthase, indole-3-glycerol phosphate synthase; FMO, flavin monooxygenase; P450, cytochrome P450.

molecular genetic studies, should lead to an improved understanding of the IPA pathway.

The IAOx pathway was initially proposed following the identification of an IAOx-forming activity in *Brassica campestris* (Chinese cabbage), a close relative of *Arabidopsis* (Ludwig-Müller and Hilgenberg, 1988). It has recently gained support with the cloning of genes that encode enzymes involved in the conversion of Trp to IAOx. Two *Arabidopsis* cytochrome P450s, CYP79B2 and CYP79B3, that can catalyse the conversion of Trp to IAOx *in vitro* have recently been discovered. CYP79B2 was first identified by its DNA sequence similarity to the tyrosine-metabolizing CYP79A1 of *Sorghum* (Bak *et al.*, 1998), although no activity was determined. The activity of CYP79B2 was discovered in a screen for *Arabidopsis* genes encoding Trp-metabolizing activities (Hull *et al.*, 2000). CYP79B3 was identified by its strong homology to CYP79B2. The results of expressing CYP79B2 in yeast and over-expressing it in plants suggested it has a role in Trp metabolism and indeed bacterially expressed CYP79B2 and CYP79B3 both metabolize Trp to IAOx (Hull *et al.*, 2000). Mikkelsen *et al.* (2000) reported the same *in vitro* activity for CYP79B2, and showed that plants over-expressing CYP79B2 have increased indole glucosinolate levels. CYP79B2 mRNA accumulates after both wounding (Mikkelsen *et al.*, 2000) and infection with a bacterial pathogen (Hull *et al.*, 2000). These expression characteristics are also seen for genes encoding Trp biosynthetic enzymes, suggesting co-regulation of genes involved in Trp metabolism.

That over-expression of CYP79B2 leads to IAOx and to increased indole glucosinolate production is not surprising, since it has been accepted for some time that the aldoxime derivative of an amino acid is the intermediate for glucosinolates (Chapple *et al.*, 1994). In addition, every characterized member of the CYP79 family has been found to N-hydroxylate specific amino acids to their aldoxime derivative (reviewed in Celenza, 2001). What is the evidence that IAOx derived from CYP79B2 and/or CYP79B3 is used for IAA biosynthesis? Both proteins have putative chloroplast transit peptides and both of these sequences can target β-glucuronidase to the chloroplast (Hull and Celenza, unpublished). A chloroplast location is consistent with their role in Trp metabolism, because Trp and IAA are synthesized in the chloroplast. Preliminary results also indicate that CYP79B2 over-expressing lines produce elevated levels of IAA (Hull and Celenza, unpublished).

In addition to cytochome P450s, several *Arabidopsis* proteins with similarity to flavin monooxygenases (FMOs) have been found that when over-expressed increase IAA production via a likely IAOx intermediate. The FMO encoded by the *YUCCA* gene was found to convert tryptamine to *N*-hydroxyl tryptamine (Zhao *et al.*, 2001) *in vitro*, although tryptamine has not yet been identified as a native compound in *Arabidopsis*. *YUCCA* was identified as an activation-tagged mutant that had elongated hypocotyls, epinastic leaves and increased apical dominance. This phenotype, which results from over-expression of the *YUCCA* mRNA, was shown to have 50% more free IAA than wild-type plants. The fact that the *yucca* mutant is resistant to toxic Trp analogues suggests that the YUCCA protein is involved in Trp-dependent IAA biosynthesis. The use of tryptamine by YUCCA, if confirmed, implies that Trp decarboxylase is present in *Arabidopsis*. The *Arabidopsis* genome contains at least two genes that appear to encode amino acid decarboxylases, but whether any of these are Trp decarboxylases is unclear. Once YUCCA has converted tryptamine to *N*-hydroxyl tryptamine, another hydroxylation is necessary to synthesize IAOx. This is presumably carried out by another FMO-like protein or a cytochrome P450, neither of which have been identified. The *Arabidopsis* genome contains nine other *YUCCA*-like genes (Zhao *et al.*, 2001). At least two of these, *YUCCA2* and *YUCCA3*, also cause an IAA-overproduction phenotype when expressed with the CaMV 35S promoter.

After the identification of both cytochrome P450s and flavin monooxygenases that can promote IAOx biosynthesis, it is worth re-evaluating the IAOx-forming activity partially purified from Chinese cabbage (Ludwig-Müller and Hilgenberg, 1988). These authors describe a plasma membrane-associated activity capable of converting Trp to IAOx. This activity is enhanced six-fold by the addition of manganese ions or H_2O_2, suggesting that this enzyme is a peroxidase. The enzyme was given the name TrpOxE, but the gene (or genes) encoding it has (have) yet to be isolated. Because neither cytochrome P450s nor flavin monooxygenase typically utilize Mn^{2+} or H_2O_2, there may indeed be a third way to make IAOx.

Indole-3-acetaldoxime: to indole-3-acetonitrile or indole-3-acetaldehyde?

Plants grown in the presence of exogenous IAOx display an IAA overproduction phenotype that those

252

grown on Trp do not (Zhao *et al.*, 2001). This suggests that conversion from Trp to IAOx is a limiting reaction for IAA production. How does IAOx get converted to IAA? The only IAOx-metabolizing gene identified so far encodes the cytochrome P450 CYP83B1 (see below). However, it most likely has a role in indole glucosinolate biosynthesis, rather than a direct role in IAA biosynthesis. Application of IAN and IAAld both mimic IAA application and thus both indole-3-acetonitrile (IAN) and indole-3-acetaldehyde (IAAld) have been proposed as intermediates between IAOx and IAA (Figure 1).

Indole-3-acetonitrile as an intermediate in IAA biosynthesis

IAOx to IAN conversion has been found in Chinese cabbage (Helminger *et al.*, 1985; Ludwig-Müller and Hilgenberg, 1990) and in banana leaves (Kumar and Mahadevan, 1963). While IAN is typically thought to come directly from IAOx, IAN can also arise from the breakdown of indole glucosinolates under acidic conditions, suggesting a more circuitous route for IAA synthesis via IAOx. Plant-associated microbes are also capable of converting IAN to IAA, and could contribute locally to increased IAA levels in nature (Normanly *et al.*, 1997). Nitrilase activity has been found in plants of the families Cruciferae, Graminae and Musaceae (Thimann and Mahadevan, 1964). While biochemical evidence supports a role for nitrilases in IAA biosynthesis, molecular genetic evidence has been somewhat ambiguous. *Arabidopsis* has four genes encoding nitrilases, *NIT1* to 4. *NIT1* to 3 are tandemly arranged on chromosome III while *NIT4* is located on chromosome V, and *NIT 1* to 3 are ca. 85% identical to each other while *NIT4* shares only 65% identity with the other *NIT* genes (Bartel and Fink, 1994). All four nitrilases have been shown to convert IAN to IAA *in vitro*, though NIT4 most likely has a role in cyanide detoxification, converting β-cyano-L-alanine to aspartic acid or asparagine and probably does not play a role in IAA metabolism (Piotrowski *et al.*, 2000). Over-expression of the nitrilases using the CaMV 35S promoter does not cause any obvious phenotypic deviations under normal growth conditions. However, plants over-expressing NIT2 are more sensitive to exogenous IAN, perhaps indicating an *in vivo* role for this enzyme in *Arabidopsis* (Normanly *et al.*, 1997). In addition, NIT2 has been expressed in tobacco using the CaMV 35S promoter and shown to convert micromolar amounts of exogenous IAN to

IAA (Schmidt *et al.*, 1996). The expression pattern of different nitrilases is also consistent with them playing a role in IAA biosynthesis. Both *NIT1* and *NIT2* are induced after infection with *Plasmodiophora brassicae*, which causes clubroot disease (Grsic-Rausch *et al.*, 2000). This suggests that nitrilases are involved in IAA biosynthesis because IAA is elevated at the site of club formation. From over-expression data, NIT2 appears likely to be involved in IAA biosynthesis. However, the only known IAN-resistant mutations are changes in the *NIT1* gene. *nit1* mutant plants, although insensitive to exogenous IAN, do not have altered levels of endogenous IAN or IAA (Normanly *et al.*, 1997). The fact that the *nit1* mutant maintains normal IAA levels can be explained in two ways. Either exogenous IAN is metabolized only by NIT1, but NIT2 and/or NIT3 can substitute functionally for NIT1 during *in planta* IAA biosynthesis or IAN is not the normal substrate of nitrilases in plants. If the latter is true, a different route for IAOx to IAA would be needed.

Indole-3-acetaldehyde as an intermediate in IAA biosynthesis

An alternative route whereby IAA could be produced from IAOx is via IAAld. Activities catalysing IAOx to IAAld conversion in both *Avena* and Chinese cabbage have been described (Helminger *et al.*, 1987; Rajagopal and Larsen, 1972). Alternatively, it has been proposed that IAOx could be non-enzymatically converted to IAAld (reduction to the corresponding imine followed by hydrolysis to IAAld at low pH) (Eckardt, 2001). IAAld would then be converted to IAA by an aldehyde oxidase (AO) (Normanly and Bartel, 1999). The strongest evidence for involvement of an AO in *Arabidopsis* IAA biosynthesis comes from analysis of the IAA-overproducing mutant *sur1*, which is allelic to *rooty*, *alf1*, *hls3* and *ivr*. The *sur1* mutant has increased levels of AO, which is able to convert IAAld to IAA (Seo *et al.*, 1998). The *SUR1* gene product is predicted to encode an aminotransferase, but its relationship to IAA metabolism is not clear (Gopalraj *et al.*, 1996). Recent analysis of the auxin-overproducing mutant *sur2* (or *rnt1*) also supports a pathway from IAOx to IAA that does not include IAN (Bak *et al.*, 2001; Barlier *et al.*, 2000). The *SUR2* gene encodes the cytochrome P450 CYP83B1 which converts IAOx to its *aci*-nitro derivative, the first step in indole glucosinolate biosynthesis, and the *sur2* mutant has decreased levels of indole glucosinolates, but increased rates of IAA production (Bak *et al.*, 2001; Barlier *et al.*, 2000). Interestingly, plants that have both *sur2* and *nit1* mutations still have increased levels of IAA production, which is consistent with a pathway from IAOx to IAA that does not involve IAN (Bak *et al.*, 2001). The *sur2* mutant accumulates high levels of IAAld, but labelling studies show that very little of this IAAld is converted to IAA (Barlier *et al.*, 2000). This means that there is evidence against both IAN and IAAld as intermediates of IAA in the *sur2* mutant unless nitrilases other than NIT1 account for the IAA production. No third pathway has been proposed and it is possible that what is seen in this mutant is not representative of the wild-type biosynthesis routes.

Tryptophan-independent IAA biosynthesis

As early as the 1950s, workers speculated that there might be a direct pathway to IAA that does not involve Trp (reviewed by Audus, 1972). The initial mass spectrometric evidence to support Trp-independent IAA biosynthesis came from labelling studies of the aquatic plant *Lemna gibba*, which grows floating on water and readily takes up [^{15}N]Trp from the medium. Baldi *et al.* (1991) showed that when the Trp pool was labelled to 98% with [^{15}N]Trp very little ^{15}N was incorporated into IAA. Calculations of the rate of Trp conversion relative to the expected rate of turnover suggested that an alternative route to IAA must have been operating. Further, crucial evidence demonstrating the activity of a Trp-independent pathway comes from the analysis of Trp-auxotrophic mutants of maize and *Arabidopsis*. Maize has two genes encoding tryptophan synthase β (TSβ). The *orange pericarp* (*orp*) double mutant lacks both of the TS enzymes and is therefore a Trp auxotroph (Wright *et al.*, 1992). Without Trp supplementation *orp* is a seedling lethal. It accumulates the Trp precursors anthranilate (as the glucoside) and indole. Interestingly, seedlings of this mutant have greatly increased (ca. 50-fold) total IAA levels compared to wild-type seedlings. Even when the plants were fed deuterium-labelled Trp, no deuterium was incorporated into IAA (Wright *et al.*, 1991). However, in parallel studies with plants grown on [^{2}H]-labelled water, the incorporation of deuterium showed that *de novo* IAA biosynthesis occurred in these seedlings. These data provide compelling evidence that maize seedlings can synthesize IAA from an indolic precursor other than Trp.

Initial studies using preparations from maize endosperm were interpreted as showing that indole was converted to IAA directly (Jensen and Bandurski, 1994; Rekoslavskaya and Bandurski, 1994). However, subsequent studies have shown that maize endosperm can synthesize IAA from Trp (Rekoslavskaya, 1995) and this appears to be the primary reaction leading to IAA in the endosperm, according to both the activity of protein preparations *in vitro* (Ilić *et al.*, 1999; Östin *et al.*, 1999) and analyses of isotopic enrichment in cultured maize kernels (Glawischnig *et al.*, 2000). These studies clearly showed that preparations from maize endosperm converted [^{14}C]indole to tryptophan and then to IAA or, alternatively, [^{14}C]tryptophan could be supplied. However, an *in vitro* system using an enzyme preparation from light-grown maize seedlings confirmed the hypothesis that seedlings use indole directly as the precursor to form IAA (Östin *et al.*, 1999). In contrast to the endosperm results, the seedling enzyme preparations converted [^{14}C]indole to IAA without labelling of tryptophan and, when supplied, [^{14}C]tryptophan was inactive as a substrate.

In maize, as well as other species, it is essential to understand plant development and to have full control over the experimental procedures in order to obtain clear, unambiguous results. As discussed above, maize endosperm has a quite different enzymatic pathway leading to IAA than that present in green tissue from seedlings. A similar difference is noted when comparing intact, dark-grown maize seedlings with coleoptile segments. Koshiba *et al.* (1995) have shown that severed dark-grown maize coleoptile tips can convert applied Trp to IAA. However, other studies, using intact coleoptiles and deuterium labelling, supply conclusive evidence that there is very little, if any, IAA biosynthesis in young maize seedlings (Jensen and Bandurski, 1996; Pengelly and Bandurski, 1983). The difference may be explained, in part, by a recent study on germinating bean axes, showing that they utilized the tryptophan-independent pathway until their cotyledons were surgically removed. In the 12 h labelling period following surgical wounding, tryptophan-dependent IAA biosynthesis could be detected. After a recovery period, tryptophan-independent IAA biosynthesis again became the predominant pathway (Sztein *et al.*, 2002). Recent studies have shown that the pathways involved in the biosynthesis of indolic compounds are closely linked to stress, wounding and pathogen attack (Figure 2), so it might be expected that IAA biosynthesis would also be affected by such environmental signals. Nevertheless, the propensity of plant biologists to use experimental systems with cut tissue sections

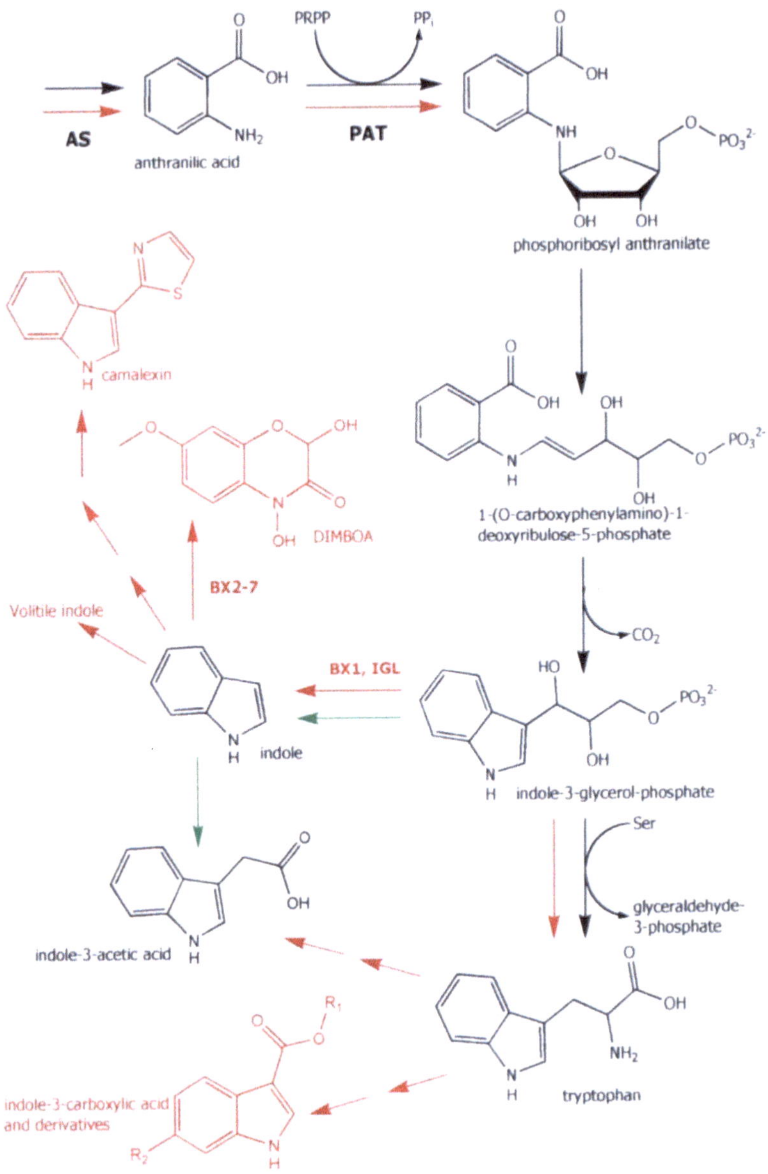

Figure 2. Indolic metabolism as determined from studies in maize and *Arabidopsis*. Reactions and intermediates in black are known to be constitutive processes. Those in red are stress, wounding or disease-related (Zhao and Last, 1996; Zook, 1998; Frey *et al.*, 1997, Frey *et al.* 2000; Hagemeir *et al.*, 2001). IAA from indole (in green) appears to occur in vegetative tissue in the absence of stress (Wright *et al.*, 1992; Normanly *et al.*, 1993; Sztein *et al.*, 2002).

makes it imperative to fully understand these relationships. Trp-independent IAA biosynthesis has also been demonstrated in protoplasts and shoot apices from tobacco (Sitbon *et al.*, 2000). In wild-type tobacco protoplasts, Trp-dependent IAA biosynthesis was estimated to account for less than 25% of total IAA biosynthesis, in contrast to the situation in IAA-overproducing protoplasts (*35S-iaaH* × *35S-iaaM*), where the Trp-dependent biosynthetic pathway was

dominant. Scots pine seedlings (Ljung *et al.*, 2001a) were shown to utilize both Trp-dependent as well as Trp-independent IAA biosynthesis during germination, and the timing of the different pathways appeared to be developmentally regulated (see below).

Trp-independent IAA biosynthesis in Arabidopsis

In *Arabidopsis*, mutations in genes encoding anthranilate phosphoribosyl transferase (*trp1-100*), tryptophan synthase α (TSα) (*trp3-1*) and tryptophan synthase β (TSβ) (*trp2-1*) have been used to further distinguish which Trp intermediates are used for Trp-independent IAA biosynthesis. *trp1-100* is a leaky allele of *trp1* that grows without supplemental Trp, but which can be distinguished by the accumulation of a blue fluorescent metabolite of anthranilate. The *trp2-1* and *trp3-1* mutants are conditional Trp auxotrophs that only require exogenous Trp when grown in high light (Last *et al.*, 1991). While the *trp1-100* mutant has normal levels of total and free IAA, the *trp2-1* and *trp3-1* mutants have dramatically increased levels of total IAA compared to wild-type plants when grown in high light. The level of free IAA is similar to wild-type levels. Furthermore, double-labelling studies with ^{15}N-anthranilate and deuterium-labelled Trp in the *trp2-1* mutant showed more ^{15}N incorporation into IAA than into Trp, and the conversion of [^{2}H$_5$]Trp to [^{2}H$_5$]IAA was only slightly faster than spontaneous, non-enzymatic conversion (Normanly *et al.*, 1993). This suggests that in Trp-independent IAA biosynthesis, IAA is derived from a precursor of Trp such as indole or indole-3-glycerol phosphate (IGP). A recent study extended these results by analysing plants expressing an anti-sense copy of the IGP synthase gene that encodes the enzyme acting immediately before TSα (Ouyang *et al.*, 2000). The anti-sense plants had nearly 50% less total IAA compared to wild-type and had phenotypes suggestive of auxin deficiency, such as small rosettes and reduced fertility (Ouyang *et al.*, 2000). In conjunction with the data obtained from *trp2* and *trp3* mutants this strongly suggests that indole and/or indole-3-glycerol phosphate (IGP) can function as (an) IAA precursor(s) in a Trp-independent pathway. Consistent with this conclusion is the discovery in maize of indole synthase, an enzyme similar to TSα that catalyses the synthesis of the indole used in biosynthesis of the defense compound DIMBOA (Frey *et al.*, 1997). *Arabidopsis* is thought to make the phytoalexin camalexin, via a Trp-independent pathway (Zook, 1998) and a TSα-like gene has been found in *Arabidopsis* based on sequence analysis (Ouyang *et al.*, 2000).

Conclusions related to the existence of the tryptophan-independent pathway for IAA biosynthesis in *Arabidopsis* have been questioned in the light of new findings (Müller *et al.*, 1998b; Müller and Weiler,

2000). However, in one of these studies (Müller *et al.*, 1998b), the rates of uptake and the effects of feeding on endogenous pools of the supplied compounds were not determined even though the levels (100 μM) of Trp and IAN used for feeding were 15–25 times higher than endogenous levels of these compounds (Tam *et al.*, 1995; Ilíc *et al.*, 1996). In a study by Rapparini *et al.* (1999) it was shown that when *Lemna gibba* plants were fed labelled Trp, the exogenous Trp was preferentially used as a substrate for IAA biosynthesis before endogenous Trp, leading to an overestimation of the Trp-dependent pathway. In the second study used to question the existence of tryptophan-independent synthesis it was determined that IGP, which co-purifies with the IAA conjugate fraction, breaks down to IAA under standard extraction and hydrolysis conditions (Müller and Weiler, 2000). Because the *trp3-1* mutant accumulates IGP, these authors suggested that the high level of IAA found in the conjugate fraction is an artefact due to the purification method used. While this might account for the entire increase in total IAA that has been reported for *trp3-1* it is not apparent how this could explain the increased IAA found in the *trp2-1* mutant. In addition, this result could not account for the isotopic enrichment data presented in Normanly *et al.* (1993), because in this study labelling of the free IAA pool was measured.

IAA metabolism

The general model for IAA metabolism describes oxidation and conjugation processes that modify the indole ring or auxin's side-chain, and thus cause loss of biological activity. Conjugation can be considered to be a catabolic process because some conjugates cannot be hydrolysed back to the free, active hormone. During conjugation the side-chain of the active hormone is modified, and two distinct groups of conjugates have been described in a variety of plant species: (1) ester-type conjugates, in which the carboxyl group of IAA is linked via the oxygen bridge to sugars (for example glucose) or cyclic poly-ols (like inositol), and (2) amide-type conjugates in which the carboxyl group forms an amide (e.g. peptide) bond with amino acids or polypeptides (reviewed in Reinecke and Bandurski, 1987; Bandurski *et al.*, 1995; Normanly, 1997; Normanly and Bartel, 1999).

Decarboxylative metabolism of IAA

Decarboxylative catabolism of IAA usually involves modification of both the side-chain and the indole ring. These complex reactions are catalysed by a variety of plant peroxidases, often referred to as 'IAA oxidases'. The major *in vitro* products are indole-3-methanol, indole-3-aldehyde (the product of indole-3-methanol reacted with free radicals), 3-methyleneoxoindole, indole-3-carboxylic acid and 3-methyloxoindole (Reinecke and Bandurski, 1987; Barcelo *et al.*, 1990; Östin, 1995). However, these compounds are very seldom identified as endogenous constituents in plants (Bandurski *et al.*, 1995; Östin, 1995), except for indole-3-methanol, which is probably a product of glucosinolate degradation (Chevolleau *et al.*, 1997). The low levels of these compounds found in plants suggest that the pathways that produce them are not the major catabolic processes for IAA *in vivo*. Studies with transgenic plants using both sense and anti-sense forms of an anionic peroxidase (defined as a peroxidase that migrates toward the anode under standard conditions of protein gel electrophoresis) confirmed that endogenous levels of free IAA are not affected by altering peroxidase levels, although the same enzyme purified from plants *in vitro* was able to degrade IAA (Lagrimini, 1991; Lagrimini *et al.*, 1997). The application of exogenous IAA or its analogues can strongly suppress the expression of anionic peroxidase genes in tobacco (Klotz and Lagrimini, 1996) and also reduce the total enzyme activity of the basic peroxidase in protein extracts from *Catharanthus roseus* cell cultures (Limam *et al.*, 1998). Not much is known about decarboxylative IAA metabolism in *Arabidopsis*, although the vast amounts of evidence accumulated from the other plants suggest it is similar between species. Although the evidence for peroxidase involvement in IAA degradation *in vivo* is no longer very strong, there are several lines of evidence suggesting there is some relationship between plant peroxidases and either IAA metabolism or activity. For example, only plant peroxidases are able to decarboxylate IAA in the absence of H_2O_2 and are able to use molecular oxygen directly. Also, several amino acid sequence similarities have been found between auxin-binding proteins and plant peroxidases (Savitsky *et al.*, 1999). Interestingly, activation of anionic peroxidase expression in *Arabidopsis* correlates with defects in auxin perception (Mayda *et al.*, 2000), which may indicate involvement of this peroxidase isoenzyme (CEVI-1) in the alteration of auxin responses when over-expressed.

Non-decarboxylative metabolism of IAA

Non-decarboxylative oxidation of the indole ring is the major IAA catabolic pathway in *Arabidopsis* (Östin *et al.*, 1998). Similar pathways have been described in other plants, but the final oxidative products differ significantly between species. Either the free hormone or its aspartic acid conjugate can be oxidized at the second carbon of the indole ring in the first committed step. In maize, the free hormone is oxidized in this manner, then oxidized at the seventh carbon of the indole ring, and finally modified by glucosylation of a newly formed hydroxyl group (Nonhebel *et al.*, 1985). While the same pathway exists in *Pinus silvestris*, other species including *Vicia faba*, tomato (fruit pericarp), carrot and hybrid aspen use IAAsp as the substrate for the first oxidation (Riov and Bangerth, 1992; Sasaki *et al.*, 1994; Tuominen *et al.*, 1994). The second oxidation in these systems involves formation of a hydroxyl group at the third position of the indole ring, and in most cases further glucosylation to form the 3-*O*-glucoside of OxIAAsp. It has been speculated that in hybrid aspen, OxIAAsp can be hydrolyzed back to OxIAA (Tuominen *et al.*, 1994) but this has not yet been confirmed. In tomato pericarp tissue, a high-molecular-weight end-product has also been described, in which OxIAAsp is linked to glucose via the N-glucoside bond from the indole ring (Östin *et al.*, 1995). It is not yet clear whether 2-oxoindole-3-acetic acid (OxIAA) or N-(2-oxoindole-3-acetyl)-aspartic acid (OxIAAsp) is the initial metabolite for this pathway in *Arabidopsis*. Both OxIAAsp and N-(3-hydroxy, 2-oxoindole-3-acetyl)-aspartic acid (DiOxIAAsp), as well as OxIAA, have been identified in *Arabidopsis* plants incubated with labeled exogenous IAA (Östin *et al.*, 1998; M. Kowalczyk and G. Sandberg, unpublished results). This suggests that the *Arabidopsis* pathway is similar to that described in hybrid aspen, tomato and carrot. However, the major difference is the presence of what has been tentatively identified as *O*-(2-oxoindole-3-acetyl)-β-glucose (OxIAGlc), an intermediate to an unidentified final product (Östin *et al.*, 1998). The *Arabidopsis* system, as presented in Figure 3, seems to be similar to that described in the orange (*Citrus sinensis*) fruit epicarp where there are two pathways that seem to operate independently: OxIAA leading to DiOxIAGlc and IAAsp leading to DiOxIAAsp (Chamarro *et al.*, 2001). Nevertheless, irrespective of the specific oxidation products, the results obtained so far for *Arabidopsis* indicate that non-decarboxylative metabolism plays the major role in maintaining auxin levels (Östin *et al.*, 1998; Kowalczyk and Sandberg, 2001).

Auxin conjugation systems

Most of our knowledge on IAA conjugation comes from studies on maize done by Bandurski and co-workers (reviewed by Reinecke and Bandurski, 1987). In this very specific system, free IAA first forms a 'high-energy' intermediate: 1-*O*-(indole-3-acetyl)-β-D-glucose (1-O-IAGlc). This reversible reaction, catalysed by a UDPG-dependent glucosyltransferase (Szerszen *et al.*, 1994 and references therein), favours the formation of 1-O-IAGlc due to the relatively high levels of UDP-glucose present in the tissue, and to the reaction being coupled to the downstream synthesis of *O*-(indole-3-acetyl)-inositol (IAInos) and its glycosides, IAInos arabinose (IAInos-Ara) and galactose (IAInos-Gal). A class of high-molecular-weight ester-linked conjugates, IAA-β-1,4-glucans, is also formed in this system, presumably from the 1-O-IAGlc. During seed germination, the levels of all of the ester-linked conjugates decreased, but only IAInos and free IAA were transported to the shoot, in a study by Chisnell and Bandurski (1988). Although the maturing endosperm does not seem to make amide-type conjugates, seedlings have an inducible capacity for the biosynthesis of IAAsp from applied IAA. Compared to maize, little is known about IAA-conjugate synthesis in other plants. It has been known for almost 50 years that vegetative tissues from the majority of dicotyledonous plants are able to form amide conjugates, mostly IAAsp and IAGlu, from exogenous IAA. Other conjugates, including 1-O-IAGlc, IAAla and IAGly, are reportedly synthesized from exogenous IAA, but several attempts to find IAA-amino acid-forming enzymes in plants have failed to uncover the mechanism responsible for their formation. A gene responsible for microbial formation of IAA-ε-N-lysine has proved useful for modification of plant IAA metabolism (Romano *et al.*, 1991), but no plant gene with homology to it, or any similar function has been identified to date.

Conjugation in Arabidopsis

Thin-layer chromatography (TLC)-based analysis reported by Sztein *et al.* (1995) indicated that *Arabidopsis* was able to form IAGlc, IAAsp and possibly IALeu from exogenous IAA. Experiments performed *in vivo* with radiolabelled IAA confirmed the synthesis

of IAAsp (especially when large amounts of the free hormone were added) and IAGlu (Östin *et al.*, 1998) as well as a novel amide conjugate, N_α-(indole-3-acetyl)-glutamine (IAGln) (Barratt *et al.*, 1999). Initially, neither IALeu nor IAGlc were observed as products in these experiments, but their presence in *Arabidopsis* was confirmed later (Barlier *et al.*, 2000; Tam *et al.*, 2000; Kowalczyk and Sandberg, 2001). Recently, the first *Arabidopsis* gene encoding a protein involved in auxin conjugation has been identified and cloned: UDPG-dependent IAA glucosyltransferase, a homologue of the maize enzyme (Jackson *et al.*, 2001). However, the *Arabidopsis* enzyme is much more specific for IAA than the maize enzyme, which can metabolize a range of carboxylic acids in addition to IAA. This suggests that the *Arabidopsis* enzyme operates in an environment where IAA levels are lower than in maize endosperm. It is not known to which downstream reaction the synthesis of IAGlc is coupled in *Arabidopsis*. However, experiments with transgenic plants over-expressing IAGlc synthase indicate that levels of all amide conjugates and OxIAA are decreased in their vegetative tissues (R. Jackson and D. Bowles, personal communication). Quantitative measurements of the concentration of two amide conjugates (IAAsp and IAGlu) and IAGlc in *Arabidopsis* indicated that they occur at very low levels (Tam *et al.*, 2000). While this is not entirely surprising, especially since both IAAsp and IAGlc can be considered as intermediates rather than end-products, the levels of conjugates measured using the 'total IAA' method are usually very high. A complete metabolic profile of indole components was recently compiled by screening vegetative tissues of *Arabidopsis* for the presence of IAA-related compounds by gas chromatography-mass spectrometry (GC-MS) (Kowalczyk and Sandberg, 2001). This experiment revealed the existence of two additional conjugates, IAAla, and IALeu, but both of these compounds were found at concentrations even lower than those of IAAsp and IAGlu, which were also detected in the metabolic screen. Importantly, the experimental procedure and the sensitivity of the instruments used make it likely that no other low-molecular-weight amide conjugates of IAA exist in vegetative tissue at significant levels. As discussed below, this may be due to the pre-dominance of high-molecular-weight IAA-protein conjugates that are difficult to analyse by TLC or GC-MS methods.

IAA bound to proteins and amino acids constitute the pool of amide-linked conjugates of IAA found in plants. The first such compound was found when

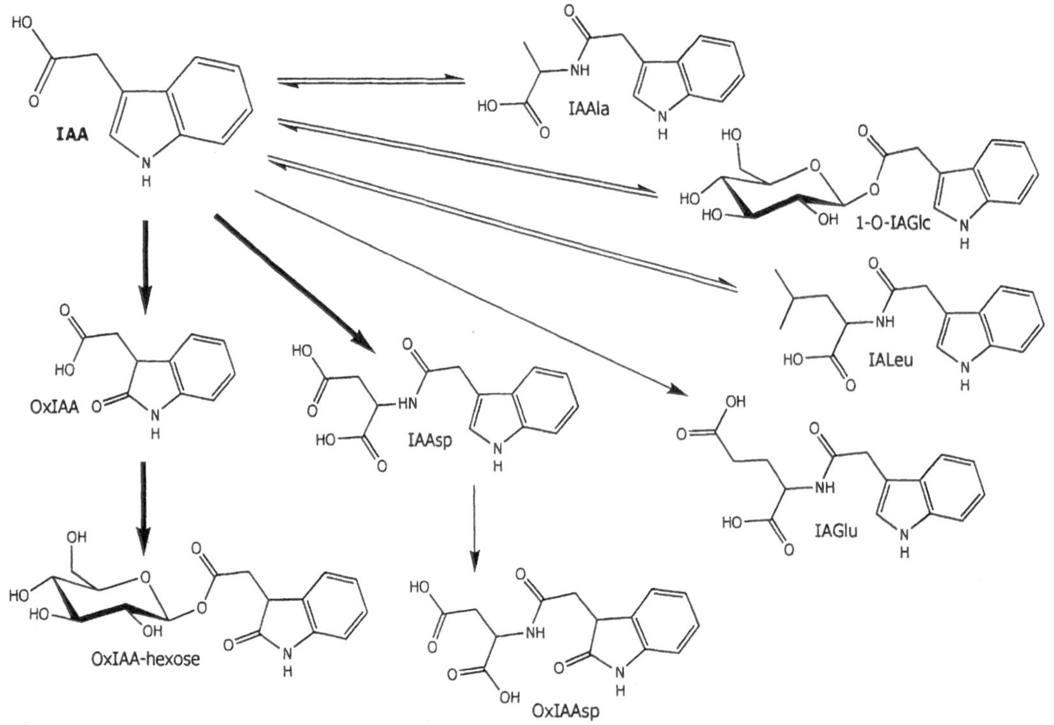

Figure 3. Pathways involved in IAA metabolism. Routes involved in catabolism and conjugation of IAA in vegetative tissues of *Arabidopsis* have been verified by labelling experiments as well as identification of endogenous constituents by mass spectrometry (Östin *et al.*, 1998; Tam *et al.*, 2000; Kowalczyk and Sandberg, 2001). Reversible products include the conjugates IAAla, IALeu and IAGlc. The major pathways regulating the IAA pool involve OxIAA and IAAsp as first products.

Bialek and Cohen (1986) isolated an IAA-containing compound from bean (*Phaseolus vulgaris*) with a very low mobility on TLC that was shown to be a conjugate consisting of two moieties of IAA attached to a hydrophobic 3.6 kDa peptide. The 3.6 kDa peptide was shown to be the smallest member of a family of proteins with IAA prosthetic groups. These protein amide conjugates accumulated during seed maturation (Bialek and Cohen, 1989) and declined as a class early in seed germination. The IAA-protein conjugates, after the initial rapid decline, began to increase in amount again early in seedling growth (Bialek and Cohen, 1992). After a few days of growth, the vegetative bean tissues contained IAA-protein conjugates that were different from those present in seeds. Stable isotope labelling studies during germination showed that a portion of the IAA produced *de novo* during rapid cell expansion was actually used for IAA-protein production, the first hint that these compounds function as active components for hormonal metabolism/regulation during a period of active IAA biosynthesis (Bialek *et al.*, 1992). The presence of the most abundant IAA-modified protein in bean seed was correlated to a developmental period of rapid growth during seed development and this specific protein was rapidly degraded during germination. The gene encoding this protein (*IAP1*) was isolated and cloned (Walz *et al.*, 2001). *IAP1* is a single-copy gene encoding a 35 kDa polypeptide. The gene has high sequence similarity to a 35 kDa late seed maturation protein from soybean that does not contain a known signalling sequence and does not appear to encode the smaller 3.6 kDa bean peptide from bean (Cohen *et al.*, 1988).

Previous studies have shown that *Arabidopsis* has both ester and amide conjugated forms of IAA (Tam *et al.*, 2000), but the smaller molecular weight conjugates previously identified did not collectively account for the bulk of the conjugate pool. A 35 kDa immunostaining protein was partially purified from *Arabidopsis* seeds and subjected to alkaline hydrolysis (Cohen *et al.*, 1986; Chen *et al.*, 1988). GC-MS analysis confirmed the presence of IAA covalently bound to this protein. The IAA-containing proteins from seeds of *Arabidopsis* appear to constitute the major portion of the solvent-insoluble (70% 2-propanol) amide-linked conjugates and account for 78% of the total IAA,

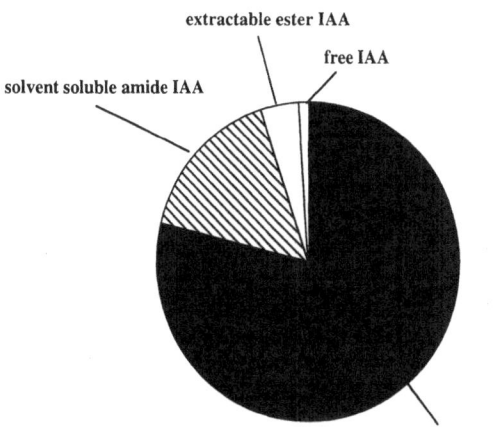

Figure 4. IAA-conjugate pools in *Arabidopsis* seeds. The 70% 2-propanol extractable amide-linked conjugates (mostly IAA-amino acids) and insoluble amide-linked conjugates (IAA-proteins) accounted for 17% and 78%, respectively, of the total IAA pool in the seed. Free IAA and ester-linked conjugates comprised only <1% and 4% of the total IAA pool. Higher–molecular-weight ester-linked conjugates were not detected (Park *et al.*, 2001).

ing that both amide- and ester-type conjugates can be hydrolysed back to the free hormone. Amide conjugates can serve as slow-release auxin sources in plant tissue cultures (Magnus *et al.*, 1992), and in high concentrations some of them can have the same phytotoxic effect as high concentrations of IAA (Slovin, 1997). Screens for plants insensitive to IAA amide conjugates led to the discovery of the first *Arabidopsis* IAA conjugate hydrolase, encoded by the *ILR1* gene (Bartel and Fink, 1995). The enzyme's amino acid sequence is similar to the procaryotic aminoacylase and hippuricase. ILR1 activity has strong specificity for IALeu and IAPhe. Using a similar approach another hydrolytic enzyme, encoded by the *IAR3* gene has been identified (Davies *et al.*, 1999). The IAR3 protein displays the highest activity for IAAla, a conjugate well known for its auxin-like activity. Overall, six different amide conjugate hydrolases have been postulated to exist in *Arabidopsis* (Davies *et al.*, 1999).

IAA homeostasis

The quantitative temporal and spatial distributions of IAA play important roles in the development of a plant throughout its life cycle. The IAA pool in a specific tissue is dependent on the combined effects of all the processes described earlier (i.e. biosynthesis, conjugation and degradation) and transport. Together, these different mechanisms maintain the IAA pool at

a level that is optimal for growth and development, i.e. preserve auxin homeostasis. Here we examine several examples of plant development where auxin homeostasis plays a crucial role.

Auxin homeostasis during embryo development

Embryogenesis in plants proceeds via a number of defined stages. The transition from globular to heart-shaped embryos marks a change from axial to bilateral symmetry and it has been shown that IAA plays an important role in this developmental process (for reviews, see Geldner *et al.*, 2000; Harada, 1999). It has also been shown using a wheat embryo system that exogenous auxin blocks the attainment of bilateral symmetry (Fischer and Neuhaus, 1996). Further studies using auxin transport inhibitors have demonstrated that auxin movement is important during normal embryo development. For example, the auxin efflux carrier inhibitor NPA disrupts the apical-basal patterning of globular carrot embryos and *Brassica juncea* embryos (Schiavone and Cooke, 1987; Hadfi *et al.*, 1998), and the fused cotyledon phenotype of the *pin1-1* mutant of *Arabidopsis* can be phenocopied using NPA (Liu *et al.*, 1993). Correct localization of auxin efflux carrier proteins is an important factor controlling embryo patterning. The PIN1 auxin efflux carrier has been observed to show a polar pattern of localization as early as the mid-globular stage. Later in embryo development PIN1 becomes localized to the basal face of the vascular precursor cells and to the opposite face of epidermal cells of the cotyledon primordia (Steinmann *et al.*, 1999). Together, these findings suggest that the distribution of IAA during embryogenesis is important in pattern formation. The aberrant cotyledon patterning seen in a proportion of mutant *pinoid (pid)* seedlings further supports the hypothesis that efflux carrier mediated IAA distribution plays an important role in embryo development (Benjamins *et al.*, 2001). The *PID* gene encodes a serine/threonine protein kinase that is proposed to function as a positive regulator of polar auxin transport and may interact either directly or indirectly with members of the auxin efflux carrier family.

Studies of auxin metabolism during somatic embryogenesis in carrot cultures (Michalczuk *et al.*, 1992b) have shown that L-tryptophan can be converted in these cells to IAA. However, the use of tryptophan for IAA biosynthesis is regulated by the chemical/hormonal environment in which the cells are grown. Carrot cell suspension cultures were grown on media containing 30% deuterated water (^{2}H$_2$O) and media containing [^{15}N]indole, both with and without 2,4-D. Label from ^{2}H$_2$O was incorporated into IAA, indicating that *de novo* biosynthesis had occurred, both in the presence and absence of 2,4-D. [^{15}N]indole studies showed the same result. However, when cells were supplied with [^{2}H$_5$]-L-tryptophan, only the cultures grown in the presence of 2,4-D incorporated label into IAA. In all cases, tryptophan pools were equally strongly labeled, with or without 2,4-D. These results demonstrate that strong developmental regulation of tryptophan-independent and tryptophan-dependent IAA biosynthesis pathways occurs, because the incorporation of label into IAA from tryptophan changed, without a corresponding change of incorporation from indole or ^{2}H$_2$O (Michalczuk *et al.*, 1992b). Excised hypocotyls from carrot seedlings produce IAA primarily by the tryptophan conversion pathway, both with and without exogenous auxin, suggesting that it is the embryogenic program rather than the presence or absence of exogenous auxin that controls the pathway for IAA biosynthesis. However, carrot hypocotyls accumulated substantially more IAA than controls when treated with 2,4-D, although they did not produce greater levels of IAA when treated with either NAA or [^{2}H$_4$]IAA. The results from the [^{2}H$_4$]IAA treatment demonstrate that external IAA can inhibit native auxin production, while NAA has little or no effect and 2,4-D actually increases IAA accumulation.

Initiation of somatic embryogenesis by the removal of 2,4-D coincides not only with a shift from IAA synthesis using a tryptophan precursor to use of the non-tryptophan pathway, but also with a dramatic 10–20-fold decrease in IAA levels (Figure 5A). This decline may result in part from changes in the rate and pathway of biosynthesis. However, it is also accompanied by a massive drop in the IAA-amide conjugate pool in the cell. It has been suggested for other plant processes that IAA turnover rates themselves may correlate with biological activity (Reinecke and Bandurski, 1987). Thus, increases in the rate of turnover and changes in the conjugate pool are implicated as important components in the process of somatic embryogenesis. While it has been shown that conjugates can be directly oxidized without release of the active hormone (Tsurumi and Wada, 1986), the decrease observed in the IAA amide conjugate pool during carrot somatic embryogenesis appears to be at least in part due to hydrolysis back to IAA. IAA-amino acid hydrolase activity increases some 5-fold during

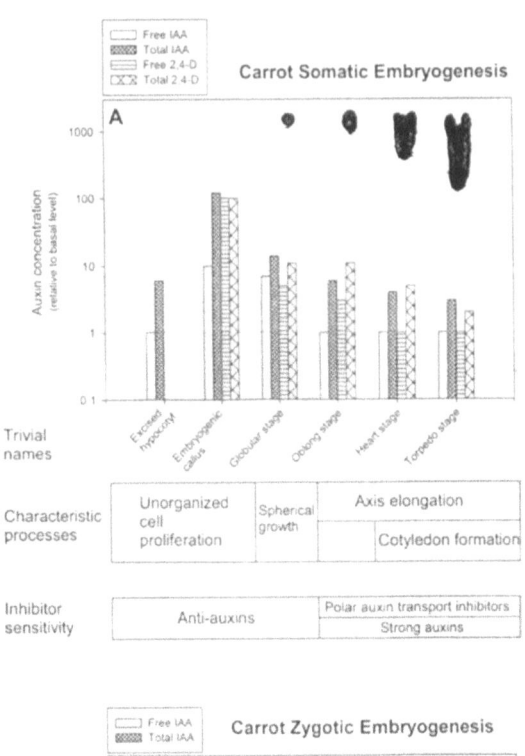

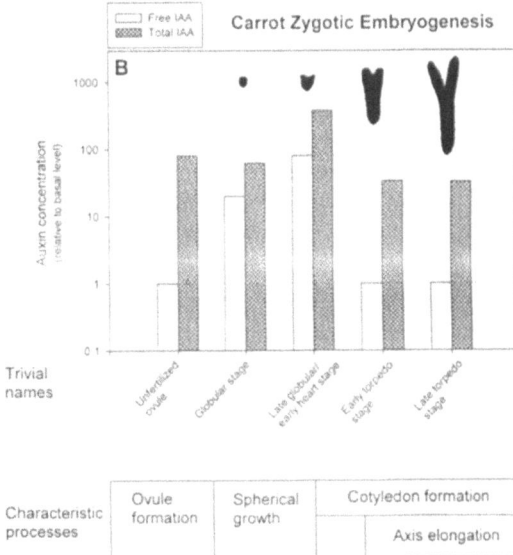

Figure 5. IAA homeostasis during embryo development. A comparison of the free and total IAA and 2,4-D induced changes observed during carrot somatic embryogenesis (A) (data calculated from Michalczuk *et al.*, 1992a) vs. the free and total IAA changes observed during carrot zygotic embryogenesis (B) (Ribnicky *et al.*, 2001). Trivial names refer to the morphological shapes observed at each embryonic stage; characteristic processes refer to the morphogenetic processes occurring at each stage; and inhibitor sensitivity refers to the auxin agonists or antagonists active at each stage. Data are expressed relative to the average basal auxin level, which for somatic embryos is the sum of the average free IAA concentration (7 ng/g FW) plus the free 2,4-D concentration (14 ng/g FW) in oblong to torpedo-stage embryos. For zygotic embryos the basal level is the average IAA level measured in unfertilized ovules and early and late torpedo stages (26 ng/g FW).

carrot somatic embryogenesis at the same time that auxin levels and *de novo* auxin biosynthesis are both decreasing (Kuleck and Cohen, 1992).

These studies of carrot somatic embryos have been followed up recently by measuring changes in IAA and conjugated IAA levels in carrot zygotic embryos (Ribnicky *et al.*, 2001). Novel methods allowed to compare the hormonal changes that occur in somatic embryos to those that occur during normal plant reproduction. The data obtained confirm that the large changes in IAA and IAA conjugate levels induced by the hormonal manipulations required to obtain embryos during somatic embryogenesis in culture (Michalczuk *et al.*, 1992a) mimic natural processes in the zygotic process where, after ertilization, a huge surge in auxin levels is seen (Figure 5B).

Auxin homeostasis during early seedling growth

As the embryo develops into a mature seed, it forms storage organs that are essential for the initial growth of the young seedling. These storage organs contain not only nutrients and other compounds needed for rapid germination and growth, but also supply different plant hormones. These stored hormones can then be rapidly released when needed and used during the initial growth phase, before *de novo* synthesis starts in the newly formed tissues. *Arabidopsis* seeds are very small and so are not ideal for investigations on how IAA is stored, released and synthesized during germination. Most knowledge on this subject comes from investigations on species such as sweet corn (*Zea mays*), rice (*Oryza sativa*), horse chestnut (*Aesculus* spp.) and, recently, Scots pine (*Pinus sylvestris*). In all these species the main stored forms of IAA are IAA-ester conjugates (Epstein *et al.*, 1980; Hall, 1980; Domagalski *et al.*, 1987; Ljung *et al.*, 2001a) such as IAA-sugar conjugates, IAA-*myo*-inositol and IAA-*myo*-inositol glucosides. In legumes like bean (*Phaseolus vulgaris*), IAA is mainly stored in the seed as amide-linked conjugates such as IAA-amino acids or IAA-peptides (Bialek *et al.*, 1992). These stored forms of IAA can then be hydrolysed to free IAA, which is used by the seedling during germination and early seedling growth. Eventually, the seedling begins to synthesize its own IAA, usually 2–7 days after germination (Bialek *et al.*, 1992; Pengelly and Bandurski, 1983; Ljung *et al.*, 2001a). The onset of IAA biosynthesis seems to vary between species, and it is probably influenced by environmental conditions such as temperature and light as well as the size

of the seed reserves of conjugated IAA. In addition, distinct IAA biosynthesis pathways are operational at different developmental stages as well as different tissues (Ljung *et al.*, 2001a; Michalczuk *et al.*, 1992b; Sitbon *et al.*, 2000). As *de novo* synthesis of IAA begins in the seedling, different mechanisms to control IAA homeostasis such as catabolism and conjugation are initiated. Studies of the developmental regulation of IAA metabolism during germination and early seedling growth in Scots pine (Ljung *et al.*, 2001a) have clearly demonstrated the differences in timing between these homeostatic mechanisms (Figure 6). IAA biosynthesis and catabolism are temporally initiated in correlation with major developmental events such as hypocotyl, root and cotyledon elongation. Trp-dependent IAA biosynthesis is initiated around four days after germination (DAG) and Trp-independent synthesis later (around 7 DAG). Catabolism and conjugation can be detected at the same time as *de novo* IAA biosynthesis begins. There are also large changes in the pools of free and conjugated IAA in the germinating seedling. A rapid increase in free IAA levels occurs during the first days of germination, and this is correlated to a rapid decrease in the stored pool of ester-linked IAA conjugates. The level of amide-linked conjugates is initially low, but starts to increase around 3 DAG, when IAA catabolism is initiated.

The major low-molecular-weight IAA-conjugates and catabolites in *Arabidopsis* have been identified (Östin *et al.*, 1998; Tam *et al.*, 2000; Kowalczyk and Sandberg, 2001). Recent work (Park *et al.*, 2001) has shown that *Arabidopsis* seeds contain primarily protein-conjugated IAA, although other types of conjugates are present in significant amounts. With methods now available to quantify these substances in small amounts of plant tissue it is possible to get a better understanding of where and when they are synthesized. Some of these IAA-conjugates can be hydrolysed back to IAA, providing a mechanism to increase the total IAA pool within the tissue, while other conjugates serve primarily as intermediates for IAA catabolism, permanently removing IAA from the pool (Bartel, 1997; Östin *et al.*, 1998; Normanly and Bartel, 1999). Kowalczyk and Sandberg (2001) performed a quantitative analysis on different tissues of an *Arabidopsis* plant at two stages of development, using heavy-labelled internal standards of OxIAA, IAAsp, IAGlu, IAAla and IALeu. In this study it was demonstrated that expanding leaves and roots generally contained high amounts of the free hormone and also the highest levels of IAAsp, IAGlu and OxIAA,

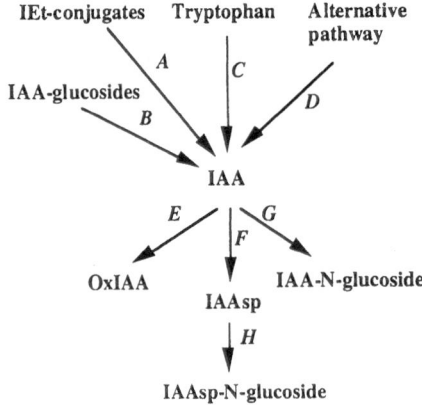

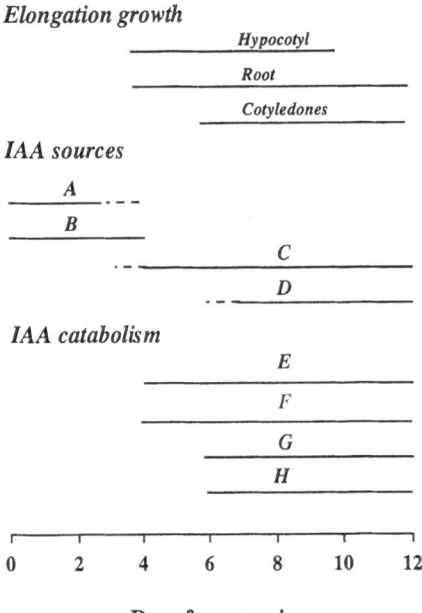

Days from sowing

Figure 6. An overview of developmental regulation of IAA homeostasis during germination and early seedling growth in Scots pine (*Pinus sylvestris*). IAA-ester conjugates were initially hydrolysed in the seed to yield a peak of free IAA prior to the initiation of root elongation. Tryptophan-dependent synthesis is initiated first, followed by tryptophan-independent synthesis, showing developmental regulation of IAA biosynthesis during germination. Induction of catabolism to yield OxIAA and irreversible conjugation to IAAsp was noticed at the same time as *de novo* synthesis was first detected. The figure is reprinted with permission from Ljung *et al.* (2001a; © American Society of Plant Biologists).

supporting the hypothesis that they are irreversible catabolic products. The levels of IALeu and IAAla did not follow the general distribution of IAA. Interestingly, the levels of IALeu and IAAla were highest in roots and aerial tissues, respectively. It is generally accepted, as mentioned above, that different conjugates play different roles in IAA metabolism. Even within the chemically similar amide conjugate family there are likely to be differences in function. It has been postulated that IAAsp is an irreversible catabolite rather than a reversible storage compound. This seems to be the case in *Arabidopsis*, where the levels of IAAsp mirror the distribution of free IAA. From a physiological standpoint, IAAsp (and possibly IAGlu) should be considered different from the IAAla and IALeu conjugate pools, since only IAAla and IALeu are likely to contribute to the free IAA pool. Therefore, experiments where all low-molecular-weight conjugates are measured as one combined pool would overestimate the relative importance of these conjugates in IAA homeostasis. This is demonstrated in Figure 7, which illustrates the relationship between the amounts of reversible conjugates and 'catabolic' conjugates in vegetative tissues of *Arabidopsis*. It is also important to note the relative contribution of the oxidation pathway, in which OxIAA is the first product. This is probably more important than 'catabolic' conjugation for maintaining the correct level of free IAA in vegetative tissues of *Arabidopsis* (Figure 7).

Regulation of IAA levels plays a role in leaf development

Leaf development is a complex process, involving differentiation of as many as twelve different cell types (Langdale, 1998). Signalling mechanisms that regulate leaf development have to be spatially and temporally coordinated in order to determine the final structure of the leaf (Tsiantis and Langdale, 1998). Auxin has been implicated in the regulation of organ initiation and phyllotaxis. Tomato shoot apices cultured in the presence of the auxin transport inhibitor NPA failed to initiate new leaf primordia, but instead formed a pin-like structure from the meristem (Reinhardt *et al.*, 2000). Local application of IAA to the pin structures induced the formation of leaf primordia at a fixed position relative to the tip of the meristem. IAA was able to influence the radial position and size of the lateral organs, but not their apical-basal positioning relative to the meristem. This suggests that the level and spatial distribution of IAA is highly regulated within shoot meristematic tissues. However, it is not clear whether this control is imposed at the level of auxin transport, biosynthesis or conjugation or a combination of all of these processes. It is known that IAA induces elongation growth in stems and coleoptiles in a dose-dependent manner, and that supra-optimal IAA concentrations inhibit expansion growth (Cle-

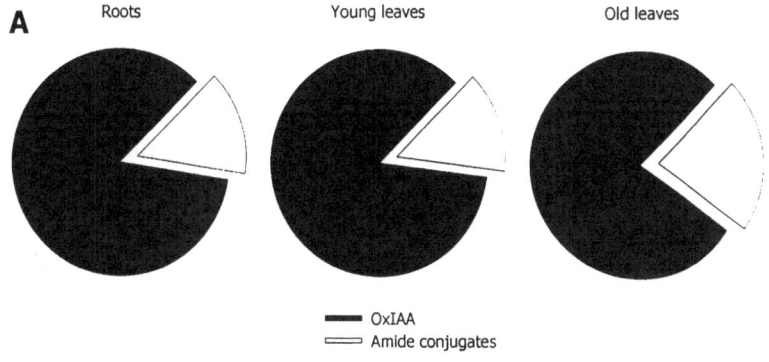

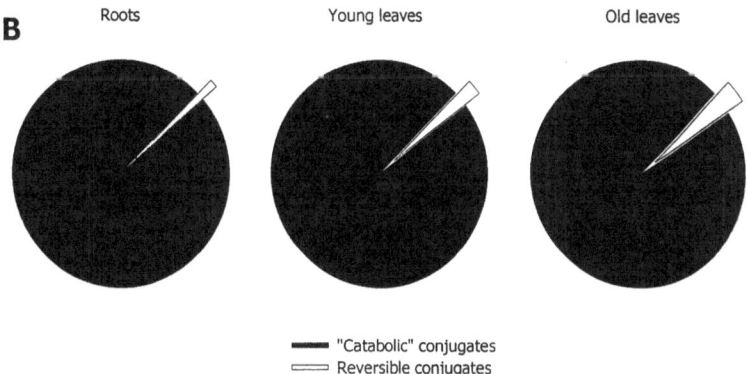

Figure 7. IAA-conjugate pools in 10-day old *Arabidopsis* seedlings. A. Relationship between the pool of OxIAA and low molecular IAA amide conjugates in different plant organs. B. Relationship between the pool of reversible IAA conjugates (IAAla, IALeu, IAGlc) and 'catabolic' conjugates (IAAsp, IAGlu) in the same tissues, demonstrating that the majority of low molecular conjugates are products of irreversible catabolism.

land, 1995). IAA is also believed to be involved in the control of leaf expansion, possibly mediated via the action of specific cell wall loosening expansin proteins (Hutchison *et al.*, 1999; Cosgrove, 2000). When the leaf expands, there is a coordinated differentiation of vascular tissue and there is strong supporting evidence for IAA playing a role in controlling this process (Nelson and Dengler, 1997). Inhibition of auxin polar transport in developing *Arabidopsis* leaves has dramatic effects on vascular tissue development (Mattsson *et al.*, 1999; Sieburth, 1999) that most likely also strongly influence the auxin transport capacity in the leaves. Leaf development was also shown to be affected in maize plants when polar auxin transport was inhibited (Tsiantis *et al.*, 1999). It is not known when the IAA transport capacity in the leaf is initiated, but it is probably closely correlated to the formation of the vascular tissue early in leaf and stem development. The parenchyma cells surrounding the vascular

strands are likely candidates for auxin transport in dicotyledonous plants (Lomax *et al.*, 1995; Gälweiler *et al.*, 1998). The role of IAA during leaf development in *Arabidopsis* has been investigated (Ljung *et al.*, 2001b) and results show that young, developing leaves contain very high amounts of IAA compared to older leaves. The levels of IAA are thus high in leaves with high rates of cell division, compared to the expanding leaves, where cell division has ceased and growth depends solely on cell expansion (Ljung *et al.*, 2001b and Figure 8). Spatial analysis of IAA levels in mesophyll tissue from expanding tobacco leaves has shown that the highest IAA levels correlate to areas where the rate of cell division is greatest (Ljung *et al.*, 2001b). Fully developed tobacco leaves have low and constant IAA levels in mesophyll tissue throughout the leaf blade (Edlund *et al.*, 1995). However, in very small leaves, where cell division occurs across the whole leaf area, IAA levels are much higher but neverthe-

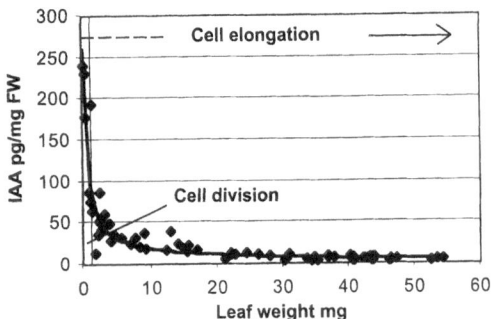

Figure 8. The relationship between IAA concentration and leaf weight in individual *Arabidopsis* leaves. The youngest leaves analysed, less than 0.5 mm in length, contained 250 pg/mg of IAA and also exhibited the highest relative capacity to synthesize this hormone. A decrease of nearly a hundred-fold in IAA content (pg/mg fresh weight) occurred as young leaves expanded to their full size, and this was accompanied by a clear shift in both pool size and IAA synthesis capacity, establishing a strict relationship between rate of cell division and the pool size of IAA (Ljung *et al.*, 2001b).

less constant in mesophyll tissue throughout the leaf blade (G. Sandberg, unpublished results). It could be concluded that the high IAA level in young leaves is essential for the intense cell division that takes place early in leaf development. Alternatively, the elevated IAA level could be a consequence of high cell division rates, and it may have some other signalling function such as in the development of leaf vascular tissue and/or IAA transport.

Recent studies have highlighted the importance of auxin homeostasis in maintaining correct leaf development. The auxin transport inhibitor NPA causes a reduction in the IAA content of expanding *Arabidopsis* leaves and this is mirrored by a decrease in leaf expansion (Ljung *et al.*, 2001b). The *sur1* and *sur2* mutants of *Arabidopsis* contain increased IAA levels compared to wild-type (see below). The first expanding leaves of *sur1* and *sur2* seedlings have significantly higher IAA levels than the wild type, and this correlates with a reduced rate of leaf expansion (Ljung *et al.*, 2001b). Together these results show that maintaining auxin homeostasis is essential for normal leaf expansion.

IAA homeostasis in root development

Although not very much is known about auxin homeostasis in the root system, it is clear that IAA has profound effects on primary and lateral root development (Reed *et al.*, 1998; Casimiro *et al.*, 2001; Bhalerao *et al.*, 2001) and root meristem patterning

(Sabatini *et al.*, 1999; Doerner, 2000). Recent work has sought to establish the source of the IAA that controls different stages in lateral root development. The cyclinB1::*uidA* marker has been used to demonstrate that the first lateral roots are initiated within 28 h after germination, and thus prior to the formation of the first pair of leaves (Casimiro *et al.*, 2001). The IAA influencing the initial cell divisions to form a lateral root primordium must therefore either be derived from the cotyledons or synthesized in the developing primary root. The onset of lateral root emergence is associated with a transient increase in the concentration of IAA in the root system 5–7 DAG (Bhalerao *et al.*, 2001). Removing the shoot from seedlings prior to the emergence of the first leaves eliminated the pulse of IAA within the root and caused the lateral roots to be arrested predominantly between stages IV and V. However, the total number of initiated lateral roots in these plants was not significantly different from that of non-excised control seedlings. In contrast, if the shoot was removed after the primary leaves had formed, and thus after the pulse of IAA had reached the root, then lateral root emergence was comparable to that of non-excised control seedlings (Bhalerao *et al.*, 2001). The pulse of IAA derived from the first leaf pair cannot therefore have a role in the initiation of lateral roots early in the development of *Arabidopsis* seedlings. The IAA pulse could however act as a regulatory mechanism in the development of the root system once the aerial tissues have emerged from the soil and begun to develop. Such communication between shoot and root tissues is likely to be important during the development of the seedling. Analysis of IAA levels in 1 mm segments from the root tip of 3-, 6- and 7-day old seedlings revealed the existence of a steep IAA concentration gradient, with the highest levels in the first millimetre of the root tip (Bhalerao *et al.*, 2001). This gradient is formed between 3 and 6 DAG, and correlates well with the expression pattern of the DR5::*uidA* reporter line, which also shows strong expression in the first millimetre of the root tip, increasing expression between 4 and 10 DAG. The high level of IAA found in the root tip supports the hypothesis that IAA has important functions during cell organization in this tissue (Sabatini *et al.*, 1999). It is also possible that the IAA gradient may be necessary for the initiation of new lateral root primordia (Bhalerao *et al.*, 2001).

Mutants that alter auxin homeostasis

Mutants defective in different steps of the putative IAA biosynthetic pathways have been identified (Bartel, 1997; Eckardt, 2001), although these pathways have yet to be fully elucidated A number of *Arabidopsis* mutants have been identified which show elevated endogenous IAA levels, and these have provided some information about the homeostatic control processes. The *sur1* and *sur2 Arabidopsis* mutants display phenotypes that are typically associated with elevated endogenous IAA-levels such as adventitious root formation, increased numbers of lateral roots and small, epinastic cotyledons (Boerjan *et al.*, 1995; Delarue *et al.*, 1998). The *sur1* mutant was found to have increased levels of both free and conjugated IAA compared to the wild type on a whole-seedling basis. It is possible that the increased levels of IAA conjugates in *sur1* reflect an increased rate of conjugation as a result of the larger amount of free IAA present. The *SUR2* gene, as discussed earlier, encodes a cytochrome P450 (CYP83B1) likely to be involved in glucosinolate biosynthesis. However, mutations in *SUR2* may lead to increased conversion of IAOx to IAA and thus severely alter IAA homeostasis in the mutant. Free IAA levels were elevated in all organs analysed in *sur2* mutant seedlings compared to the wild type (Barlier *et al.*, 2000). The expression of this gene was induced by stress, by overproduction of auxin and by treatment with exogenous auxin. The *sur2* phenotype reverts to the wild type 12–15 DAG. This reversion to the wild-type phenotype is correlated to a reduction in the level of free IAA in both the aerial parts of the plant and the root system. The wild-type phenotype was also restored by growing the seedling in the presence of exogenous IAA or on low-pH medium. Elevated levels of IAA catabolites and IAA conjugates (OxIAA, IAAsp, IAGlu) were found in *sur2* seedlings compared to the wild type. Similar levels of hydrolysable conjugates such as IALeu and IAAla were found in *sur2* seedlings compared to wild-type seedlings in early stages of germination, but the capacity to synthesize these conjugates was apparently lost in *sur2* seedlings around 6 DAG.

Recently two other *Arabidopsis* mutants have been described that have elevated endogenous IAA levels. The *bus1-1* mutant shows a bushy habit with crinkly leaves and altered vascular development. The *BUS1* gene encodes a cytochrome P450 (CYP79F1), which is thought to catalyse the formation of glucosinolates derived from methionine. Mutant null *bus1* plants lacked short-chain glucosinolates and *BUS1* over-expressing plants showed increased levels of these compounds (Reintanz *et al.*, 2001). Interestingly, levels of IAA and the IAA precursor indole-3-acetonitrile were increased in the *bus1-1* mutants, indicating that the homeostatic control of IAA biosynthesis had been disrupted in these plants. This was also shown in the recently described mutant *supershoot (sps)*, which is allelic to *bus1-1* (Tantikanjana *et al.*, 2001). The *sps* mutant displayed elevated IAA levels as well as a 3- to 9- fold increase in the levels of zeatin-like cytokinins. An activation tagging approach was used in *Arabidopsis* to isolate a dominant mutant called *yucca* that showed elevated levels of free IAA (Zhao *et al.*, 2001). The mutant phenotype was shown to be a result of over-expression of a flavin monooxygenase gene. Feeding experiments indicate that *in vitro* YUCCA catalyses hydroxylation of the amino group of tryptamine, a step in Trp-dependent IAA biosynthesis (see discussion earlier).

Homeostatic control of IAA levels via biosynthesis

In order to preserve IAA homeostasis, the plant has several mechanisms that cooperatively keep the IAA concentration at the correct level. One of the factors affecting the IAA level in a specific tissue is the rate of *de novo* IAA biosynthesis. Plants have multiple IAA biosynthetic pathways, as earlier described, that can be developmentally controlled both temporally and spatially. As first described in studies of amino acid biosynthesis, *de novo* biosynthesis of a variety of compounds can be studied by incubating the plants with deuterated water (Mitra *et al.*, 1976). In this way it has been possible to measure the total IAA biosynthetic rate (see Pengelly and Bandurski, 1983; Wright *et al.*, 1992; Bialek *et al.*, 1992; Jensen and Bandurski, 1996). This technique yields valuable information, regardless of the pathway(s) being used by the plant. It also permits estimation of the size of the pool of IAA synthesized *de novo* in relation to the total IAA pool present in different plant organs. It does not, however, lend itself easily to analysis of the route of biosynthesis used to produce the accumulating IAA. Therefore, feeding experiments with labelled IAA precursors have been used to provide information on the relative importance of different biosynthetic pathways during plant development. There are, however, potential problems with this approach since different substances are taken up at different rates by the plant during incubation. It is also important to show that the substance under study reaches critical parts of the plant, i.e. cells within a specific tissue or all compartments within a specific cell. Failure to account for these possibilities can lead to significant errors in the estimation of biosynthetic rates and metabolic pathways.

Very little is still known about the actual mechanisms of IAA biosynthetic regulation, and various control mechanisms may cooperate in this process. Two recent experiments have demonstrated that there might be a mechanism that controls IAA levels by feedback inhibition of IAA biosynthesis (Ribnicky *et al.*, 1996, Ljung *et al.*, 2001b). In studies by Ribnicky *et al.* (1996), $[^2H_4]IAA$ fed to plants in culture reduced the formation of unlabelled IAA significantly more than either 2,4-D or NAA, suggesting that a direct biochemical rather than hormonal feedback mechanism controls biosynthesis of IAA. Ljung *et al.* (2001b) used NPA, a substance known to block polar IAA transport, to study IAA regulation by altering sites of IAA accumulation, thereby trapping IAA in synthesizing source tissues such as leaves. In their experiments, treatment of plants with NPA initially resulted in increased IAA levels in expanding leaves, as would be expected. However, after reaching a maximum 16 h after NPA treatment, the IAA levels began to drop, indicating the initiation of feedback inhibition of IAA synthesis in this tissue. This was further verified in experiments where the rates of IAA biosynthesis were measured in plants incubated in deuterated water or in deuterated water supplemented with NPA.

Young developing leaves are believed to be an important source of IAA for plant growth and development, yet very little research has been performed to verify this hypothesis by actually measuring IAA levels and IAA biosynthetic rates in different tissues. Experiments designed to identify sites of IAA biosynthesis within different plant organs used the developing *Arabidiopsis* seedling as a model system. Figure 9 shows the IAA distribution and IAA biosynthetic capacity in different organs of 10-day old *Arabidopsis* seedlings incubated for 24 hours with 30% deuterated water and 40 μM NPA. There is, of course, a large difference in size between the different organs (Figure 9A). Comparison of the total IAA pools in different tissues (Figure 9B) shows that roots and expanding leaves contain significantly higher proportions (38% and 33%, respectively) of the pool compared to the youngest leaves and cotyledons (16% and 13%). The relative IAA distribution between old and young leaves is explained by the large difference in

A
Relative weight of different tissues

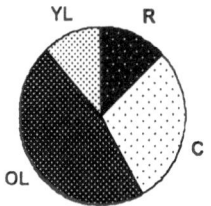

B
IAA pool size in different tissues

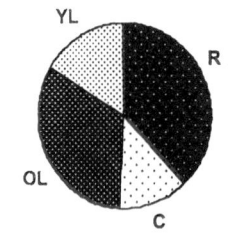

C
IAA biosynthesis in different tissues

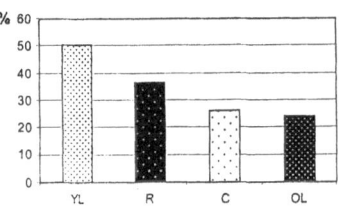

Figure 9. IAA pool size and IAA biosynthesis capacity in different parts of 10-day old *Arabidopsis* seedlings. Young leaves (YL), old leaves (OL), cotyledons (C) and the root system (R). A. Relative weight of different plant organs. B. The relationship between the IAA pool size in different plant organs. C. IAA biosynthesis capacity in different plant organs (Ljung *et al.*, 2001b).

size between these organs, which balances the very high concentration of free IAA found in young leaves (see Figure 8). When IAA synthesis capacity was compared between tissues, youngest leaves showed the highest synthesis rates followed by the root, cotyledons and expanding leaves (Figure 9C). Calculations of the proportions of the total IAA pool in different tissues that was newly synthesized after a 24 h labelling incubation period revealed that about half of the pool in the young leaves consisted of newly synthesized IAA. This was in contrast to proportions of newly-synthesized IAA amounting to 35% in the root and only 25% in cotyledons and old leaves. Even though the plants had been treated with NPA to block polar IAA transport, the possibility that IAA could have been transported from the shoot to the root system via a non-NPA sensitive route could not be discounted. However, experiments with dissected roots incubated with 30% deuterated water showed that there was still major synthesis capacity in the root, suggesting that the root is a major source of IAA (Ljung et al., 2001b). These data clearly demonstrate that all parts of the plant can potentially act as important sites of IAA biosynthesis, and that there is a complex, interacting network of different IAA sources and transport routes. Both polar transport of IAA as well as transport in the phloem (Baker, 2000; Swarup et al., 2001) can potentially serve as routes for relocalization of newly synthesized IAA from the different sources. It has yet to be determined when and where IAA biosynthesis is initiated in the root, and if root versus shoot derived IAA serves different purposes during root development. One hypothesis is that IAA synthesis is initiated in the developing lateral roots after they have reached a certain size, and that the root system then gradually becomes independent of shoot-derived IAA for lateral root development (Bhalerao et al., 2001).

Auxin transport from source to sink tissues as a homeostatic regulatory process

Auxin, uniquely amongst plant hormones, has been shown to exhibit polar transport. Therefore the sites of IAA biosynthesis and action may be spatially distinct. Polar auxin transport (PAT) is thought to be mediated by the auxin efflux carrier proteins (Gälweiler et al., 1998). However, in addition to long distance PAT, IAA can also be moved from cell to cell via the action of both influx and efflux carriers. Identification of components of both the influx and efflux carriers has allowed detailed analysis of expression profiles (Gälweiler et al., 1998; Müller et al., 1998a; Swarup et al., 2001). The AUX1 protein is a putative component of the influx carrier machinery (Bennett et al., 1996; Marchant et al., 1999) and is one of four proteins in *Arabidopsis* that share extensive sequence homology. Mutations in *AUX1* lead to auxin-resistant agravitropic roots and reduced numbers of lateral roots. An epitope tagging approach has been adopted to localize AUX1 in *Arabidopsis* root tissues. AUX1 is asymmetrically localized to the plasma membrane of protophloem cells and is thought to facilitate the acropetal movement of IAA to the root apex. AUX1 was also found to be localized to a subset of columella and lateral root cap cells, where it is proposed to facilitate gravitropic responses. The first

efflux carrier protein was identified by a number of independent groups and has been termed PIN2 (Müller et al., 1998a), AGR1 (Chen et al., 1998; Utsuno et al., 1998) and EIR1 (Luschnig et al., 1998). There are eight known members of this family of proteins, which are likely to perform distinct spatial or developmental roles in the plant (Gälweiler et al., 1998; Müller et al., 1998a; Steinmann et al., 1999). It is likely that the control of polar auxin transport as well as local IAA redistribution will be important in maintaining correct delivery of IAA to target organs. A number of genes have been isolated which are proposed to encode regulators of auxin transport. The *pis1* mutant was shown to be hypersensitive to NPA, and it has been proposed that PIS1 acts as a negative regulator of the action of the efflux carrier inhibitors NPA and TIBA (Fujita and Syono, 1997). The *PINOID (PID)* gene, which encodes a serine/threonine protein kinase, has been proposed to function as a positive regulator of polar auxin transport (Benjamins et al., 2001). A second study resulted in isolation of the same gene, but concluded that it functioned as a negative regulator of auxin signaling (Christensen et al., 2000). Currently it is not clear how PID could regulate PAT, although it has been speculatively proposed that there could be an interaction between PID and the IAA efflux carrier proteins. Maintenance and correct regulation of PAT may be required to prevent IAA accumulation in source tissues resulting in feedback inhibition of biosynthesis and/or increased conjugation. Though our knowledge of auxin transport and its regulation has increased in recent years, further work is required to fully understand how the process is regulated and the how the influx and efflux carriers interact to control auxin movement throughout the plant.

Future prospects

In auxin biology, redundancy is a theme that consistently arises. We are beginning to understand that Trp-dependent and Trp-independent IAA biosynthetic pathways have distinct developmental roles and differ in their environmental responses. Whether their roles overlap is unclear. In addition there is evidence for at least two Trp-dependent pathways that use IAOx as an intermediate, though their roles are not known. In catabolism and conjugation the existence of multiple pathways indicate that tissue- and time-specific resolution may occur in the regulation of the IAA pool. The newly discovered IAA-conjugated peptides in

268

Arabidopsis most likely have a role in IAA homeostasis, possibly as an auxin source during germination and early seedling growth. Recent developments in metabolic screens to identify all IAA-related molecules give hope for the future. These analyses have to be done not only during vegetative growth, but also during seed maturation and germination, where it is expected that other forms of conjugated IAA will be of importance. The development of quantitative mass spectrometry has now reached the level where we can analyse the IAA content in root tips and leaf primordia. This technology has recently contributed data that will provide the basis for an extensive future map of IAA distribution in *Arabidopsis*. This will be important for the entire auxin field, since the auxin markers that have become available lately sometimes correlate well to the IAA level in specific experiments and tissues, but not in general terms. Quantitative and qualitative mass spectrometric analysis of precursors and catabolites/conjugates in a tissue-specific manner together with measurements of biosynthesis rates in specific tissues will be useful tools for increasing our understanding of auxin metabolism and homeostasis. The recent completion of the *Arabidopsis* genome provides more useful tools. Reverse genetic methods applied to genes with predicted functions in auxin biology will certainly yield interesting results. However, as exemplified by identification of the genes *YUCCA* and *PINOID* using activation tagging, new sophisticated screens may be necessary to identify genes in these pathways. Genetic screens are only as good as the phenotype being screened for. Mutations that affect auxin levels several-fold might not have significant morphological phenotypes, and high-throughput biochemical screens (rather than phenotypic screens) may be needed to find many of the genes involved in auxin biology. Nevertheless, over the next few years it is certain that we will be able to identify many more genes involved in auxin metabolism, and to determine how they influence plant development.

References

Audus, L.J. 1972. Plant Growth Substances, vol. 1: Chemistry and Physiology. Leonard Hill, London.

Bak, S., Kahn, R.A., Nielsen, H.L., Møller, B.L. and Halkier, B.A. 1998. Cloning of three A-type cytochromes P450, CYP71E1, CYP98, and CYP99 from *Sorghum bicolor* (L.) Moench by a PCR approach and identification by expression in *Escherichia coli* of CYP71E1 as a multifunctional cytochrome P450 in the biosynthesis of the cyanogenic glucoside dhurrin. Plant. Mol. Biol. 36: 393–405.

Bak, S., Tax, F.E., Feldmann, K.A., Galbraith, D.W. and Feyereisen, R. 2001. CYP83B1, a cytochrome P450 at the metabolic branch point in auxin and indole glucosinolate biosynthesis in *Arabidopsis thaliana*. Plant Cell 13: 101–111.

Baker, D.A. 2000. Long-distance vascular transport of endogenous hormones in plants and their role in source : sink regulation. Israel J. Plant Sci. 48: 199–203.

Baldi, B.G., Maher, B.R., Slovin, J.P. and Cohen, J.D. 1991. Stable isotope labeling, *in vitro*, of D- and L-tryptophan pools in *Lemna gibba* and the low incorporation of label into IAA. Plant Physiol. 95: 1203–1208.

Bandurski, R.S., Cohen, J.D., Slovin, J.P. and Reinecke, D.M. 1995. Auxin biosynthesis and metabolism. In: P.J. Davies (Ed.) Plant Hormones: Physiology, Biochemistry and Molecular Biology, Kluwer Academic Publishers, Dordrecht, Netherlands, pp. 39–65.

Barcelo, A.R., Pedreno, M.A., Ferrer, M.A., Sabater, F. and Munoz, R. 1990. Indole-3-methanol is the main product of the oxidation of indole-3-acetic-acid catalyzed by 2 cytosolic basic isoperoxidases from *Lupinus*. Planta 181: 448–450.

Barlier, I., Kowalczyk, M., Marchant, A., Ljung, K., Bhalerao, R., Bennett, M., Sandberg, G. and Bellini, C. 2000. The *SUR2* gene of *Arabidopsis thaliana* encodes the cytochrome P450 CYP83B1, a modulator of auxin homeostasis. Proc. Natl. Acad. Sci. USA. 97: 14819–14824.

Barratt, N.M., Dong, W.Q., Gage, D.A., Magnus, V. and Town, C.D. 1999. Metabolism of exogenous auxin by *Arabidopsis thaliana*: Identification of the conjugate N-α-(indole-3-ylacetyl)-glutamine and initiation of a mutant screen. Physiol. Plant. 105: 207–217.

Bartel, B. 1997. Auxin biosynthesis. Annu. Rev. Plant Physiol. 48: 51–67.

Bartel, B. and Fink, G.R. 1994. Differential regulation of an auxin-producing nitrilase gene family in *Arabidopsis thaliana*. Proc. Natl. Acad. Sci. USA 91: 6649–6653.

Bartel, B. and Fink, G.R. 1995. ILR1, an amidohydrolase that releases active indole-3-acetic acid from conjugates. Science 268: 1745–1748.

Benjamins. R., Quint, A., Weijers, D., Hooykaas, P. and Offringa, R. 2001. The PINOID protein kinase regulates organ development in *Arabidopsis* by enhancing polar auxin transport. Development, 128: 4057–4067.

Bennett, M.J., Marchant, A., Green, H.G., May, S.T., Ward, S.P., Millner, P.A., Walker, A.R., Schultz, B. and Feldmann, K.A. 1996. *Arabidopsis AUX1* gene: a permease-like regulator of root gravitropism. Science 273: 948–950.

Bhalerao, R.P., Eklöf, J., Ljung, K., Marchant, A., Bennett, M. and Sandberg, G. 2002. Shoot derived auxin is essential for early lateral root emergence in *Arabidopsis* seedlings. Plant J. 29: 325–332.

Bialek, K. and Cohen, J.D. 1986. Isolation and partial characterisation of the major amide-linked conjugate of indole-3-acetic acid from *Phaseolus vulgaris* L. Plant Physiol. 80: 99–104.

Bialek, K. and Cohen, J.D. 1989. Free and conjugated indole-3-acetic acid in developing bean seeds. Plant Physiol. 91: 775–779.

Bialek, K. and Cohen, J.D. 1992. Amide-linked indoleacetic acid conjugates may control levels of indoleacetic acid in germinating seedlings of *Phaseolus vulgaris*. Plant Physiol. 100: 2002–2007.

Bialek, L., Michalczuk, L. and Cohen, J.D. 1992. Auxin biosynthesis during seed germination in *Phaseolus vulgaris*. Plant Physiol. 100: 509–517.

Boerjan, W., Cervera, M., Delarue, M., Beeckman, T., Dewitte, W., Bellini, C., Caboche, M., Van Onckelen, H., Van Montagu, M. and Inzé, D. 1995. *superroot*, a recessive mutation in *Arabidopsis*, confers auxin overproduction. Plant Cell 7: 1405–1419.

Casimiro, I., Marchant, A., Bhalerao, R.P., Beeckman, T., Dhooge, S., Swarup, R., Graham, N., Inzé, D., Sandberg, G., Casero, P.J. and Bennett, M. 2001. Auxin transport promotes *Arabidopsis* lateral root initiation. Plant Cell 13: 843–852.

Celenza, J.L. 2001. Metabolism of tyrosine and tryptophan: new genes for old pathways. Curr. Opin. Plant Biol. 4: 234–240.

Chamarro, J., Östin, A. and Sandberg, G. 2001. Metabolism of indole-3-acetic acid by orange (*Citrus sinensis*) flavedo tissue during fruit development. Phytochemistry 57: 179–187.

Chapple, C.C.S., Shirley, B.W., Zook, M., Hammerschmidt, R. and Somerville, S.C. 1994. Secondary metabolism in *Arabidopsis*. In: E.M. Meyerowitz and C.R. Somerville (Eds.) *Arabidopsis*, Cold Spring Harbor Laboratory Press, Plainview, NY, pp. 989–1030.

Chen, R., Hilson, P., Sedbrook, J., Rosen, E., Caspar, T. and Masson, P.H. 1998. The *Arabidopsis thaliana AGRAVITROPIC 1* gene encodes a component of the polar-auxin transport efflux carrier. Proc. Natl. Acad. Sci. USA 95: 15112–15117.

Chen, K.-H., Miller A.N., Patterson, G.W. and Cohen, J.D. 1988. A rapid and simple procedure for purification of indole-3-acetic acid prior to GC-SIM-MS analysis. Plant Physiol. 86: 822–825.

Chevolleau, S., Gasc, N., Rollin, P. and Tulliez, J. 1997. Enzymatic, chemical, and thermal breakdown of H-3-labeled glucobrassicin, the parent indole glucosinolate. J. Agric. Food Chem. 45: 4290–4296.

Chisnell, J.R. and Bandurski, R.S. 1988. Translocation of radiolabeled indole-3-acetic-acid and indole-3-acetyl-myo-inositol from kernel to shoot of *Zea mays* L. Plant Physiol. 86: 79–84.

Christensen, S.K., Dagenais, N., Chory, J. and Weigel, D. 2000. Regulation of auxin response by the protein kinase PINOID. Cell 100: 469–478.

Cleland, R.S. 1995. Auxin and Cell Elongation. In: P.J. Davies (Ed.) Plant Hormones: Physiology, Biochemistry and Molecular Biology, Kluwer Academic Publishers, Dordrecht, Netherlands, pp. 214–227.

Cohen, J.D., Baldi B.G. and Slovin J.P. 1986. 13C(6)[benzene ring]-indole-3-acetic acid – a new internal standard for quantitative mass-spectral analysis of indole-3-acetic acid in plants. Plant Physiol. 80: 14–19.

Cohen, J.D., Slovin, J.P., Bialek, K., Chen, K.-H. and Derbyshire, M. 1988. Mass spectrometry, genetics and biochemistry: Understanding the metabolism of indole-3-acetic acid. In: G.L. Steffens and T.S. Rumsey (Eds.) Beltsville Symposia on Agricultural Research 12, Biomechanisms Regulating Growth and Development, Kluwer Academic Publishers, Dordrecht, Netherlands, pp. 229–241.

Cosgrove, D.J. 2000. New genes and new biological roles for expansins. Curr. Opin. Plant Biol. 3: 73–78.

Davies, R.T., Goetz, D.H., Lasswell, J., Anderson, M.N. and Bartel, B. 1999. IAR3 encodes an auxin conjugate hydrolase from *Arabidopsis*. Plant Cell 11: 365–376.

Delarue, M., Prinsen, E., Van Onckelen, H, Caboche, M. and Bellini, C. 1998. *Sur2* mutations of *Arabidopsis thaliana* define a new locus involved in the control of auxin homeostasis. Plant J. 14: 603–611.

Doerner, P. 2000. Root patterning: does auxin provide positional cues? Curr. Biol. 10: R201–R203.

Domalgaski, W., Schulze, A. and Bandurski, R.S. 1987. Isolation and characterisation of esters of indole-3-acetic acid from the liquid endosperm of the horse chestnut (*Aesculus* species). Plant Physiol. 84: 1107–1113.

Eckardt, N.A. 2001. New insights into auxin biosynthesis. Plant Cell 13: 1–3.

Edlund, A., Eklöf, S., Sundberg, B., Moritz, T. and Sandberg, G. 1995. A microscale technique for gas chromatography-mass spectrometry measurements of picogram amounts of indole-3-acetic acid in plant tissues. Plant Physiol. 108: 1043–1047.

Epstein, E., Cohen, J.D. and Bandurski, R.S. 1980. Concentration and metabolic turnover of indoles in germinating kernels of *Zea mays* L. Plant Physiol. 65: 415–421.

Fischer, C. and Neuhaus, G. 1996. Influence of auxin on the establishment of bilateral symmetry in monocots. Plant J. 9: 659–669.

Frey, M., Chomet, P., Glawischnig, E., Stettner, C., Grun, S., Winklmair, A., Eisenreich, W., Bacher, A., Meeley, R.B., Briggs, S.P., Simcox, K. and Gierl, A. 1997. Analysis of a chemical plant defense mechanism in grasses. Science 277: 696–699.

Frey, M., Stettner, C., Paré, P.W., Schmelz, E.A., Tumlinson, J.H. and Gierl, A. 2000. An herbivore elicitor activates the gene for indole emission in maize. Proc. Natl. Acad. Sci. USA 97: 14801–14806.

Fujita, H. and Syono, K. 1997. PIS1, a negative regulator of the action of auxin transport inhibitors in *Arabidopsis thaliana*. Plant J. 12: 583–595.

Gälweiler, L., Guan, L., Müller, A., Wisman, E., Mendgen, K., Yephremov, A. and Palme, K. 1998. Regulation of polar auxin transport by AtPIN1 in *Arabidopsis* vascular tissue. Science 282: 2226–2230.

Geldner, N., Hamann, T. and Jürgens, G. 2000. Is there a role for auxin in early embryogenesis? Plant Growth Reg. 32: 187–191.

Glawischnig, E., Toams, A., Eisenreich, W., Spiteller, P., Bacher, A. and Gierl, A. 2000. Auxin biosynthesis in maize kernels. Plant Physiol. 123: 1109–1119.

Gopalraj, M., Tseng, T.-S. and Olszewski, N. 1996. The *Rooty* gene of *Arabidopsis* encodes a protein with highest similarity to aminotransferases. Plant Physiol. 111 (suppl.): 114.

Grsic-Rausch, S., Kobelt, P., Siemens, J. M., Bischoff, M. and Ludwig-Müller, J. 2000. Expression and localization of nitrilase during symptom development of the clubroot disease in *Arabidopsis*. Plant Physiol. 122: 369–378.

Hadfi, K., Speth, V. and Neuhaus, G. 1998. Auxin-induced developmental patterns in *Brassica juncea* embryos. Development 125: 879–887.

Hagemeir, J., Schneider, B., Oldham, N.J. and Hahlbrock, K. 2001. Accumulation of soluble and wall-bound indolic metabolites in *Arabidopsis thaliana* leaves infected with virulent or avirulent *Pseudomonas syringae* pathovar tomato strains. Proc. Natl. Acad. Sci. USA 98: 753–758.

Hall, P.J. 1980. The occurrence of indole-3-acetyl-*myo*-inositol in kernels of *Oryza sativa*. Phytochemistry 19: 22121–22123.

Harada, J.J. 1999. Signalling in plant embryogenesis. Curr. Opin. Plant Biol. 2: 23–27.

Helminger, J., Rausch, T. and Hilgenberg, W. 1985. Metabolism of [14C]-indole-3-acetaldoxime by hypocotyls of Chinese cabbage. Phytochemistry 24: 2497–2502.

Helminger, J., Rausch, T. and Hilgenberg, W. 1987. A soluble protein factor from chinese cabbage converts indole-3-acetaldoxime to IAA. Phytochemistry 26: 615–618.

Huchison, K.W., Singer, P.B., McInnis, S., Diaz-Sala, C. and Greenwood, M.S. 1999. Expansins are conserved in conifers and expressed in hypocotyls in response to exogenous auxin. Plant Physiol. 120: 827–831.

Hull, A.K., Vij, R. and Celenza, J.L. 2000. *Arabidopsis* cytochrome P450s that catalyze the first step of tryptophan-dependent indole-

270

3-acetic acid biosynthesis. Proc. Natl. Acad. Sci. USA 97: 2379–2384.

Ilíc, N., Normanly, J. and Cohen, J.D. 1996. Quantification of free plus conjugated indole-3-acetic acid in *Arabidopsis* requires correction for the non-enzymatic conversion of indolic nitriles. Plant Physiol. 111: 781–788.

Ilíc, N., Östin, A. and Cohen, J.D. 1999. Differential inhibition of IAA and tryptophan biosynthesis by indole analogues. I. Tryptophan dependent IAA biosynthesis. Plant Growth Regul. 27: 57–62.

Jackson, R.G., Lim, E.-K., Li, Y., Kowalczyk, M., Sandberg, G., Hoggett, J., Ashford, D.A. and Bowles, D.J. 2001. Identification and biochemical characterization of an *Arabidopsis* indole-3-acetic acid glucosyltransferase. J. Biol. Chem. 276: 4350–4349.

Jensen, P.J. and Bandurski, R.S. 1994. Metabolism and synthesis of indole-3-acetic acid (IAA) in *Zea mays*. Levels of IAA during kernal development and the use of *in vitro* endosperm systems for studying IAA biosynthesis. Plant Physiol. 106: 343–351.

Jensen, P.J. and Bandurski, R.S. 1996. Incorporation of deuterium into indole-3-acetic acid and tryptophan in *Zea mays* seedlings grown on 30% deuterium oxide. Plant Physiol. 147: 679–702.

Klotz, K.L. and Lagrimini, L.M. 1996. Phytohormone control of the tobacco anionic peroxidase promoter. Plant Mol. Biol. 31: 565–573.

Koshiba, T., Kamiya, Y. and Iino, M. 1995. Biosynthesis of indole-3-acetic acid from L-tryptophan in coleoptile tips of maize (*Zea mays* L.). Plant Cell Physiol. 36: 1503–1510.

Kowalczyk, M. and Sandberg, G. 2001. Quantitative analysis of indole-3acetic acid metabolites in *Arabidopsis thaliana*. Plant Physiol., in press.

Kuleck, G.A. and Cohen, J.D. 1992. The partial purification and characterization of IAA-alanine hydrolase from *Daucus carota*. Plant Physiol. 99 (suppl): 18.

Kumar, S.A. and Mahadevan, S. 1963. 3-indoleacetaldoxime hydro-lyase: a pyridoxal-5'-phosphate activated enzyme. Arch. Biochem. Biophys. 103: 516–518.

Lagrimini, L.M. 1991. Peroxidase, IAA oxidase and auxin metabolism in transformed tobacco plants. Plant Physiol. 96 (suppl): 77.

Lagrimini, L.M., Joly, R.J., Dunlap, J.R. and Liu, T.T. 1997. The consequence of peroxidase overexpression in transgenic plants on root growth and development. Plant Mol. Biol. 33: 887–895.

Langdale, J.A. 1998. Cellular differentiation in the leaf. Curr. Biol. 10: 734–738.

Last, R.L., Bissinger, P.H., Mahoney, D.J., Radwanski, E.R. and Fink, G.R. 1991. Tryptophan mutants in *Arabidopsis*: the consequences of duplicated tryptophan synthase genes. Plant Cell 3: 345–358.

Limam, F., Chahed, K., Ouelhazi, N., Ghrir, R. and Ouelhazi, L. 1998. Phytohormone regulation of isoperoxidases in *Catharanthus roseus* suspension cultures. Phytochemistry 49: 1219–1225.

Liu, C.-M., Xu, Z.-H. and Chua, N.-H. 1993. Auxin polar transport is essential for the establishment of bilateral symmetry during early plant embryogenesis. Plant Cell 5: 621–630.

Ljung, K., Östin, A., Lioussanne, L. and Sandberg, G. 2001a. Developmental regulation of indole-3-acetic acid turnover in scots pine seedlings. Plant Physiol. 125: 464–475.

Ljung, K., Bhalerao, R.P. and Sandberg, G. 2001b. Sites and homeostatic control of auxin biosynthesis in *Arabidopsis* during vegetative growth. Plant J., in press.

Lomax, T.L., Muday, G.K. and Rubery, P.H. 1995. Auxin Transport. In: P.J. Davies (Ed.) Plant Hormones: Physiology, Bio-

chemistry and Molecular Biology, Kluwer Academic Publishers, Dordrecht, Netherlands, pp. 509–530.

Ludwig-Müller, J. and Hilgenberg, W. 1988. A plasma membrane-bound enzyme oxidizes L-tryptophan to indole-3-acetaldoxime. Physiol. Plant. 74: 240–250.

Ludwig-Müller, J. and Hilgenberg, W. 1990. Conversion of indole-3-acetaldoxime to indole-3-acetonitrile by plasma membranes from Chinese cabbage. Physiol. Plant. 79: 311–318.

Luschnig, C., Gaxiola, R., Grisafi, P. and Fink, G. 1998. EIR1, a root-specific protein involved in auxin transport, is required for gravitropism in *Arabidopsis thaliana*. Genes Dev. 12: 2175–2187.

Magnus, V., Nigovic, B., Hangarter, R.P. and Good, N.E. 1992. N-(indole-3-ylacetyl)amino acids as sources of auxin in plant tissue culture. Plant Growth Reg. 11: 19–28.

Marchant, A., Kargul, J., May, S. T., Muller, P., Delbarre, A., Perrot-Rechenmann, C. and Bennett, M. J. 1999. AUX1 regulates root gravitropism in *Arabidopsis* by facilitating auxin uptake within root apical tissues. EMBO J. 18: 2066–2073.

Mattsson, J., Sung, Z.R. and Berleth, T. 1999. Responses of plant vascular systems to auxin transport inhibition. Development 126: 2979–2991.

Mayda, E., Marques, C., Conejero, V. and Vera, P. 2000. Expression of a pathogen-induced gene can be mimicked by auxin insensitivity. Mol. Plant-Microbe Interact. 13: 23–31.

Michalczuk, L., Cooke, T. and Cohen, J.D. 1992a. Auxin levels at different stages of carrot somatic embryogenesis. Phytochemistry 31: 1097–1103.

Michalczuk, L., Ribnicky, D.M., Cooke, T.J. and Cohen, J.D. 1992b. Regulation of indole-3-acetic acid biosynthetic pathways in carrot cell cultures. Plant Physiol. 100: 1346–1353.

Mikkelsen, M.D., Hansen, C.H., Wittstock, U. and Halkier, B.A. 2000. Cytochrome P450 CYP79B2 from *Arabidopsis* catalyzes the conversion of tryptophan to indole-3-acetaldoxime, a precursor of indole glucosinolates and indole-3-acetic acid. J. Biol. Chem. 275: 33712–33717.

Mitra, R., Burton, J. and Varner, J.E. 1976. Deuterium oxide as a tool for the study of amino acid metabolism. Analyt. Biochem. 70: 1–17.

Müller, A. and Weiler, E. W. 2000. Indolic constituents and indole-3-acetic acid biosynthesis in the wild-type and a tryptophan auxotroph mutant of *Arabidopsis thaliana*. Planta 211: 855–863.

Müller, A., Guan, C., Gälweiler, L., Tänzler, P., Huijser, P., Marchant, A., Parry, G., Bennett, M., Wisman, E. and Palme, K. 1998a. *AtPIN2* defines a locus of *Arabidopsis* for root gravitropism control. EMBO J. 17: 6903–6911.

Müller, A., Hillebrand, H. and Weiler, E.W. 1998b. Indole-3-acetic acid is synthesised from L-tryptophan in roots of *Arabidopsis thaliana*. Planta 206: 362–369.

Nelson, T. and Dengler, N. 1997. Leaf vascular pattern formation. Plant Cell 9: 1121–1135.

Nonhebel, H.M., Kruse, L.I. and Bandurski, R.S. 1985. Indole-3-acetic acid catabolism in *Zea mays* seedlings. Metabolic conversion of oxindole-3-acetic acid to 7-hydroxy-2-oxindole-3-acetic acid 7'-O-β -D-glucopyranoside. J. Biol. Chem. 260: 12685–12689.

Normanly, J. 1997. Auxin metabolism. Physiol. Plant. 100: 431–442.

Normanly, J. and Bartel, B. 1999. Redundancy as a way of life: IAA metabolism. Curr. Opin. Plant Biol. 2: 207–213.

Normanly, J., Cohen, J.D. and Fink, G.R. 1993. *Arabidopsis thaliana* auxotrophs reveal a tryptophan-independent biosynthetic pathway for indole-3-acetic acid. Proc. Natl. Acad. Sci. USA 90: 10355–10359.

Normanly, J., Grisafi, P., Fink, G.R. and Bartel, B. 1997. *Arabidopsis* mutants resistant to the auxin effects of indole-3-acetonitrile are defective in the nitrilase encoded by the *NIT1* gene. Plant Cell 9: 1781–1790.

Östin, A. 1995. Metabolism of indole-3-acetic acid in plants with emphasis on non-decarboxylative catabolism. Ph.D. dissertation, Department of Forest Genetics and Plant Physiology, SLU, Umeå, Sweden.

Östin, A., Ilíc, N. and Cohen, J.D. 1999. An *in vitro* system from maize seedlings for tryptophan-independent indole-3-acetic acid biosynthesis. Plant Physiol. 119: 173–178.

Östin, A., Kowalczyk, M., Bhalerao, R.P. and Sandberg, G. 1998. Metabolism of indole-3-acetic acid in *Arabidopsis*. Plant Physiol. 118: 285–296.

Ouyang, J., Shao, X. and Li, J. 2000. Indole-3-glycerol phosphate, a branchpoint of indole-3-acetic acid biosynthesis from the tryptophan biosynthetic pathway in *Arabidopsis thaliana*. Plant J. 24: 327–334.

Park, S., Walz, A., Momonoki, Y.S., Slovin, J.P., Ludwig-Müller, J. and Cohen, J.D. 2001. Partial characterization of major amide-linked conjugates of IAA in *Arabidopsis* seed (Abstract 321). Final Program July 2001, American Society of Plant Biologists/Canadian Society of Plant Physiologist meeting, Providence, RI, pp. 81–82.

Pengelly, W.L. and Bandurski, R.S. 1983. Analysis of indole-3-acetic acid metabolism in *Zea mays* using deuterium oxide as a tracer. Plant Physiol. 73: 445–449.

Piotrowski, M., Schönfelder, S. and Weiler, E.W. 2001. The *Arabidopsis thaliana* isogene *NIT4* and its orthologs in tobacco encode β-cyano-L-alanine hydratase/nitrilase. J. Biol. Chem. 27: 2616–2621.

Rajagopal, R. and Larsen, P. 1972. Metabolism of indole-3-acetaldoxime in plants. Planta 103: 45–54.

Rapparini, F., Cohen, J.D. and Slovin, J.P. 1999. Indole-3-acetic acid biosynthesis in *Lemna gibba* studied using stable isotope labeled anthranilate and tryptophan. Plant Growth Regul. 27: 139–144.

Reed, R.C., Brady, S.R. and Muday, G.M. 1998. Inhibition of auxin movement from the shoot into the root inhibits lateral root development in *Arabidopsis*. Plant Physiol. 118: 1369–1378.

Reinecke, D.M. and Bandurski, R.S. 1987. Auxin biosynthesis and metabolism. In: P.J. Davies (Ed) Plant Growth and Development, Martinus Nijhoff, Dordrecht, Netherlands, pp. 24–42.

Reinhardt, D., Mandel, T. and Kuhlemeier, C. 2000. Auxin regulates the initiation and radial position of plant lateral organs. Plant Cell 12: 507–518.

Reintanz, B., Lehnen, M., Reichelt, M., Gershenzon, J., Kowalczyk, M., Sandberg, G., Godde, M., Uhl, R. and Palme, K. 2001. *Bus*, a bushy *Arabidopsis* CYP79F1 knockout mutant with abolished synthesis of short-chain aliphatic glucosinolates. Plant Cell 13: 351–367.

Rekoslavskaya, N.I. 1995. Pathways of indoleacetic acid and tryptophan synthesis in developing maize endosperm: studies *in vitro*. Russ. J. Plant Physiol. 42: 143–151.

Rekoslavskaya, N.I. and Bandurski, R.S. 1994. Indole as a precursor of indole-3-acetic acid in *Zea mays*. Phytochemistry 35: 905–909.

Ribnicky, D.M., Ilíc, N., Cohen, J.D. and Cooke, T.J. 1996. The effect of exogenous auxins on endogenous indole-3-acetic acid metabolism: Implications for somatic embryogenesis in carrot. Plant Physiol. 112: 549–558.

Ribnicky, D.M., Cohen, J.D., Hu, W.-S. and Cooke, T.J. 2001. An auxin surge following fertilization in carrots: a mechanism for regulating plant totipotency. Planta, in press.

Riov, J. and Bangerth, F. 1992. Metabolism of auxin in tomato fruit tissue: formation of high molecular weight conjugates of oxindole-3-acetic acid via the oxidation of indole-3-acetylaspartic acid. Plant Physiol. 100: 1396–1402.

Romano, C.P., Hein, M.B. and Klee, H.J. 1991. Inactivation of auxin in tobacco transformed with the indoleacetic acid-lysine synthetase gene of *Pseudomonas savastanoi*. Genes Dev. 5: 438–446

Sabatini, S., Beis, D., Wolkenfeldt, H., Murfett, J., Guilfoyle, T., Malamy, J., Benfey, P., Leyser, O., Bechtold, N., Weisbeek, P. and Scheres, B. 1999. An auxin-dependent distal organiser of pattern and polarity in the *Arabidopsis* root. Cell 99: 463–472.

Sasaki, K., Shimomura, K., Kamada, H. and Harada, H. 1994. IAA metabolism in embryogenic and nonembryogenic carrot cells. Plant Cell Physiol. 35: 1159–1164.

Savitsky, P.A., Gazaryan, I.G., Tishkov, V.I., Lagrimini, L.M., Ruzgas, T. and Gorton, L. 1999. Oxidation of indole-3-acetic acid by dioxygen catalysed by plant peroxidases: specificity for the enzyme structure. Biochem. J. 340: 579–583.

Schiavone, F.M. and Cooke, T.J. 1987. Unusual patterns of somatic embryogenesis in the domesticated carrot: developmental effects of exogenous auxins and auxin transport inhibitors. Cell Differ. 21: 53–62.

Schmidt, R.C., Müller, A., Hain, R., Bartling, D. and Weiler, E.W. 1996. Transgenic tobacco plants expressing *Arabidopsis thaliana* nitrilase II enzyme. Plant J. 9: 683–691.

Seo, M., Akaba, S., Oritani, T., Delarue, M., Bellini, C., Caboche, M. and Koshiba, T. 1998. Higher activity of an aldehyde oxidase in the auxin-overproducing *superroot1* mutant of *Arabidopsis thaliana*. Plant Physiol. 116: 687–693.

Sieburth, L.E. 1999. Auxin is required for leaf vein patterning in *Arabidopsis*. Plant Physiol. 121: 1179–1190.

Sitbon, F., Åstot, C., Edlund, E., Crozier, A. and Sandberg, G. 2000. The relative importance of tryptophan-dependent and tryptophan-independent biosynthesis of indole-3-acetic acid in tobacco during vegetative growth. Planta 211: 715–721.

Slovin, J.P. 1997. Phytotoxic conjugates of indole-3-acetic acid. Potential agents for biochemical selection of mutants in conjugate hydrolysis. Plant Growth Regul. 21: 215–221.

Steinmann, T., Geldner, N., Grebe, M., Mangold, S., Jackson, C.L., Paris, S., Gälweiler, L., Palme, K. and Jürgens, G. 1999. Coordinated polar localisation of auxin efflux carrier PIN1 by GNOM ARF GEF. Science 286: 316–318.

Swarup, R., Friml, J., Marchant, A., Ljung, K., Sandberg, G., Palme, K. and Bennett, M. 2001. Localisation of the auxin permease AUX1 in the *Arabidopsis* root apex reveals two novel functionally distinct hormone transport pathways. Genes Dev., in press.

Szerszen, J.B., Szczyglowski, K. and Bandurski, R.S. 1994. *iaglu*, a gene from *Zea mays* involved in conjugation of growth hormone indole-3-acetic acid. Science 265: 1699–1701.

Sztein, A.E., Cohen, J.D., Slovin, J.P. and Cooke, T.J. 1995. Auxin metabolism in representative land plants. Am. J. Bot. 82: 1514–1521.

Sztein, A.E., Ilíc, N, Cohen, J.D. and Cooke, T.J. 2001. Indole-3-acetic acid biosynthesis in isolated axes from germinating bean seeds: the effect of wounding on the biosynthetic pathway. Plant Growth Regul., in press.

Tam, Y.Y., Epstein, E. and Normanly, J. 2000. Characterisation of auxin conjugates in *Arabidopsis*. Low steady-state levels of indole-3-acetyl apartate, indole-3-acetyl-glutamate, and indole-3-acetyl-glucose. Plant Physiol. 123: 589–595.

Tam, Y.Y. and Normanly, J. 1998. Determination of indole-3-pyruvic acid levels in *Arabidopsis thaliana* by gas chromatography-selected ion monitoring-mass spectrometry. J. Chromatogr. A 800: 101–108.

Tam, Y.Y., Slovin, J.P. and Cohen J.D. 1995. Selection and characterization of methyltryptophan resistant lines of *Lemna gibba* showing a rapid rate of indole-3-acetic acid turnover. Plant Physiol. 107: 77–85.

Tantikanjana, T., Yong, J.W.H., Letham, D.S., Griffith, M., Hussain, M., Ljung, K., Sandberg, G. and Sundaresan, V. 2001. Control of axillary bud initiation in *Arabidopsis* through the SUPERSHOOT gene. Genes Dev. 15: 1577–1588.

Thimann, K.V. and Mahadevan, S. 1964. Nitrilase. I. Occurrence, preparation, and general properties of the enzyme. Arch. Biochem. Biophys. 105: 133–141.

Tsiantis, M., Brown, M.I.N., Skibinski, G. and Langdale, J.A. 1999. Disruption of auxin transport is associated with aberrant leaf development in maize. Plant Physiol. 121: 1163–1168.

Tsiantis, M. and Langdale, J.A. 1998. The formation of leaves. Curr. Opin. Plant Biol. 1: 43–48.

Tsurumi, S. and Wada, S. 1986. Dioxindole-3-acetic acid conjugates formation from indole-3-acetylaspartic acid in *Vicia* seedlings. Plant Cell Physiol. 27: 1513–1522.

Tuominen, H., Östin, A., Sandberg, G. and Sundberg, B. 1994. A novel metabolic pathway for indole-3-acetic-acid in apical shoots of *Populus tremula* (L.) × *Populus tremuloides* (Michx). Plant Physiol. 106: 1511–1520.

Utsuno, K., Shikanai, T., Yamada, Y. and Hashimoto, T. 1998. *AGR*, an agravitropic locus of *Arabidopsis thaliana*, encodes a novel membrane-protein family member. Plant Cell Physiol. 39: 1111–1118.

Walz, A., Park, S., Momonoki, Y.S., Slovin, J.P., Ludwig-Müller, L., and Cohen, J.D. 2001. A gene encoding a protein modified by the phytohormone indoleacetic acid. (Abstract 326). Final Program July 2001, American Society of Plant Biologists/Canadian Society of Plant Physiologist meeting, Providence, RI, p. 82.

Wright, A.D., Moehlenkamp, C.A., Perrot, G.H., Neuffer, M.G. and Cone, K.C. 1992. The maize auxotrophic mutant *orange pericarp* is defective in duplicate genes for tryptophan synthase ?. Plant Cell 4: 711–719.

Zhao, J. and Last, R.L. 1996. Coordinate regulation of the tryptophan biosynthetic pathway and indolic phytoalexin accumulation in *Arabidopsis*. Plant Cell 8: 2235–2244.

Zhao, Y., Christensen, S.K., Fankhauser, C., Cashman, J.R., Cohen, J.D., Weigel, D. and Chory J. 2001. A role for flavin monooxygenase-like enzymes in auxin biosynthesis. Science 291: 306–309.

Zook, M. 1998. Biosynthesis of camalexin from tryptophan pathway intermediates in cell-suspension cultures of *Arabidopsis*. Plant Physiol. 118: 1389–1398.

Plant Molecular Biology **49**: 273–284, 2002.
Perrot-Rechenmann and Hagen (Eds.), Auxin Molecular Biology.
© 2002 *Kluwer Academic Publishers.*

Polar auxin transport – old questions and new concepts?

Jiří Friml[1,2,*] and Klaus Palme[3]
[1]*Zentrum für Molekularbiologie der Pflanzen, Universität Tübingen, Auf der Morgenstelle 3, 72076 Tübingen, Germany (*author for correspondence; e-mail jiri.friml@zmbp.uni-tuebingen.de);* [2]*Department of Biochemistry, Faculty of Science, Masaryk University, Kotlářská 2, 611 37 Brno, Czech Republic (e-mail friml@chemi.muni.cz);* [3]*Max-Delbrück-Laboratorium in der Max-Planck-Gesellschaft, Carl-von-Linné-Weg 10, 50829 Köln, Germany*

Received 13 September 2001; accepted 22 October 2001

Key words: auxin, polar auxin transport

Abstract

Polar auxin transport controls multiple aspects of plant development including differential growth, embryo and root patterning and vascular tissue differentiation. Identification of proteins involved in this process and availability of new tools enabling 'visualization' of auxin and auxin routes *in planta* largely contributed to the significant progress that has recently been made. New data support classical concepts, but several recent findings are likely to challenge our view on the mechanism of auxin transport. The aim of this review is to provide a comprehensive overview of the polar auxin transport field. It starts with classical models resulting from physiological studies, describes the genetic contributions and discusses the molecular basis of auxin influx and efflux. Finally, selected questions are presented in the context of developmental biology, integrating available data from different fields.

Abbreviations: AEI, auxin efflux inhibitors; *agr*, agravitropic; ARF, auxin response factor; ARF GEF, guanine-nucleotide exchange factor on ADP-ribosylation factor G protein; BFA, Brefeldin A; *eir*, ethylene-insensitive root; IAA, indole-3-acetic acid; 1-NAA, 1-naphtylacetic acid; NPA, 1-*N*-naphthylphthalamic acid; PAT, polar auxin transport; *rcn1*, roots curl in NPA; TIBA, 2,3,5'-triiodobenzoic acid; *tir*, transport inhibitor response

Historical concepts

Polar auxin transport represents a unique process specific to plants and the phytohormone auxin. Already in the 19th century classical experiments studying the phototropism of canary grass coleoptiles indicated the existence of a transmissible signal, which was later identified and termed auxin (Darwin and Darwin, 1881). Further research in this field led to the formulation of the Cholodny-Went hypothesis, which proposed that differential auxin distribution represents the mechanism for tropic growth (summarized in Went, 1974; Firn *et al.*, 2000). Auxin transport experiments using chemical inhibitors revealed the physiological importance of auxin transport and led to the formulation of the chemiosmotic model that explains auxin transport at the cellular level through the action of specific influx and efflux carriers (Rubery and Shel-

drake, 1974; Raven, 1975). Joint efforts of genetics and molecular biology resulted in identification of genes encoding putative auxin influx and efflux carrier proteins represented by *AUX* and *PIN* gene families respectively (Palme and Gälweiler, 2000). Although direct biochemical evidence establishing their function as carriers is still lacking, it is undisputable that AUX and PIN proteins participate in the auxin transport process. Today the availability of new tools for localization of these proteins (Palme and Gälweiler, 2000; Friml *et al.*, 2001b; Swarup *et al.*, 2001) and *in situ* monitoring of auxin accumulation (Sabatini *et al.*, 1999) as well as availability of mutants impaired in auxin transport (Palme and Gälweiler, 2000; Friml *et al.*, 2001b) enable us to give the old physiological theories new molecular meaning and precisely address the role of auxin transport in many developmental processes. So far these studies largely support the widely accepted classical concepts and give a fresh input to old discussions about auxin as a plant morphogen (Sabatini *et al.*, 1999; Friml *et al.*, 2001b). However recent data suggest that auxin efflux carriers may not solely act at the plasma membrane. Based on these data a new view on the mode of auxin transport inhibitors action is proposed (Geldner *et al.*, 2001). These findings are likely to change our long-held views about auxin transport and, therefore, further advances in this field are eagerly awaited.

Physiology of auxin transport

Auxin is thought to be synthesized in plants locally, in young growing regions, predominantly in the shoot apex, young leaves and developing seeds (Normanly *et al.*, 1991; Ljung *et al.*, 2001), but it seems that almost any plant tissue can at certain times be responsive for auxin (Davies, 1995). Already in the 1920s, Cholodny and Went were independently trying to hypothesize how auxin moves from the apex into the elongation zone (Went, 1974).

Two main pathways describe the transport of auxin, a fast, non-directional transport in the phloem and a slower, directional, so-called polar auxin transport (PAT) in various tissues. The evidence for the existence of phloem transport was established through experiments with radioactively labelled auxin (Morris and Thomas, 1978). This transport occurs in both basipetal and acropetal directions, proceeds relatively fast (5–20 cm/h) and seems to correlate well with the transport of assimilates and inactive auxin conjugates (Nowacki and Bandurski, 1980). Direct auxin analysis revealed physiological relevant amounts of free indole-3-acetic acids (IAA) within the phloem exudate (Baker, 2000). Experiments in pea showed that the labeled auxin transported within the phloem was later detected in the PAT system indicating that both transport pathways may be linked (Cambridge and Morris, 1996).

In contrast to phloem transport, PAT is specific for active free auxins, occurs in a cell-to-cell manner and has a strictly unidirectional character. The main PAT stream runs from the apex basipetally with a velocity of 5–20 mm/h towards the base of the plant (Lomax *et al.*, 1995). Using radioactively labelled auxin this kind of transport was mainly detected in the cambium and adjacent, partially differentiated xylem elements (Morris and Thomas, 1978). In roots the auxin stream continues acropetally towards the root

tip, where part of the auxin is redirected backwards and transported basipetally through the root epidermis to the elongation zone (Rashotte *et al.*, 2000). In contrast, PAT in shoots occurs in lateral direction (Morris and Thomas, 1978). Auxin transport assays also revealed that PAT requires energy, is saturable and sensitive to protein synthesis inhibitors. These results suggested the existence of specific auxin transport proteins and led in the middle of the 1970s to the formulation of a coherent model for auxin transport, termed the chemiosmotic hypothesis (Rubery and Sheldrake, 1974; Raven, 1975). In the relatively acidic environment of the cell wall (pH around 5.5) about 15% of IAA exists in its protonated form (IAAH). This non-charged, lipophilic molecule passes easily through the plasma membrane by diffusion. In the more basic cytoplasm (pH around 7) IAAH dissociates and hence the resulting IAA$^-$anion is 'trapped' in the cell due to its poor membrane permeability. Therefore, a specific efflux carrier was postulated and the polarity of the auxin flux was explained by its asymmetric distribution in cells. In addition, the existence of specific auxin influx carriers was hypothesized (Goldsmith, 1977) and later saturable auxin influx, probably working as an IAA$^-$/2H$^+$ co-transporter, was physiologically demonstrated (Lomax *et al.*, 1985; Benning, 1986). Auxin influx and efflux pathways can be physiologically distinguished using auxin efflux inhibitors (AEI). These pharmacological tools arose from correlative exploration of structure-activity profiles of chemicals with auxin-like activity (Katekar and Geissler, 1977) and were demonstrated to inhibit efflux of auxin from cells and hypocotyl segments (reviewed in Rubery, 1990). Most widely recognized AEI such as 1-*N*-naphthylphthalamic acid (NPA) share a benzoic acid ortho-linked with an aromatic ring system. However other very effective inhibitors such as 2,3,5-triiodobenzoic acid (TIBA) do not share this structural pattern. In order to explain the mechanism of inhibition of auxin efflux by AEI, the existence of an NPA-binding protein forming part of the auxin efflux carrier complex was postulated. Additional experiments with auxin efflux and NPA binding studies in the presence of inhibitors of protein synthesis suggested the existence of a third unstable component of the auxin efflux carrier complex likely coupling the NPA-binding protein to auxin transport protein (Morris *et al.*, 1991). Attempts to localize NPA-binding protein within the cell through NPA binding studies led to controversial results favouring either association with cytoskeleton (Cox and Mudday, 1994) or integral membrane localization (Bernasconi *et al.*, 1996). Despite very limited data on the molecular mechanism of AEI function, they have shown to be valuable tools for the establishment of the role of the auxin efflux carrier in plant development. Due to the lack of similar inhibitors of auxin influx the characterization of the corresponding protein complex and elucidation of the role of auxin influx in plant development was retarded. Nevertheless, the recent isolation of compounds such as 1-naphthoxyacetic acid (1-NOA) and 3-chloro-4-hydroxyphenylacetic acid (CHPAA) specifically inhibiting auxin uptake in tobacco culture cells has re-established the balance (Imhoff *et al.*, 2000).

Genetic and molecular analysis of auxin transport

The biochemical attempts to isolate the components of auxin influx and efflux carrier complexes and other proteins involved in PAT have been so far not fully successful. The main contributions concerning their identity come from the genetic field, especially from the analysis of *Arabidopsis* mutants. Different screening strategies have been successfully applied to identify mutants affected in PAT. Some mutants have been selected on the basis of abnormal responses to auxin transport inhibitors or were identified fortuitously in screens for developmental alterations and only later the connection to PAT was discovered. A mutant called *rcn1* was isolated, whose roots curl in the presence of NPA in contrast to straight root growth in wild type (Garbers *et al.*, 1996). This mutant shows a reduction in root and hypocotyl elongation and is also defective for apical hook formation. The *RCN1* gene was cloned and shown to encode a protein phosphatase II A subunit. *RCN1* may control the level of phosphorylation and thereby the activity of a component involved in PAT (Garbers *et al.*, 1996). This hypothesis was corroborated by recent findings that the *rcn* mutant displays enhanced basipetal auxin transport, a phenotype feature, which was also observed in plants treated with the phosphatase inhibitor cantharidin (Rashotte *et al.*, 2001). The *pinoid* (*pid*) mutants display defects in the formation of flowers and cotyledons resembling plants grown on AEI. Moreover, auxin transport in *pid* inflorescences is reduced (Okada *et al.*, 1991). The *PID* gene has been recently cloned and shown to encode a serine-threonine protein kinase, which was proposed to act as a negative regulator of auxin signaling (Christensen *et al.*, 2000). However, another

analyses demonstrate that PID action is sensitive to AEI thus favouring the hypothesis that PID functions as a positive regulator of auxin transport (Benjamins *et al.*, 2001). Thus analysis of both RCN and PID highlights a role of phosphorylation in the regulation of PAT. The *tir* (transport inhibitor response) mutants were isolated on their ability to allow root elongation in the presence of AEI (Ruegger *et al.*, 1997). The mutant called *tir3* displays a variety of morphological defects including reduced elongation of root and inflorescences, decreased apical dominance and reduced lateral root formation. Both auxin transport and NPA-binding activity are reduced in *tir3* mutants (Ruegger *et al.*, 1997), suggesting that the *TIR3* gene may encode the NPA-binding protein or some closely related protein (Hobbie, 1998). The corresponding gene designated as *BIG* has been recently cloned (Gil *et al.*, 2001) and shown to encode a protein with several putative Zn-finger domains homologous to the *Drosophila* Calossin (CalO)/Pushover protein. A defect in CalO interferes with neurotransmitter release in *Drosophila*. The *big* mutation interferes with an effect of auxin efflux inhibition on putative auxin efflux barrier AtPIN1 cycling (Gil *et al.*, 2001) supporting the function of *BIG* in vesicle trafficking, although the mechanism of action has not yet been not clarified.

AUX1 protein: an influx carrier?

Another mutant called *aux1*, which confers a root agravitropic and auxin-resistant phenotype, was instructive for identification of a gene possibly encoding an auxin influx carrier (Bennett *et al.*, 1996). The *aux1* phenotype is consistent with a defect in auxin influx, but similar phenotypes have been observed also in mutants defective in auxin response (Lincoln *et al.*, 1990). The *AUX1* gene encodes a 485 amino acid protein sharing significant similarity with plant amino acid permeases favouring the role for AUX1 in the uptake of the tryptophan-like IAA (Bennett *et al.*, 1996). Despite the fact that final biochemical proof of AUX1 function as an auxin uptake carrier is still lacking, several lines of evidence strongly support the involvement of AUX1 in auxin influx. The strongest support came from a detailed analysis of the *aux1* phenotype. It has been demonstrated that the membrane permeable 1-NAA rescues the *aux1* root agravitropic phenotype much more efficiently than membrane less permeable IAA or 2,4-D and that this rescue coincides with restoration of basipetal auxin transport (Yamamoto and Yamamoto, 1998; Marchant *et al.*, 1999). Moreover, this phenotype including its specific NAA rescue can be mimicked by growing seedlings on recently isolated inhibitors of auxin influx (Parry *et al.*, 2001). The most direct support that AUX1 participates in auxin influx came from the comparison of auxin transport properties of *aux1* and wild-type roots. Uptake assays using radioactively labelled auxins and auxin analogues revealed that *aux1* roots accumulated significantly less 2,4-D than wild-type roots (Marchant *et al.*, 1999). Interestingly, such a difference was not found when the membrane-permeable 1-NAA or the IAA-like amino acid tryptophan were assayed. Recently the AUX1 protein was localized within *Arabidopsis* root tissue by an epitope tagging approach (Swarup *et al.*, 2001). The AUX1 protein was detected in a remarkable pattern in a subset of stele, columella, lateral root cap and epidermal cells exclusively in the root tips (Figure 1). Disruption of AUX1 causes changes in cell-specific auxin accumulation associated with tissues mediating basipetal auxin transport (Swarup *et al.*, 2001; Rashotte *et al.*, 2001). However, *aux1* mutants are also defective in auxin supply to the root tip, since mutant root tips contain less free auxin than those of wild type. This paradox taken together with localization of AUX1 at the upper side of protophloem cells suggest a role of AUX1 protein in unloading of the bulk flow via the protophloem to the root apical meristem (Swarup *et al.*, 2001). Thus AUX1 would appear to provide the first molecular connection between polar and non-polar auxin transport routes.

PIN proteins: efflux carriers?

The knitting needle-like *pin1* mutant phenotype strongly resembles plants treated with inhibitors of auxin efflux. In addition, this mutant displays a strong reduction in basipetal auxin transport (Okada *et al.*, 1991). The *AtPIN1* gene was cloned by transposon tagging and found to encode a 622 amino acid protein with up to 12 putative transmembrane segments. AtPIN1 similarity to a group of transporters from bacteria supports a role of this protein as an auxin efflux carrier (Gälweiler *et al.*, 1998). Alternatively and equally likely on the basis of currently available genetic evidence, it could act as a regulator of auxin transport. Almost simultaneously a homologous gene was found by several laboratories, *At-PIN2/EIR1/AGR1* (Chen *et al.*, 1998; Luschnig *et al.*, 1998; Müller *et al.*, 1998; Utsuno *et al.*, 1998), and cloning and analysis of additional sequences (*AtPIN3*

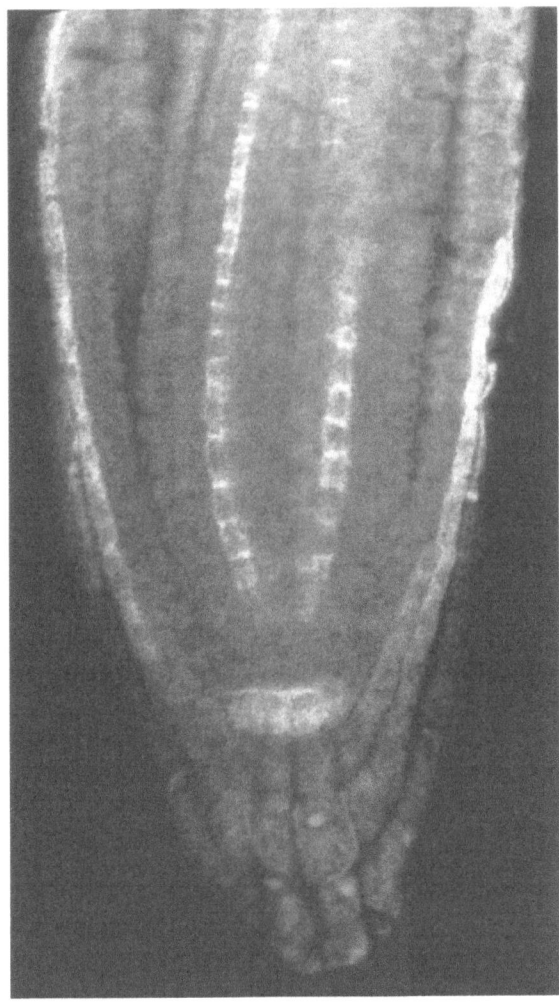

Figure 1. The AUX1 protein localizes to the apical surface of protophloem cells, to the columella and to the lateral root cap.

and *AtPIN4*) followed (Friml *et al.*, 2001a, b). In total, the *Arabidopsis PIN* gene family consists of eight members and homologous genes were found in other plant species such as maize, rice and soybean. As for the role of AUX1 in auxin influx, the proposed function for PIN proteins as efflux carriers has not been biochemically demonstrated to date. However, there are several lines of evidence strongly supporting a role of PIN proteins in auxin transport and particularly in auxin efflux.

AtPIN protein sequences and topology suggest transport function

The AtPIN proteins share more than 70% similarity and have an identical topology – two highly hydrophobic domains with five to six transmembrane segments linked by a hydrophilic region. Transporters of the major facilitator class display similar topology. Moreover, PIN proteins were demonstrated to share limited sequence similarity with prokaryotic and eukaryotic transporters (Palme and Gälweiler, 1999, Müller *et al.*, 1998; Luschnig *et al.*, 1998; Chen *et al.*, 1998; Utsuno *et al.*, 1998) supporting a proposed transport function for PIN proteins.

Yeast over-expressing AtPIN2 are resistant to fluoroindole

To date, the only experimental system used for to address AtPIN transport activity were yeast cells carrying a mutation in the *GEF1* gene resulting in altered ion homeostasis (Gaxiola *et al.*, 1998; Luschnig *et al.*, 1998). Yeast cells defective in this gene and over-expressing AtPIN2 (in these experiments designated EIR1 and AGR1) showed enhanced resistance to the yeast toxin fluoroindole, a substance with some albeit limited structural similarity to auxin (Luschnig *et al.*, 1998). Other experiments demonstrated that AtPIN2 over-expressing yeast retain less radioactively labelled auxin than control yeast (Chen *et al.*, 1998). The decreased accumulation of labelled auxins or toxic auxin analogues may indicate an auxin efflux function for AtPIN2 protein in yeast. Nevertheless, direct measurements of auxin efflux instead of auxin retention have not been demonstrated, leaving this issue unsolved.

AtPIN proteins are localized in a polar manner in auxin transport competent cells

The classical concept of chemiosmotic hypothesis predicted that auxin efflux carriers adopt a polar localization (Rubery and Sheldrake, 1974; Raven, 1975). Such a remarkable localization in accordance with the known direction of auxin flux has been demonstrated for several AtPIN proteins in PAT-competent cells. AtPIN1 protein is localized at the lower side of elongated parenchymatous xylem and cambial cells of *Arabidopsis* inflorescence axes in accordance with the basipetal direction of PAT (Gälweiler *et al.*, 1998; Palme and Gälweiler, 2000). In contrast, the AtPIN2 protein was polarly localized at the upper side of the lateral root cap and epidermis cells (Müller *et al.*, 1998). The AtPIN3 protein localizes predominantly to the lateral side of shoot endodermis cells (Friml *et al.*, 2002a) and the polar localization of AtPIN4 in root directs towards columella initials, the site of auxin accumulation in the root apex (Sabatini *et al.*, 1999; Friml *et al.*, 2002b).

Atpin mutants are defective in PAT

One of the strongest arguments for the involvement of PIN proteins in auxin transport is a reduction of PAT in *Atpin* mutants, which directly correlates with loss of AtPIN expression in corresponding tissue. This was demonstrated for basipetal auxin transport in stem of *Atpin1* mutant or in root of *Atpin2* mutant (Okada *et al.*, 1991; Rashotte *et al.*, 2000).

Disruption of AtPIN function cause changes in cell-specific auxin accumulation

Auxin accumulation has been indirectly monitored by the activity of an auxin-responsive construct (e.g. *DR5::GUS* Sabatini *et al.*, 1999), which seems to correlate very well with direct free IAA measurements (Casimiro *et al.*, 2001; Friml *et al.*, 2002b). Using this approach changes in cell-specific auxin accumulation were found in several *Atpin* mutants which correlated with loss of AtPIN expression in corresponding tissues. These changes were demonstrated for *Atpin2* (in these experiments called *eir1*) mutant roots after changes in gravistimulation (Luschnig *et al.*, 1998) and for *Atpin4* mutant roots and embryos where the defects in establishment and maintenance of the characteristic auxin response maximum were observed (Friml *et al.*, 2002b). The use of radioactively labelled auxins enables more direct determination of auxin content than the visualization by auxin-responsive promotors, but does not provide the cellular resolution. These experiments in root tips of wild type and *Atpin2* (in these experiments called *agr1*) mutants revealed that with radioactive IAA-preloaded *Atpin2* root tips retain more radioactivity than similarly treated wild-type roots (Chen *et al.*, 1998).

Atpin mutants are hypersensitive to 1-NAA

Loss of auxin efflux carrier activity would be predicted to result in an accumulation of its substrate 1-NAA in plant cells, leading to greater sensitivity towards this auxin analogue. Indeed, root growth assays have revealed that several *Atpin2* mutant alleles exhibit an 1-NAA-hypersensitive phenotype, consistent with our predictions (Müller *et al.*, 1998; Parry and Bennett, personal communication).

Phenotypes of Atpin mutants can be phenocopied by auxin efflux inhibitors

Most of the defects observed in *Atpin* mutants are in processes known to be regulated by PAT and can be phenocopied by AEI. These include:

- the *Atpin1* aerial phenotype with defects in stem vasculature (Gälweiler *et al.*, 1998);
- the affected root and shoot gravitropism in *Atpin2* and *Atpin3* mutants, respectively (Maher and Martindale, 1980, Friml *et al.*, 2002a);
- the defect in hypocotyl and root elongation in light, in apical hook opening and in lateral root initiation, which have been reported for *Atpin3* mutants (our unpublished data);
- the *Atpin4* root meristem pattern aberrations, which can be also found in seedlings germinated on low concentrations of AEI (Friml *et al.,* 2002b).

The data accumulated so far, especially from *Atpin* mutants analyses, provides an extensive body of evidence to argue that AtPIN proteins are involved in some important aspects of auxin transport. However, the central question still remains whether PIN proteins represent either transport or a regulatory components. To answer this question, auxin transport assays have to be developed to establish directly carrier functions of different PIN proteins and to determine their substrate specificities, affinities and kinetic properties.

PIN protein dynamics - turning over the concept

To date the identification of polarly localized PIN proteins as putative efflux carriers and the analysis of corresponding knockout mutants appear to agree with the chemiosmotic hypothesis (Rubery and Sheldrake, 1974; Raven, 1975). However, recent studies focused on cell biological requirements of the polar localization of AtPIN1 protein revealed surprising facts, which are difficult to reconcile with this model. These studies even question the specificity of AEI, which represents one of the most valuable tools to study PAT which have been developed in the past decades. Previous studies have revealed that interference with vesicle trafficking either genetically by disrupting the *GNOM* gene or chemically by Brefeldin A (BFA), a fungal toxin, interferes with correct plasma membrane localization of the AtPIN1 protein (Steinmann *et al.*, 1999). More detailed studies in *Arabidopsis* roots have shown that in the presence of BFA the AtPIN1 protein is rapidly and reversibly internalized from the plasma membrane into the endosomal compartment (Figure 2; Geldner *et al.*, 2001). This also occurred in the presence of a protein synthesis inhibitor thereby demonstrating that internalized AtPIN1 originated from the plasma membrane and that AtPIN1 is rapidly cycling between an endosomal compartment and the plasma membrane. The combinative treatment with BFA and Cytochalasin D or Latrunculin B, which are known to disrupt actin, led to the inhibition of this cycling, suggesting that AtPIN1 vesicles are transported along the actin cytoskeleton. The analysis of effects of AEI on AtPIN1 localization revealed that they leave AtPIN1 at the plasma membrane, but block AtPIN1 cycling. Surprisingly, AEI block also the cycling of other functionally unrelated membrane proteins. This suggested a much more general mode for AEI, namely inhibition of membrane protein trafficking.

It is reasonable to assume that the rapidly cycling AtPIN proteins are components very much sensitive to trafficking inhibition. This taken together with importance of AtPIN-dependent PAT in plant development would explain seemingly specific physiological effects of AEI. Most interestingly, these studies show that the very rapid cycling of the AtPIN1 protein seems to be an essential part of the PAT process and AtPIN1 function, since vesicle trafficking inhibitor BFA is also known to rapidly interfere with auxin efflux and could phenocopy the effect of AEI (Morris and Robinson 1998; Delbarre *et al.*, 1998; Geldner *et al.*, 2001). This surprising finding can be hardly incorporated in the old static models in which influx and efflux carrier complexes sit at the plasma membrane and mediate their transport function. The remaining crucial question is the functional relevance of cycling? Different scenarios can be conceived: (1) a high turnover of auxin efflux complexes would provide important flexibility for rapid changes in polarity of plasma membrane localization and thereby for redirection of an auxin flux; (2) if the hypothesis proposed by Hertel would be true and a component of auxin efflux has a dual receptor/transporter function as was proposed for sugar carriers (Lalonde *et al.*, 1999), the cycling might be part of signal transduction and receptor regeneration, as is known for other kinds of receptors (Knutson, 1991); (3) the most exciting scenario would be that the vesicle trafficking itself is part of the auxin transport machinery and that, in analogy with neurotransmitter release, auxin would be a vesicle cargo, released from cells by polar exocytosis. Regardless whether one or more of these scenarios are true, uncovering of cell mechanisms controlling the subcellular dynamics of the auxin carriers seems to be central for our understanding of auxin transport.

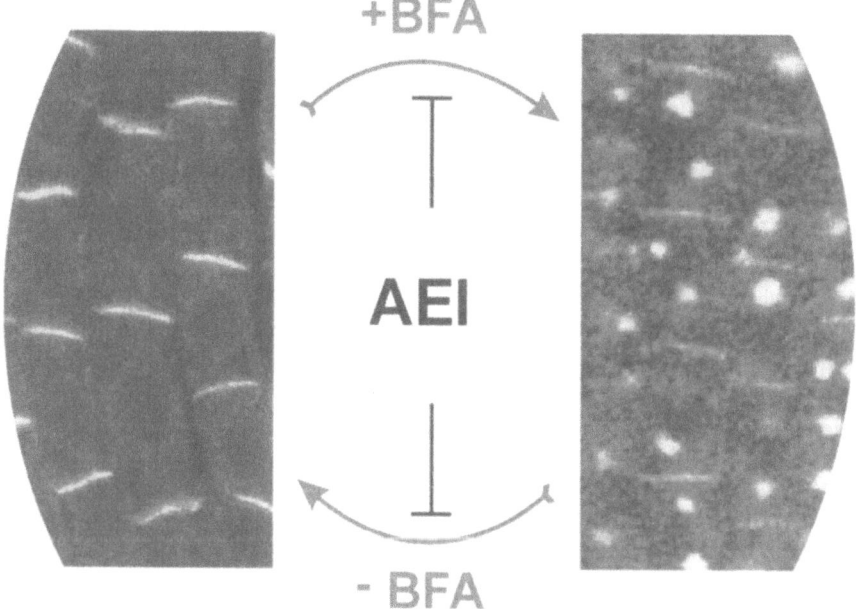

Figure 2. AtPIN1 cycling. The reversible internalization of AtPIN1 label upon Brefeldin A (BFA) treatment suggests that AtPIN1 cycles between the plasma membrane and an endosomal compartment. Polar auxin transport inhibitors (AEI) such as TIBA block AtPIN1 cycling.

The role of polar auxin transport in plant development

The central role of PAT in plant development, especially in tropic growth, vascular development, lateral root initiation, embryo and root patterning as well as apical dominance was initially established by physiological studies using chemical inhibitors of PAT (Lomax *et al.*, 1995). The recent availability of genetic and molecular tools now allows to address more specifically the role of auxin transport in each of these developmental processes.

Vascular tissue development and canalization hypothesis

The manipulation of plants with auxin transport inhibitors drew attention to the role of auxin in vascular development (Camus, 1949). Early grafting experiments demonstrated that buds induced the differentiation of new vascular tissues and the replacement of buds by auxin indicated its importance as the major inductive signal (Wetmore and Rier, 1963). Several reports discussed auxin as a correlative and morphogenetic signal, which influences vascular maturation through a partially self-organizing mechanism (Newcomb *et al.*, 1970; Sachs, 1989). This so-called canalization hypothesis assumes that centres with el-

evated auxin levels gradually gain more competency for auxin conduction and narrowing auxin flow which finally results in differentiation of vascular strands. Auxin transport inhibition experiments have shown that vascular development in leaves and cotyledons of *Arabidopsis* plants grown on different concentration of NPA was severely affected (Mattsson *et al.*, 1999): secondary veins were not developed and, using higher NPA concentrations, the differentiation of the whole leaf vasculature was completely abolished. These studies together with previous studies on vascular regeneration after wounding (Wetmore and Rier, 1963) seemed to be in very good agreement with the canalization hypothesis and provided a strong link between PAT and vascular development. However, recent results of extensive genetic screens in *Arabidopsis* for mutants defective in vascular patterning led to isolation of seven different loci designated *van1–van7* (Koizumi *et al.*, 2000). Most of these mutations led to fragmentation of otherwise fully differentiated cotyledon vasculature without largely affecting the overall architecture of the vascular network thereby favouring a pre-patterning mechanism for vasculature differentiation (Koizumi *et al.*, 2000). The molecular characterization of these mutants and use of molecular markers for visualization of auxin transport routes and early vascular fate identity will certainly contribute to the further clarification of the role of auxin canalization versus prepatterning in vascular tissue differentiation.

Differential growth in shoot: a role for AtPIN3-regulated lateral auxin transport?

Differential growth in shoot (gravitropism, phototropism and apical hook formation) have been proposed to result from the asymmetric redistribution of auxin, which subsequently promotes or inhibits cell growth and elongation resulting in bending (Went, 1974). This model gained strong support by visualization of differentially distributed auxin response reporters in gravity-stimulated tobacco shoots (Li et al., 1991), in Arabidopsis light and gravity stimulated hypocotyls and newly formed apical hooks (Friml et al., 2002a). In this system PAT inhibitors blocked both the asymmetric distribution of auxin response and differential growth, implicating PAT as the process underlying differential auxin distribution and thereby regulating differential growth (Friml et al., 2002a; Lehman et al., 1996). The lateral auxin transport system has been proposed to facilitate the exchange of auxin between the main basipetal stream in vasculature and peripheral regions of shoot, where control of elongation occurs (Epel et al., 1992). The recent characterization of the putative auxin efflux carrier, AtPIN3, recently provided experimental data for this concept (Friml et al., 2002a). The Atpin3 mutants display defects in hypocotyl differential growth. Moreover, AtPIN3 expression and localization studies revealed that the AtPIN3 protein is predominantly localized at the lateral side of endodermis cells further supporting the model of lateral auxin transport.

Root gravitropism: an influx and efflux matter

In the root the sites of gravity perception are columella cells and the site of response is the elongation zone, where elevated auxin levels on the lower side lead to the inhibition of growth and thereby downward bending of the root (for an overview, see Chen et al., this issue). The main contribution of PAT to the root gravitropic response was supposed to be the basipetal transport from columella region through epidermis towards the elongation zone. The identification of several root agravitropic mutants (agr1/eir1/wav6/Atpin2) in a gene encoding the root-specific putative auxin efflux carrier AtPIN2 supports this hypothesis (Chen et al., 1998; Müller et al., 1998; Luschnig et al., 1998; Utsuno et al., 1998). The AtPIN2 protein was localized to the cortex, epidermis and lateral root cap,

predominantly at the upper side of cells, consistent with the model (Figure 3; Müller et al., 1998). The role of AtPIN2 in basipetal auxin transport was further supported by the finding that the auxin response maximum at the lower side of the root after a change of gravity vector is not established and basipetal auxin transport is reduced in Atpin2 mutants (Luschnig et al., 1998; Rashotte et al., 2000). AUX1 is also involved in both regulation of root gravitropism and auxin transport. Aux1 mutants display defects in basipetal auxin transport correlating with root agravitropic phenotype (Swarup et al., 2001). The aux1 mutant phenotype was rationalized by the visualization of AUX1 protein at the plasma membrane of lateral root cap and epidermis cells (Figure 1; Swarup et al., 2001). Thus it seems that both auxin influx and efflux as well as its molecular components are required for root gravitropism. Nonetheless it is still not clear if PAT plays a permissive or regulatory role in root gravitropism and where and how the asymmetry of the auxin distribution is initially established. These questions have recently been addressed by visualization of auxin distribution and auxin carriers in gravistimulated roots. The presence of both putative auxin influx and efflux carriers in columella cells (AUX1, Figure 1, Swarup et al., 2001; AtPIN3,4, Figure 3; Friml et al., 2002a,b) suggests that the asymmetry of auxin distribution may be established there and not during the basipetal transport to the responsive tissue. Moreover, the actin dependent cycling of AtPIN3 and its rapid lateral relocation after gravity stimulation in columella cells provide a likely mechanism for establishment of this asymmetry (Friml et al., 2002a). In this model the sedimentation of statoliths upon gravity stimulation leads to rearrangement of the actin cytoskeleton resulting in relocation of the efflux carrier components. This causes the asymmetric flux of auxin from columella cells and auxin is translocated by basipetal transport to the responsive tissue, where it elicits differential growth.

Root meristem patterning: climbing to the sink

The role of auxin and PAT in roots is not restricted to growth responses. Exogenous manipulation of auxin levels as well as analysis of mutants impaired in auxin signalling demonstrated a role for auxin in regulating the pattern of cell division and differentiation (Kerk et al., 1994; Ruegger et al., 1997) and raised a debate about a possible role of auxin as a morphogen (Sabatini et al., 1999). Similarly to animal morphogens, auxin concentration gradients exist as

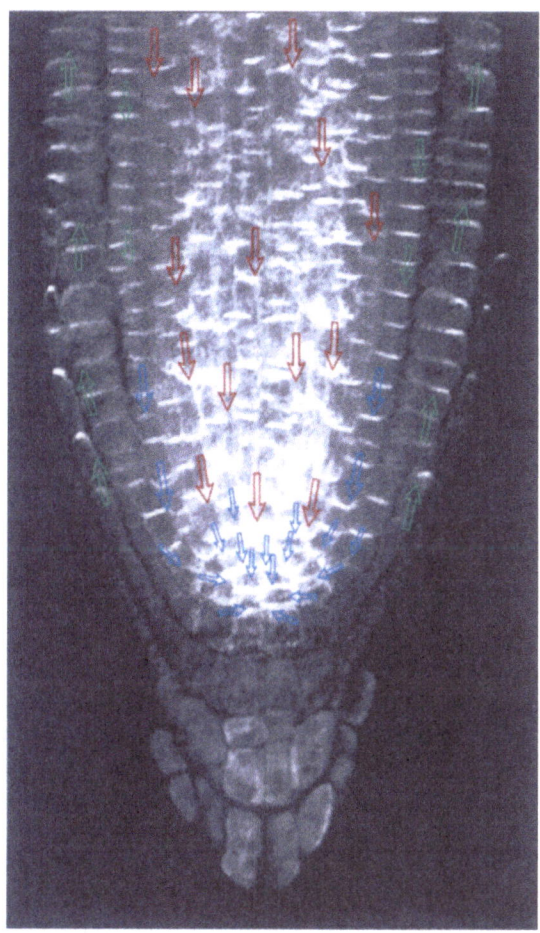

Figure 3. Immunolocalization of AtPIN proteins with presumptive auxin routes in the *Arabidopsis* root tip. AtPIN1 localization suggests auxin flux from apical tissues towards the central root meristem (red arrows), AtPIN4 regulates the focusing of auxin flux into columella establishing the auxin maximum (blue arrows). Part of the auxin is rerouted through outer cell layers backward to the elongation zone regulating root bending (AtPIN2, green arrows).

has been demonstrated by direct auxin measurements in Scots pine (Uggla *et al.*, 1998) as well as by indirect visualization using auxin responsive promoter *DR5* in *Arabidopsis* roots and embryos (Sabatini *et al.*, 1999). In the *Arabidopsis* root the maximum of auxin response was detected in the columella initial cells and first columella layer (Sabatini *et al.*, 1999). Localization studies of the recently characterized AtPIN4 (Figure 3) revealed polar localization pointing towards the same area, suggesting involvement of AtPIN4 in establishment of the auxin gradient with a maximum in columella initials and columella cells. Both chemical inhibition of PAT and the *Atpin4* mutation change the auxin gradient and relocalize its maximum.

These changes in auxin gradients were accompanied with various patterning defects and correlated well with changes in cell fate specification thus providing a genetic link between PAT, auxin gradients and patterning. Morphogen gradients known from animal systems are source-driven. Surprisingly, it appears that the data emerging from an analysis of the role of AtPIN4 and PAT in the root meristem can be best integrated into a model involving an active AtPIN4-dependent sink for auxin in the columella region, implying that auxin gradients are sink-driven. According to this model the At-PIN4 protein promotes auxin transport within the central root meristem tissues and actively increases auxin concentration thereby creating maximum in columella cells. The establishing and proper positioning of this maximum seems to be necessary for local down-regulation of auxin, since both NPA-treated roots and *Atpin4* mutant roots display elevated auxin levels in the root tip and fail to canalize exogenously applied auxin properly (Friml *et al.*, 2002b). However, which mechanisms are involved in auxin down-regulation is unclear. Experiments in maize demonstrated high levels of ascorbate oxidase, an enzyme degrading auxin *in vitro*, in the root tip (Kerk *et al.*, 2000). However, there are no data available favouring oxidative decarboxylation as a means for auxin degradation *in vivo*. Another possibility to remove auxin would be active export or passive leaking of the auxin from the root or by pealing off columella cells loaded with auxin. An interesting mechanism would be the continual trafficking of auxin in the apoplast between columella initial and columella cells which is suggested by AtPIN proteins localization. (Friml *et al.*, 2002b). The longer presence of auxin in the oxidative environment of the apoplast would greatly facilitate auxin degradation (G. Sandberg, personal communication).

Embryonic development: establishment of axis

Exogenous manipulations of auxin levels and PAT in carrot callus-derived somatic embryos as well as *in vitro* cultured Indian mustard (*Brassica juncea* L.) zygotic embryos established a role for PAT in the initiation and maintenance of polarized growth in developing embryos (Schiavone *et al.*, 1987; Hadfi *et al.*, 1998). AEI interfered with cotyledon separation highlighting the important role of auxin and its transport in the progression from the radially symmetric to the bilaterally symmetric embryo in analogy to the regulation of radial position of lateral organs in later development (Reinhardt *et al.*, 2000). Genetic studies using

282

Arabidopsis mutants support a role for PAT in embryo development and suggest a possible role in the establishment of the apical basal axis. The *monopteros (mp)* mutant was isolated as an embryo mutant with defects in apical-basal pattern formation (Berleth and Jürgens, 1993), but adult *mp* plants display a strong reduction in PAT (Przemeck *et al.*, 1996). The *MP* gene encodes an auxin-regulated transcription factor from the ARF family raising the possibility that the MP protein could regulate the expression of components of the PAT machinery (Hardtke *et al.*, 1998). Another mutant called *gnom (gn)* displays defects strongly resembling AEI-treated embryos (Mayer *et al.*, 1993). The *GN* gene encodes a guanine nucleotide exchange factor of ADP-ribosylation factors (ARF GEF), a protein regulating vesicle trafficking to the plasma membrane (Steinmann *et al.*, 1999). It has been demonstrated that gradual polarity establishment of AtPIN1 localization in *gn* embryos is disturbed suggesting an involvement of GN in proper localization of AtPIN1 (Steinmann *et al.*, 1999). However, in view of recent reports suggesting that vesicle trafficking represents an essential part of PAT (Geldner *et al.*, 2001), *gn* may by disrupting vesicle trafficking directly interfere with this process. Despite the physiologically and genetically well established role of PAT in embryogenesis, little is known about auxin transport routes and auxin accumulation in embryos. Due to technical difficulties especially in tiny *Arabidopsis* embryos, we can deduce these only indirectly from the polarity of AtPIN protein localizations and activity of auxin response reporters. Two *AtPIN* genes, *AtPIN1* and *AtPIN4*, are expressed during embryogenesis and can be found localized in both polar and non-polar fashion (Steinmann *et al.*, 1999; Friml *et al.*, 2002b). AtPIN1 was detected already in the 8-cell stage at the inner cell boundaries. The polarity of the AtPIN1 localization is then gradually established at the globular stage at the basal side of provascular cells, where also AtPIN4 protein is localized. The establishment of AtPIN1 and AtPIN4 polarity in provascular initials coincides with that establishment of a basally positioned *DR5* auxin response maximum (Friml *et al.*, 2002b). At the late globular and triangular stage the polar localization of AtPIN1 is also established in epidermis cells towards the position of future outgrowing cotyledons. During further progression towards the torpedo stage the vascular precursor cells become marked at their basal sides forming a typical Y-shaped pattern. These AtPIN1 and AtPIN4 localization patterns suggest auxin flux towards the quiescent centre, below

which an auxin response maximum is formed. The AtPIN4 protein was also detected at the basal side of upper suspensor cells (Friml *et al.*, 2002b), suggesting that auxin homeostasis may be regulated via the suspensor. However, such deduced auxin routes are hypothetical and there are no data available yet on a role of auxin influx carriers during this early stage of development. Thus, the role of PAT in embryogenesis remains still far from clarified. It seems that the auxin homeostasis regulation system during embryogenesis is rather robust, since mutations in *AtPIN1* or *AtPIN4* genes results only in less pronounced and not fully penetrating defects in embryo development (Okada *et al.*, 1991; Friml *et al.*, 2002b).

Outlook

It seems that PAT is now a process very well characterized from a physiological point of view. The use of genetic approaches resulted in an extensive collection of mutants defective in PAT. The main recent breakthrough in the field happened when some of these mutants were characterized at the molecular level, and *AUX1* and *PIN* gene families encoding putative auxin influx and efflux carrier components have been identified. The main task for the near future is the biochemical establishment of auxin transport function and transport properties of these proteins and the identification of their interaction partners. Also the isolation and characterization of remaining members of *AUX* and *PIN* families in *Arabidopsis* is awaited in order to describe the whole network of these carriers in plants, their possible functional redundance and their expression and activity regulation in a developmental context. However, one of the most interesting areas of PAT research lies probably in their cellular analysis. Here the polarly localized PIN and AUX proteins provide ideal tools to study cellular polarity establishment, a process poorly understood in higher plants. For PAT research, however, we are awaiting with curiosity further studies on the relationship between the vesicle trafficking pathway and auxin efflux.

Acknowledgements

We are very grateful to Eva Benková, Malcolm Bennett, Niko Geldner, Matthias Godde, Thorsten Hamann and Gerd Jürgens for helpful discussions and critical reading of the manuscript. We

[34]

also acknowledge support by DAAD (J.F.) and the Schwerpunktprogramme 'Phytohormone' by the Deutsche Forschungsgemeinschaft.

References

Baker, D.A. 2000. Long-distance vascular transport of endogenous hormones in plants and their role in source: sink regulation. Israel J. Plant Sci. 48: 199–203.

Benjamins, R., Quint, A., Weijers, D., Hooykaas, P. and Offringa, R. 2001. The PINOID protein kinase regulates organ development in *Arabidopsis* by enhancing polar auxin transport. Development, 28: 4057–4067.

Bennett, M.J., Marchant, A., Green, H.G., May, S.T., Ward, S.P., Millner, P.A., Walker, A.R., Schulz, B. and Feldmann, K.A. 1996. *Arabidopsis AUX1* gene: a permease-like regulator of root gravitropism. Science 273: 948–950.

Benning, C. 1986. Evidence supporting a model of voltage-dependent uptake of auxin into cucurbita-pepo vesicles. Planta 169: 228–237.

Berleth, T. and Jürgens, G. 1993. The role of the MONOPTEROS gene in organizing the basal body region of the *Arabidopsis* embryo. Development 118: 575–587.

Bernasconi, P., Bhavesh, P.C., Reagan, J.D. and Subramanian, M.V. 1996. The N-naphthylphthalamic acid-binding protein is an integral membrane protein. Plant Physiol. 111: 427–432.

Cambridge, A.P. and Morris, D.A. 1996. Transfer of exogenous auxin from the phloem to the polar auxin transport pathway in pea (*Pisum sativum* L.). Planta 199: 583–588.

Camus, G. 1949. Recherches sur le role de bourgeons dans les phénomènes de morphogènes. Rev. Cytol. Biol. Vég. 11: 1–195.

Casimiro, I., Marchant, A., Bhalerao, R.P., Beeckman, T., Dhooge, S., Swarup, R., Graham, N., Inzé, D., Sandberg, G., Casero, P.J. and Bennett, M. 2001. Auxin transport promotes *Arabidopsis* lateral root initiation. Plant Cell.13: 843–852.

Chen, R., Hilson, P., Sedbrook, J., Rosen, E., Caspar, T. and Masson, P.H. 1998. The *Arabidopsis thaliana AGRAVITROPIC1* gene encodes a component of polar auxin-transport efflux carrier. Proc. Natl. Acad. Sci. USA 95: 15112–15117.

Christensen, S.K., Dagenais, N., Chory, J. and Weigel, D. 2000. Regulation of auxin response by the protein kinase PINOID. Cell 100: 469–478.

Cox, D.N. and Muday, G.K. 1994. NPA binding-activity is peripheral to the plasma membrane and is associated with the cytoskeleton. Plant Cell 6: 1941–1953.

Darwin, C. and Darwin, F. 1881. The power of movement in plants (Deutsche Übersetzung: Das Bewegungsvermögen der Planze). Darwins gesammelte Werke, Bd. 13, Schweizer-bart'sche Verlagsbuchhandlung, Stuttgart, Germany.

Davies, P.J. 1995. Plant Hormones: Physiology, Biochemistry and Molecular Biology. Martinus Nijhoff, Dordrecht, Netherlands.

Delbarre, A., Muller, P. and Guern, J. 1998. Short-lived and phosphorylated proteins contribute to carrier-mediated efflux, but not to influx, of auxin in suspension-cultured tobacco cells. Plant Physiol. 116: 833–844.

Epel, B.L., Warmbrodt, R.P. and Bandurski., R.S. 1992. The ethylene signal transduction pathway in plants. Science 268: 667–675.

Firn, R.D., Wagstaff, C. and Digby, J. 2000. The use of mutants to probe models of gravitropism. J. Exp. Bot. 51: 1323–1340.

Friml, J., Wisniewska, J., Benková, E., Mendger, K. and Palme, K. 2002a. Lateral relocation of auxin efflux regulator AtPIN3 mediates tropism in *Arabidopsis*. Nature: in press.

Friml, J., Benková, E., Blilou, I., Wisniewska, J., Hamann, T., Ljung, K., Woody, S., Sandberg, G., Scheres, B., Jürgens, G. and Palme, K. 2002b. AtPIN4 mediates sink driven auxin gradients and patterning in *Arabidopsis* roots. Submitted.

Gälweiler, L., Guan, C., Müller, A., Wisman, E., Mendgen, K., Yephremov, A. and Palme, K. 1998. Regulation of polar auxin transport by AtPIN1 in *Arabidopsis* vascular tissue. Science 282: 2226–2230.

Garbers, C., DeLong, A., Deruere, J., Bernasconi, P. and Soll, D. 1996. A mutation in protein phosphatase 2A regulatory subunit A affects auxin transport in *Arabidopsis*. EMBO J. 15: 2115–2124.

Gaxiola, R.A., Juan, D.S., Klausner, R.D. and Fink, G.R. 1998. The yeast CLC chloride channel functions in cation homeostasis. Proc. Natl Acad. Sci USA 95: 4046–4050.

Geldner, N., Friml, J., Stierhof, Y.D., Jurgens, G. and Palme, K. 2001. Auxin transport inhibitors block PIN1 cycling and vesicle trafficking. Nature 413: 425–428.

Gil, P., Dewey, E., Friml, J., Zhao, Y., Snowden, K.C., Putterill, J., Palme, K., Estelle, M. and Chory, J. 2001. BIG: a calossin-like protein required for polar auxin transport in *Arabidopsis*. Genes Dev. 15: 1985–1997.

Goldsmith, M.H.M. 1977. The polar transport of auxin. Annu. Rev. Plant. Physiol. 28: 439–478.

Hadfi, K., Speth, V. and Neuhaus, G. 1998. Auxin-induced developmental patterns in *Brassica juncea* embryos. Development 125: 879–887.

Hardtke, C.S. and Berleth, T. 1998. The *Arabidopsis* gene *MONOPTEROS* encodes a transcription factor mediating embryo axis formation and vascular development. EMBO J. 17: 1405–1411.

Hertel, R. 1983. The mechanism of auxin transport as a model for auxin action. Z. Pflanzenphysiol. 112: 53–67.

Hobbie, L.J. 1998. Auxin: molecular genetic approaches in *Arabidopsis*. Plant Physiol. Biochem. 36: 91–102.

Imhoff, V., Muller, P., Guern, J. and Delbarre, A. 2000. Inhibitors of the carrier-mediated influx of auxin in suspension-cultured tobacco cells. Planta 210: 580–588.

Katekar, G.F. and Geisler, A.E. 1977. Auxin transport inhibitors. Plant Physiol. 60: 826–829.

Kerk, N. and Feldman, L. 1994. The quiescent center in roots of maize: initiation, maintenance and role in organization of the root apical meristem. Protoplasma 183: 100–106.

Kerk, N.M., Jiang, K. and Feldman L.J. 2000. Auxin metabolism in the root apical meristem. Plant Physiol. 122: 925–932.

Knutson V.P. 1991. Cellular trafficking and processing of the insulin receptor. FASEB J. 5: 2130–2138.

Koizumi, K., Sugiyama, M. and Fukuda, H. 2000. A series of novel mutants of *Arabidopsis thaliana* that are defective in the formation of continuous vascular network: calling the auxin signal flow canalization hypothesis into question. Development 127: 3197–3204.

Lalonde, S., Boles, E., Hellmann, H., Barker, L., Patrick, J.W., Frommer, W.B. and Ward, J. 1999. The dual function of sugar carriers: transport and sugar sensing. Plant Cell 11: 707–726.

Lehman, A. Black, R. and Ecker, J.R. 1996. *HOOKLESS1*, an ethylene response gene, is required for differential cell elongation in the *Arabidopsis* hypocotyl. Cell 85: 183–194.

Li, Y., Hagen, G. and Guilfoyle, T.J. 1991. An auxin-responsive promoter is differentially induced by auxin gradients during tropisms. Plant Cell 3: 1167–1176.

Lincoln, C., Britton, J.H. and Estelle, M. 1990. Growth and development of the *axr1* mutants of *Arabidopsis*. Plant Cell 2: 1071–1080.

Ljung, K., Ostin, A., Lioussanne, L. and Sandberg, G. 2001. Developmental regulation of indole-3-acetic acid turnover in Scots pine seedlings. Plant Physiol. 125: 464–475.

Lomax, T.L., Mehlhorn, R.J. and Briggs, W.R. 1985. Active auxin uptake by zucchini membrane vesicles: quantitation using ESR volume and ΔpH determinations. Proc. Natl. Acad. Sci USA 82 1986: 6541–6545.

Lomax, T.L., Muday, G.K. and Rubery, P.H. 1995. Auxin transport. In: P.J. Davies (Ed.) Plant Hormones: Physiology, Biochemistry and Molecular Biology, Kluwer, Dordrecht, Netherlands, pp. 509–530.

Luschnig, C., Gaxiola, R., Grisafi, P. and Fink, G. 1998. EIR1, a root specific protein involved in auxin transport, is required for gravitropism in *Arabidopsis thaliana*. Genes Dev. 12: 2175–2187.

Maher, E.P. and Martindale, S.J. 1980. Mutants of *Arabidopsis thaliana* with altered responses to auxins and gravity. Biochem. Genet. 18: 1041–1053.

Marchant, A., Kargul, J., May, S.T., Muller, P., Delbarre, A., Perrot-Rechenmann, C. and Bennett, M.J. 1999. AUX1 regulates root gravitropism in *Arabidopsis* by facilitating uptake within root apical tissue. EMBO J. 18: 2066–2073.

Mattsson, J., Sung, Z.R. and Berleth, T. 1999. Responses of plant vascular system to auxin transport inhibition. Development 126: 2979–2991.

Mayer, U., Buettner, G. and Jürgens, G. 1993. Apical-basal pattern formation in the *Arabidopsis* embryo: studies on the role of the *gnom* gene. Development 117: 149–162.

Morris, D. A. and Robinson, J. 1998. Targeting of auxin carriers to the plasma membrane: differential effects of brefeldin A on the traffic of auxin uptake and efflux carriers. Planta 205: 606–612.

Morris, D.A. and Thomas, 1978. A microautoradiographic study of auxin transport in the stem of intact pea seedlings (*Pisum sativum* L.). J. Exp. Bot. 29: 147–157.

Morris, D.A., Rubery, P.H., Jarman, J. and Sabater, M. 1991. Effects of inhibitors of protein synthesis on transmembrane auxin transport in *Cucurbita pepo* L. hypocotyl segments. J. Exp. Bot. 42: 773–783.

Müller, A., Guan, C., Gälweiler, L., Tänzler, P., Huijser, P., Marchant, A., Parry, G., Bennet, M., Wisman, E. and Palme, K. 1998. AtPIN2 defines a locus of *Arabidopsis* for root gravitropism control. EMBO J. 17: 6903–6911.

Newcomb, W. and Wetherell, D.F. 1970. The effects of 2,4,6-trichlorophenoxyacetic acid on embryogenesis in wild type carrot tissue cultures. Bot. Gaz. 131: 242–245.

Normanly, J., Cohen, J.D. and Fink, G.R. 1991. *Arabidopsis thaliana* auxotrophs reveal a tryptophan-independent biosynthetic pathway for indole-3-acetic acid. Proc. Natl. Acad. Sci. USA 90: 10355–10359.

Nowacki, J. and Bandurski, R.S. 1980. Myo-inositol esters of indole-3-acetic acid as seed auxin precursors of *Zea mays* L. Plant Physiol. 65: 422–427.

Okada, K., Ueda, J., Komaki, M.K., Bell, C.J. and Shimura, Y. 1991. Requirement of the auxin polar transport system in early stages of *Arabidopsis* floral bud formation. Plant Cell 3: 677–684.

Palme, K. and Galweiler, L. 1999. PIN-pointing the molecular basis of auxin transport. Curr. Opin. Plant Biol. 2: 375–381.

Parry, G., Delbarre, A., Marchant, A., Swarup, R., Napier, R., Perrot-Rechenmann, C. and Bennett, M.J. 2001. Novel auxin transport inhibitors phenocopy the auxin influx carrier mutation *aux1*. Plant J. 25: 399–406.

Przemeck, G.K., Mattsson, J., Hardtke, C.S., Sung, Z.R. and Berleth T. 1996. Studies on the role of the *Arabidopsis* gene *MONOPTEROS* in vascular development and plant cell axialization. Planta 200: 229–237.

Raven, J.A. 1975. Transport of indolacetic acid in plant cells in relation to pH and electrical potential gradients, and its significance for polar IAA transport. New Phytol. 74: 163–172.

Rashotte, A.M., Brady, S., Reed, R., Ante, S. and Muday, G.K. 2000. Basipetal auxin transport is required for gravitropism in roots of *Arabidopsis*. Plant Physiol. 122: 481–490.

Rashotte, A.M., DeLong, A. and Muday, G.K. 2001. Genetic and chemical reductions in protein phosphatase activity alter auxin transport, gravity response, and lateral root growth. Plant Cell 13: 1683–1697.

Reinhardt, D., Mandel, T. and Kuhlemeier, C. 2000. Auxin regulates the initiation and radial position of plant lateral organs. Plant Cell 12: 507–518.

Rubery, P.H. 1990. Phytotropins: receptors and endogenous ligands. Symp. Soc. Exp. Biol. 44: 119–146.

Rubery, P.H. und Sheldrake, A.R. 1974. Carrier-mediated auxin transport. Planta 118: 101–121.

Ruegger, M., Dewey, E., Hobbie, L., Brown, D., Bernasconi, P., Turner, J., Muday, G.K. and Estelle, M. 1997. Reduced naphthylphthalamic acid binding in the *tir3* mutant of *Arabidopsis* is associated with a reduction in polar auxin transport and diverse morphological defects. Plant Cell 9: 745–757.

Sabatini, S., Beis, D., Wolkenfelt, H., Murfett, J., Guilfoyle ,T., Malamy, J., Benfey, P., Leyser, O., Bechtold, N., Weisbeek, P. and Scheres, B. 1999. An auxin-dependent distal organizer of pattern and polarity in the *Arabidopsis* root. Cell 99: 463–472.

Sachs, T. 1989. The development of vascular networks during leaf development. Curr. Top. Plant Biochem. Physiol. 18: 168–183.

Schiavone, F.M. and Cooke, T.J. 1987. Unusual patterns of somatic embryogenesis in the domesticated carrot developmental effects of exogenous auxins and auxin transport inhibitors. Cell Differ. 21: 53–62.

Steinmann, T., Geldner, N., Grebe, M., Mangold, S., Jackson, C.L., Paris, S., Gälweiler, L., Palme, K. and Jürgens, G. 1999. Coordinated polar localization of auxin efflux carrier PIN1 by GNOM ARF GEF. Science 286: 316–318.

Swarup, R., Friml, J., Marchant, A., Ljung, K., Sandberg, G., Palme, K. and Bennett, M. 2001. Localisation of the auxin permease AUX1 suggests two functionally distinct hormone transport pathways operate in the *Arabidopsis* root apex. Genes Dev., in press.

Uggla, C., Mellerowicz, E.J. and Sundberg, B. 1998. Indole-3-acetic acid controls cambial growth in scots pine by positional signalling. Plant Physiol. 117: 113–121.

Utsuno, K., Shikanai, T., Yamada, Y. and Hashimoto, T. 1998. AGR, an Agravitropic locus of *Arabidopsis thaliana*, encodes a novel membrane protein family member. Plant Cell Physiol. 39: 1111–1118.

Went, F.W. 1974. Reflections and speculations. Annu. Rev. Plant Physiol. 25: 1–26.

Wetmore, R.H. and Rier, J.P. 1963. Experimental induction of vascular tissues in callus of angiosperms. Am. J. Bot. 50: 418–430.

Yamamoto, M. and Yamamoto, K.T. 1999. Effects of natural and synthetic auxins on the gravitropic growth habit of roots in two auxin-resistant mutants of *Arabidopsis*, *axr1* and *axr4*: evidence for defects in the auxin influx mechanism of *axr4*. J. Plant Res. 112: 391–396.

Plant Molecular Biology **49**: 285–303, 2002.
Perrot-Rechenmann and Hagen (Eds.), Auxin Molecular Biology.
© 2002 *Kluwer Academic Publishers.*

Protein phosphorylation in the delivery of and response to auxin signals

Alison DeLong[1],*, Keithanne Mockaitis[2] and Sioux Christensen[3]
[1]*Department of Molecular Biology, Cell Biology and Biochemistry, Brown University, Providence, RI 02912, USA*
(*author for correspondence; e-mail Alison_DeLong@Brown.edu); [2]Laboratoire de Biologie du Développement*
des Plantes, UMR CNRS 8618, IBP, Bâtiment 630, Université de Paris-Sud, 91405 Orsay Cedex, France;
[3]*Department of Molecular, Cell and Developmental Biology, University of California Los Angeles, Los Angeles,*
CA

Received 25 May 2001; accepted 12 July 2001

Key words: floral development, kinase cascade, MAPK pathway, *PINOID* kinase, polar auxin transport, *rcn1*
mutant

Abstract

The importance of reversible protein phosphorylation in regulation of plant growth and development has been
amply demonstrated by decades of research. Here we discuss recent studies that suggest roles for protein phosphorylation in regulation of both auxin responses and polar auxin transport. Specific kinases act at auxin-requiring
steps in floral and embryonic development, and at the junction(s) between light and auxin signaling pathways
in hypocotyl elongation and phototropism responses. New evidence for rapid mitogen-activated protein kinase
(MAPK) activation by auxin treatment suggests that MAPK cascade(s) might mediate cellular responses to auxin.
Protein phosphorylation also may play a crucial role in regulating the activity or turnover of auxin-responsive
transcription factors. Auxin transport is modulated by phosphorylation, and protein phosphatase activity is involved
in regulation of auxin transport streams in roots. Although the regulatory circuits have not been fully elucidated,
these studies suggest that protein phosphorylating and dephosphorylating enzymes perform key functions in auxin
biology. In some cases, these enzymes act at the intersections between auxin signaling and other signaling
pathways.

Introduction

Reversible protein phosphorylation is a fundamental
post-translational regulatory mechanism, and is involved in controlling a vast array of cellular events
and processes. The genetic, biochemical and molecular analyses of plant growth and development have
yielded a lengthy and rapidly growing list of phenomena controlled by protein phosphorylation, and have
revealed some surprises, such as the importance of
histidine-aspartate phosphorelay systems in plant signal transduction (reviewed in Hardie, 1999; Thomason and Kay, 2000). Genetic screens have identified
crucial kinase and phosphatase activities that are required for perception and/or transduction of ethylene,
cytokinin, abscisic acid and brassinosteroid signals.
Against this backdrop, it is perhaps remarkable that

genetic screens for auxin sensitivity mutants have
identified no protein kinase- or phosphatase-encoding
genes. Several factors may explain this disparity. Kinases and phosphatases crucial to auxin action may
play essential roles or may not be 'dedicated' to auxin-
related functions, leading to lethal or pleiotropic mutant phenotypes that obscure specific roles in auxin
biology. Alternatively, partial functional redundancy
of members of a gene family may impede isolation of
loss-of-function mutants with dramatic alterations in
hormone sensitivity. Finally, protein phosphorylation
enzymes may provide rheostatic regulation, modulating auxin action rather than functioning as on/off
switches.

Nonetheless, recent work suggests that protein
phosphorylation does play important regulatory roles
in the delivery of and response to auxin signals. The

emerging evidence indicates that several kinases and phosphatases perform tissue-specific functions, and may have partners or homologues that perform partially redundant functions. Here we focus on the possible roles of the PINOID kinase, of kinases acting at the interface between light and auxin signaling systems, and of protein phosphatase 2A. In each of these cases, analysis of mutant phenotypes has provided important evidence on the impact of protein phosphorylation in auxin biology. We also discuss evidence for the role of mitogen-activated protein kinase (MAPK) cascades in auxin responses. Although mutations in auxin-activated MAPK components are not yet available, the relevance of this cascade to auxin effects has been demonstrated using an auxin response mutant and an auxin-responsive reporter gene.

Photoreceptor serine/threonine protein kinases and auxin-mediated processes

Although experimental evidence suggesting that auxin signaling plays a fundamental role in a plant's ability to respond to both the direction and the quality of incident light has existed for over a century (Darwin, 1880), until recently the mechanism by which auxin potentiates light-signaling events, and the point of convergence between the light- and auxin signaling pathways has remained largely speculative. Recent molecular genetic and biochemical analyses of photomorphogenic processes, including hypocotyl elongation, phototropism and shade avoidance, have now begun to identify proteins that integrate the two signaling systems (Huala et al., 1997; Tian and Reed, 1999; Colón-Carmona et al., 2000; Harper et al., 2000; Hsieh et al, 2000; Morelli and Ruberti, 2000). Not surprisingly, serine/threonine protein kinases, which figure prominently in responses to external stimuli in other eukaryotes (Hanks et al., 1988), also appear to mediate light-dependent differential growth responses in plants.

While a comprehensive understanding of the interactions between the auxin- and light-signaling pathways has not yet emerged, recent studies indicate that one role of photoreceptor serine/threonine protein kinases involves the regulation of cell- or tissue-specific gene transcription. Semidominant mutations in the *SHY2/IAA3* gene, which encodes a member of the auxin-up-regulated Aux/IAA transcription factor family, result in photomorphogenic development and the expression of light-inducible genes in dark-grown *Arabidopsis* seedlings (Kim et al., 1996; Tian and Reed, 1999). Furthermore, these mutations suppress the hypocotyl elongation phenotypes of plants that carry defects in the phytochrome class of serine/threonine kinase photoreceptor proteins (Kim et al., 1996; Reed et al., 1998a). Analogous gain-of-function mutations in *AXR2/IAA7* and *AXR3/IAA17* also result in similar seedling phenotypes (Nagpal et al., 2000). These observations suggest that for certain photomorphogenic responses, the phytochromes function, directly or indirectly, through the regulation of Aux/IAA activity.

The genetic argument for interaction between the phytochromes and Aux/IAA transcription factors has been corroborated by biochemical data showing that these proteins interact *in vitro*. Co-sedimentation assays using bacterially expressed Aux/IAA proteins from *Arabidopsis* (AXR3/IAA17) and pea (PsIAA4) show that these proteins interact with recombinant oat phytochrome A (PHYA) *in vitro* (Colón-Carmona et al., 2000). Similarly, *in vitro* kinase assays have demonstrated that a number of *Arabidopsis* Aux/IAA proteins are also substrates for PHYA phosphorylation *in vitro* (Colón-Carmona et al., 2000; reviewed by Liscum and Reed, this issue). The possibility that the phytochromes interact directly with a family of proposed transcription factors is consistent with the observation that phytochromes contain nuclear localization sequences and a GUS-PHYB fusion protein exhibits light-dependent nuclear accumulation (Sakamoto and Nagatani, 1996).

The biological relevance of phosphorylation to Aux/IAA activity is unclear, but it is tempting to speculate that the proposed interaction between the light- and auxin-regulated signaling pathways may affect the post-translational regulation of the Aux/IAA family. However, the precise intermolecular interactions that define the convergence of the light- and auxin-signaling pathways are not yet firmly established. For example, because both the Pr and Pfr forms of PHYA are equally active in the *in vitro* phosphorylation assays (Colon-Carmona et al., 2000), a correlation between the observed phosphorylation of Aux/IAA proteins and light activation of the phytochromes has not yet been observed.

The recognition that the ubiquitin/RUB pathway is a key component of the auxin signaling apparatus has identified additional steps in which the regulation of protein stability by phosphorylation is likely to be important. It has been suggested that the predicted repressor of the early response genes, or the short-lived Aux/IAA proteins themselves, may be targeted for ubiquitin-mediated degradation. Furthermore, the recruitment of these proteins into the ubiquitin ligase complex, a step known to require substrate phosphorylation, may be controlled by an auxin up-regulated kinase (Gray and Estelle, 2000; reviewed by Dharmasiri and Estelle, this issue).

New experimental evidence indicates that post-translational modification by RUB conjugation and ubiquitin-mediated proteolysis may regulate other aspects of auxin transport and signaling. *Arabidopsis* seedlings expressing a GUS reporter fused to the promoter and coding region of the root-specific auxin efflux carrier *AGR1/EIR1/PIN2* (EIF1-GUS) exhibited a transcription-independent decrease in GUS activity after treatment with either the synthetic auxin analogue 1-NAA or the auxin transport inhibitor TIBA (Luschnig *et al.*, 1998; Müller *et al.*, 1998; Sieberer *et al.*, 2000). Quantitative GUS assays performed in the presence of cycloheximide showed that the decrease in GUS activity resulted from decreased protein stability, and that protein destabilization required the expression of the *AXR1* gene, which encodes a subunit of the RUB-activating enzyme (Leyser *et al.*, 1993; Seiberer *et al.*, 2000; reviewed by Dharmasiri and Estelle, this issue). Mutations in a second auxin-resistant gene, *AXR4*, which may be genetically linked to the RUB pathway, suppress the activation of MAPK activity in response to auxin application (Hobbie and Estelle, 1995; Mockaitis and Howell, 2000). These data suggest that targeted phosphorylation of proteins destined for ubiquitin-mediated degradation is likely to be an important regulatory step in the response to changes in auxin homeostasis.

A second phenomenon implicating auxin signaling in the response to light perception involves the asymmetric growth of plants in response to directional light. The first molecular evidence linking gradients of auxin distribution to differential gene expression utilized a reporter construct driven by an auxin-responsive soybean promoter derived from the SAUR family of auxin up-regulated genes (Li *et al.*, 1991). Tobacco seedlings exposed to a directional light source or changes in the gravitational field exhibited differential expression of the SAUR::GUS reporter that correlated with stem curvature. In these experiments cells in the more rapidly elongating region of the stem showed 2–5-fold higher levels of GUS activity than cells on the opposing side. This differential expression was blocked when plants were treated with the polar auxin transport inhibitors 1-naphthylphthalamic acid (NPA) or 2,3,5-tri-iodobenzoic acid (TIBA), which also abolish the gravitropic and phototropic responses (Li *et al.*, 1991).

In *Arabidopsis*, an important receptor for blue light-stimulated stem elongation is encoded by the *non-phototropic hypocotyl1* (*NPH1*) locus. The *NPH1* gene has been shown to encode a membrane-associated photoreceptor serine/threonine protein kinase with properties distinct from those of the phytochromes (Liscum and Briggs, 1995; Huala *et al.*, 1997; Christie *et al.*, 1998). In addition to the conserved kinase domain, NPH1 contains two LOV domains, each of which binds a flavin chromophore (reviewed in Briggs *et al.*, 2001). A similar photoreceptor encoded by the *NPL1* gene mediates phototropic responses to blue light at higher fluence doses (Sakai *et al.*, 2001). Although NPH1 undergoes light-dependent autophosphorylation, no other substrates for NPH1 kinase activity have as yet been identified. However, biochemical analyses have shown that NPH1 interacts with the protein encoded by a second phototropic gene, *NPH3* (Motchoulski and Liscum, 1999). Significantly, NPH3 co-localizes with NPH1 to the plasma membrane and appears to undergo post-translational modification in response to blue light irradiation (Motchoulski and Liscum, 1999). Analysis of NPH3 structure reveals that this protein contains multiple protein-protein interaction domains. Liscum and colleagues have suggested that NPH3 may function as a scaffolding protein that anchors an NPH1-activated phosphorelay complex (Motchoulski and Liscum, 1999). A third phototropic mutant, NPH4, that is conditionally required for phototropic responses, encodes a member of the auxin response factor (ARF) protein family (Harper *et al.*, 2000; reviewed by Liscum and Reed, this issue). While NPH4 activity has not been molecularly linked to NPH1 or NPH3, it is likely that NPH4 acts downstream of the proposed membrane-associated phosphorylation complex.

Changes in *PINOID* expression results in auxin-related defects

At least one non-receptor serine/threonine protein kinase, PINOID (PID), is required for the transmission or propagation of auxin signals in the aerial parts of the plant (Figure 1). While mutants carrying weak *pid* alleles exhibit indeterminate inflorescence growth and produce floral meristems, strong *pid* alleles are characterized by a dramatic loss of lateral primordium

A

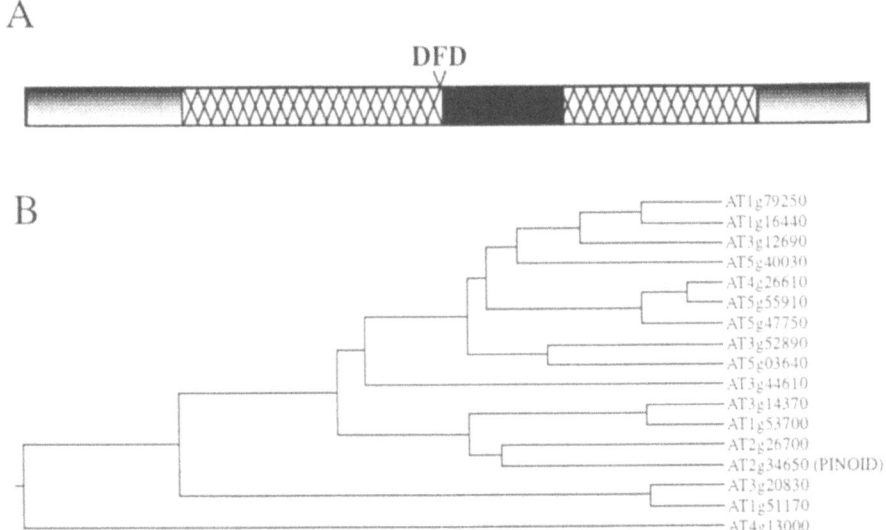

Figure 1. The PINOID-like family of serine/threonine protein kinases. A. Schematic diagram of the protein structure of PINOID and the PID-like kinases. Shaded regions denote the non-conserved amino and carboxy termini. Hatched regions represent the conserved serine/threonine catalytic domains, including subdomains I–XI. Solid black box indicates the non-conserved sequences found between catalytic subdomains VII and VIII in all DFD serine/threonine protein kinases. The position of the plant-specific DFD motif is indicated. B. Cladogram of PID-like kinases. Accession numbers are those of The Institute for Genomic Research (www.tigr.org). Predicted amino acid sequences were aligned by the clustal method using the MegAlign program (DNASTAR).

formation, resulting in a pin-like inflorescence (Bennett *et al.*, 1995). In both weak and strong *pid* mutants, the precise spiral phyllotaxy of wild-type plants is replaced by an apparently random distribution of lateral structures (Bennett *et al.*, 1995). Similarly, both weak and strong *pid* mutants display characteristic defects in floral organ number and patterning (Bennett *et al.*, 1995; Christensen *et al.*, 2000). The mutant flowers have supernumerary petals that are occasionally fused, have fewer stamens than wild type, and exhibit gross morphological defects in gynoecium formation in which the paired valves that make up the outer walls of the ovaries are missing or displaced (Bennett *et al.*, 1995; Christensen *et al.*, 2000). These floral defects are accompanied by a characteristic disruption of the underlying vascular elements similar to that found in *ETT/ARF3* mutants (Sessions and Zambryski, 1995; Sessions *et al.*, 1997; Christensen *et al.*, 2000). Finally, *pid* loss-of-function mutants are also deficient in the ability to establish bilateral symmetry during late embryogenesis (Bennett *et al.*, 1995).

The *pid* phenotypes are similar to but not identical with those associated with mutations at second *Arabidopsis* locus, *PINFORMED1* (*PIN1*). Loss-of-function *pin1* alleles phenocopy the effects of growing wild-type plants on media containing polar auxin transport inhibitors (Okada *et al.*, 1991). Subse-

quently, a third gene, *MONOPTEROS* (*MP*), was identified with inflorescence phenotypes similar to those of the *pid* and *pin1* mutants (Berleth and Jürgens, 1993; Przemeck *et al.*, 1996). *PIN1* and *MP* are now cloned and encode a putative auxin efflux transport protein and a member of the ARF family of transcriptional regulators, respectively (Gälweiler *et al.*, 1998; Hardtke and Berleth, 1998). The similarity of the *pid*, *pin1*, and *mp* phenotypes suggests that PID also regulates steps in either the transport of or response to auxin.

Constitutive expression of *PINOID* in wild-type plants using the CaMV 35S promoter also results in phenotypes consistent with the disruption of auxin transport or signaling. Hemizygous *35S::PID* plants are dwarfs with small, dark green, crinkled leaves decreased internode elongation and decreased apical dominance. While the flowers of *35S::PID* plants are also reduced in size, organ numbers and organ boundary formation appear normal. By contrast, the underlying vasculature patterns of *35S::PID* flowers are highly disorganized (S. Christensen, J. Chory and D. Weigel, unpublished data).

Although *PID* is not required for normal root growth, *35S::PID* plants have severe root defects (Christensen *et al.*, 2000); the roots of hemizygous *35S::PID* seedlings are agravitropic and lack lateral roots. Treatment of *35S::PID* seedlings with either IAA or the membrane diffusible synthetic auxin analogue NAA fails to rescue the primary and lateral root phenotypes. These data indicate that PID is not involved in auxin influx. In the absence of exogenous hormone, *35S::PID* roots are approximately half the length of the wild type. However, increasing NAA concentrations progressively inhibit primary root elongation in both wild-type and *35S::PID* plants (Christensen *et al.*, 2000). At NAA concentrations below 0.5 μM, wild-type and *35S::PID* roots show a proportional decrease in primary root length. At higher NAA concentrations, the root phenotype of these plants is indistinguishable from that of the wild type. Expression of a GUS reporter construct under the control of the DR5 promoter, which contains a multimerized synthetic auxin-response element (Ulmasov *et al.*, 1997b; Sabatini *et al.*, 1999), is reduced in *35S::PID* plants in response to both endogenous and exogenously applied auxin when compared to expression levels in wild-type plants (Christensen *et al.*, 2000).

The phenotypes observed in plants over-expressing *PID* are consistent with the proposed role of PID as a regulator of auxin function. There are at least two explanations for the observation that *35S::PID*, but not *pid*, plants exhibit root phenotypes. First, PID may have a previously undetected role in root development that is masked in *pid* mutants by the redundant activity of an analogous kinase in the root. Alternatively, the ectopic expression of PID in the root may interfere with the normal function of a root-specific PID homologue. Both hypotheses raise the possibility that tissue-specific *PID* homologues may exist.

PINOID encodes a member of a plant-specific serine/threonine protein kinase family

PID encodes a protein with homology to several classes of serine/threonine protein kinases (S/TPKs). PID contains the eleven subdomains typical of the protein kinase catalytic domain (Hanks *et al.*, 1988). Within the plant kinases PID falls into the PVPK-1 subfamily, which includes soybean and rice PID homologues of unknown function and is closely related by sequence to the calcium-dependent protein kinases (Hardie, 1999). Despite the high level of sequence identity with known S/TPKs, the PID sequence differs from its fungal and metazoan counterparts in two respects. Comparison of the PID amino acid sequence with a consensus sequence compiled from more than

65 yeast and metazoan protein kinases (Hanks *et al.*, 1988) revealed that PID contains only 13 of the 14 invariant amino acids that define the classic protein kinase catalytic domain (Christensen *et al.*, 2000). The single exception is the replacement of a glycine at position 225 with aspartate, changing the conserved DFG motif to DFD. However, the substitution in this highly conserved amino acid triplet, which has been implicated in ATP binding (Hanks *et al.*, 1988), does not abolish kinase activity as PID is able to autophosphorylate in *in vitro* kinase assays (Christensen *et al.*, 2000). A second difference highlighted by the alignment of the predicted PID amino acid sequence with the protein kinase consensus sequence is the insertion of a stretch of ca. 50 amino acids between the conserved subdomains VII and VIII, which roughly bisects the conserved catalytic region (Christensen *et al.*, 2000). This region does not contain similarity to known protein motifs, and the significance of this region is currently unknown.

BLAST analysis of sequences deposited in GenBank and the *Arabidopsis thaliana* database reveals that the DFD motif is present in a subset of S/TPKs from most or all plant families represented, including rice, maize, soybean and *Brassica* (S. Christensen, unpublished data) as well as in the green algae *Chlamydomonas* (P. Lefebvre, personal communication). Although widespread in plants, the DFD motif is not found in fungal or metazoan S/TPKs, indicating that this family of protein kinases probably arose shortly after the plant and animal lineages diverged.

In *Arabidopsis* the predicted DFD proteins can be divided into at least two major sub-families. The first class consists of putative membrane-associated receptor kinases, only two of which, NPH1 and its paralogue NPH1-like (NPL1), have been associated with mutant phenotypes (reviewed in Briggs *et al.*, 2001). In addition to the NPH1-like subfamily, at least 17 members of a second highly conserved DFD gene family that includes PID are predicted to encode soluble proteins. Members of this DFD kinase class all contain an insertion between subdomains VII and VIII analogous to that found in PID. Interestingly, although present in each of the PID-like kinases, the sequences within this domain are not conserved, even between closely related family members. Significantly, none of the *PID*-like genes have been identified genetically, supporting the idea that the functions of these proteins may be overlapping.

Interpreting the *pid* phenotype

The primary defect in *pid* loss-of-function mutants appears to be the failure to properly specify lateral meristem position and promote lateral meristem proliferation. Consistent with this hypothesis, those lateral meristems that do form develop flowers with the wild-type complement of cell types and organs in the appropriate spatial context, indicating that PID is not required for determining tissue or organ identity. However, these flowers exhibit defects in establishing organ boundaries within individual floral whorls and in the positioning of specific tissues within specific organs. The stochastic occurrence of meristem and organ defects in *pid* mutants suggests that slight fluctuations in the local cellular environment may dramatically affect PID activity. Consistent with this idea, the *PID* promoter contains a consensus auxin response element, TGTCTC, 240 bp upstream of the transcription start site (Ulmasov *et al.*, 1997a; S. Christensen, unpublished data), indicating that the induction or derepression of *PID* expression may involve ARF binding.

Auxin is transported basipetally from its site of synthesis in or near the plant apex towards the roots through cell files containing asymmetrically distributed transport proteins (reviewed by Friml and Palme, this issue). Transport assays following the movement of ^{14}C-labeled IAA through excised stem fragments have been used to demonstrate decreased polar auxin flow in auxin transport mutants such as the *pin1* mutant of *Arabidopsis* (Okada *et al*, 1991; Bennett *et al.*, 1995). Bennett and associates found that in young plants (19–26 days) carrying strong *pid* alleles auxin transport was decreased to levels ca. 75% of those recorded for wild-type plants, a decrease significantly less that that observed in *pin1* mutants, where transport levels fell to 40–50% of those of wild type. However, polar transport levels in older *pid* plants (35 days), in which inflorescence elongation was complete, dropped to only 7% of wild-type controls (Bennett *et al.*, 1995). Because this dramatic decrease in polar auxin transport was also induced by decapitating wild-type plants, the authors concluded that the observed reduction in transport might be an indirect consequence of reduced growth at the inflorescence apex of *pid* mutants (Bennett *et al.*, 1995). Recently, Offringa and colleagues have found a 50% reduction in auxin transport levels in plants carrying an allele of *pid* disrupted by a transposon insertion. In addition, they also observed a 10–20% increase in polar auxin transport in lines constitutively expressing the PID kinase (R. Offringa, personal communication). This reciprocal relationship between PID expression levels and polar transport suggests that PID may directly or indirectly regulate the rate or extent of auxin transport.

Thus, the current genetic and molecular data involving the role of PID in auxin activity have not allowed an unambiguous interpretation of PID function. The phenotypic similarities between the *pid*, *pin* and *mp* loss-of-function phenotypes suggest that these proteins may affect a common process. A more detailed understanding of the relationship between auxin transport and the activity of transcription factors, such as MP, will be required to unravel the causal nature of the mutant phenotypes. Similarly, over-expression of the PID protein has failed to distinguish between alternative models for PID function in auxin-mediated processes. The constitutive expression of *PID* in roots results in phenotypes most similar to those of auxin-insensitive mutants such as *axr1*, *axr2* and *axr4* (Lincoln *et al.*, 1990; Timpte *et al.*, 1994, 1995; Hobbie and Estelle, 1995; Nagpal *et al.*, 2000), or to the effects elicited by growing plants on auxin transport inhibitors (Reed *et al.*, 1998b). Furthermore, the down-regulation of *DR5::GUS* expression in *35S::PID* plants indicates that auxin concentrations may be dramatically reduced in the root apex (Christensen *et al.*, 2000; Offringa *et al.*, personal communication). PID may normally act to repress auxin transport or responses in the root. A second possibility is that constitutive PID expression may result in the inappropriate spatial localization of auxin pools within root tissues. Finally, ectopic expression of *PID* in the root might interfere with the activity of a root-specific PID-like kinase. The identification of proteins that interact directly with PID and a careful characterization of the effects of phosphorylation on those interacting proteins will be required to precisely define the role of *PID* in regulating auxin activity.

Reversible protein phosphorylation in regulation of auxin transport

Our understanding of the machinery involved in polar auxin transport has improved dramatically, with the recent identification of genes encoding a putative auxin influx carrier (AUX1; Bennett *et al.*, 1996; Marchant *et al.*, 1999) and a family of putative auxin efflux carriers including the PIN1 and AGR1/EIR1/PIN2 proteins (Chen *et al.*, 1998; Gälweiler *et al.*, 1998;

Luschnig *et al.*, 1998; Müller *et al.*, 1998; Utsuno *et al.*, 1998; reviewed by Friml and Palme, this issue). These carriers are believed to facilitate entry of membrane-permeant protonated auxin molecules into cells (AUX1), and to transport membrane-impermeant auxin anions outward across the plasma membrane (efflux carriers PIN1 and AGR1/EIR1/PIN2, etc.). Existing data indicate that the efflux step confers polarity on auxin transport (reviewed in Lomax *et al.*, 1995), and efflux may be the more highly regulated component of the transport process. Studies employing polar auxin transport inhibitors led to the proposal that auxin efflux is mediated by a complex of two or more proteins, with separate auxin-carrier and regulatory components (reviewed in Muday, 2000). Several synthetic compounds such as NPA block polar auxin transport and auxin-dependent growth responses by inhibiting efflux, although auxins do not compete with these inhibitors for binding sites (reviewed in Rubery, 1990). Endogenous metabolites such as flavonoids have been hypothesized to act as natural auxin transport regulators through interaction with the NPA-binding component (Jacobs and Rubery, 1988), and recent data suggest that endogenous flavonoid compounds do regulate auxin transport (Murphy *et al.*, 2000; Brown *et al.*, 2001). Efflux is sensitive to inhibitors of protein synthesis and trafficking, indicating that at least one component of the efflux apparatus is short-lived (Morris *et al.*, 1991; Delbarre *et al.*, 1998; Morris and Robinson, 1998). Specific inhibitors of the influx carrier have been identified recently, and may allow a more detailed analysis of influx carrier function (Parry *et al.*, 2001; Rahman *et al.*, 2001).

Additional data suggest that protein phosphorylation regulates auxin efflux. Noting that three flavonoid compounds identified as possible endogenous regulators inhibit the activity of protein tyrosine kinases, Bernasconi (1996) hypothesized that the NPA-binding protein may have tyrosine kinase activity. Both the flavonoid genistein and the tyrosine kinase inhibitor tyrphostin A47 reduce NPA binding to isolated plasma membranes, and antagonize NPA effects in an auxin accumulation assay in zucchini hypocotyls. Neither compound dramatically affects auxin accumulation directly, suggesting that tyrosine phosphorylation may affect the NPA-binding protein rather than the auxin carrier itself (Bernasconi, 1996). In another study, the effects of kinase and phosphatase inhibitors were assayed on auxin influx and efflux carrier activities in suspension-cultured tobacco cells. The broad-spectrum kinase inhibitors staurosporine and K252a

inhibit auxin efflux, while an inhibitor of protein kinase C and a calmodulin antagonist are ineffective in this assay (Delbarre *et al.*, 1998). These studies clearly suggest roles for both tyrosine-specific and serine/threonine-specific kinase activities in regulation of efflux and its sensitivity to NPA. The pharmacological evidence for protein phosphatase involvement has been more equivocal. The phosphatase inhibitors calyculin A and cantharidin inhibit efflux and also reduce influx slightly. However, two other phosphatase inhibitors, microcystin-LR and okadaic acid, have negligible effects on influx and efflux (Delbarre *et al.*, 1998). The basis for this differential sensitivity to phosphatase inhibitors is unclear. All four phosphatase inhibitors affect protein phosphatase 1 (PP1) and protein phosphatase 2A (PP2A) activities, but microcystin-LR and calyculin A are equally potent inhibitors of both phosphatases, while cantharidin and okadaic acid preferentially inhibit PP2A (MacKintosh and MacKintosh, 1994). Thus phosphatase inhibitor effectiveness in these influx and efflux assays does not correlate with effectiveness against specific phosphatase targets.

Given the prevalence of phosphorylation-based regulatory mechanisms, it hardly seems surprising that auxin transport may be modulated by phosphorylation. To date, though, few specific connections between components of the auxin transport machinery and protein phosphorylating or dephosphorylating activities have been demonstrated. Future studies of auxin carriers may be expected to reveal whether carrier complex function is directly regulated by phosphorylation or through an indirect connection. In either case, regulation by reversible phosphorylation may provide a ready means to integrate signals from many other pathways to provide appropriate changes in auxin transport, in response to specific developmental events, hormonal signals, wounding, etc. For instance, ethylene treatment can affect polar auxin transport (Suttle, 1988; Lee *et al.*, 1990) while wounding may disrupt the normal transport stream, and both ethylene treatment and wounding initiate kinase cascades (reviewed in Johnson and Ecker, 1998). Relatively few experiments have explored the impact of reversible protein phosphorylation on polar auxin transport in whole plants. Inhibitors were used to implicate protein phosphorylation in the gravitropic response of oat pulvini (Chang and Kaufman, 2000), and a calcium/calmodulin-dependent protein kinase activity in light-regulated root gravitropism in maize

(Lu and Feldman, 1997). However, the mechanistic roles of the inhibitor targets are entirely unknown.

PP2A and auxin transport: growth phenotypes of the *rcn1* mutant

Phenotypic analysis of the *Arabidopsis rcn1* mutant indicates a role for PP2A activity in the regulation of polar auxin transport. The *rcn1* mutant was isolated in a screen for mutants that maintain a differential cell elongation response (root curling) in the presence of NPA. In the absence of NPA, mutant seedlings exhibit defects in apical hypocotyl hook formation, root waving and root curling, three growth responses requiring polar auxin transport (Garbers *et al.*, 1996). Although mutant seedlings show reduced organ elongation in the absence of growth regulators or inhibitors, they exhibit normal sensitivity to NPA and TIBA, and to the synthetic auxin 2,4-dichlorophenoxyacetic acid in root elongation assays. These data suggest that regulation of auxin transport, rather than auxin response or NPA sensitivity, is altered in the mutant. The *rcn1* mutant carries a T-DNA insertion in a gene encoding a regulatory A subunit of PP2A. Function of the *RCN1* gene product as a PP2A subunit was initially tested in a yeast complementation assay, and the effect of *rcn1* on PP2A activity in *Arabidopsis* was subsequently demonstrated (Garbers *et al.*, 1996; Deruère *et al.*, 1999; see below).

PP2A is highly conserved among eukaryotes, and is a heterotrimeric enzyme in which the catalytic (C) subunit is bound by 'dedicated' A and B regulatory subunits. The A subunit acts as a scaffold for the holoenzyme complex (Ruediger *et al.*, 1994; Groves *et al.*, 1999), and also has direct regulatory effects on the activity of the catalytic subunit (Price and Mumby, 2000). B subunits are far more heterogeneous than A or C subunits, and are encoded by three unrelated gene families with protein products ranging from 54 to 130 kDa. The binding of different B subunits alters PP2A activity towards specific substrates *in vitro* and *in vivo* (Mayer-Jaekel and Hemmings, 1994; Shu *et al.*, 1997). The A, B and C subunits of PP2A are each encoded by gene families in *Arabidopsis* (Ariño *et al.*, 1993; Casamayor *et al.*, 1994; Rundle *et al.*, 1995; Corum *et al.*, 1996; Stamey *et al.*, 1996; Latorre *et al.*, 1997; Haynes *et al.*, 1999). There are three A, five C, and at least twelve different B subunit genes, in principle allowing the production of a dizzying number of different PP2A holoenzymes.

Some isoforms may be expressed in a cell-, tissue- or stage-specific manner; reporter gene fusions reveal that promoters for *RCN1* (Aα) and the genes encoding the C2, B55α and B55β subunits exhibit tissue specificity in seedling and floral expression patterns (Deruère *et al.*, 1999; Thakore *et al.*, 1999). The *rcn1* mutant phenotype clearly indicates that *RCN1* function is not supplied by the two other A subunit genes, but expression of those genes has been assayed only by RNA blot analysis. Thus the extent of overlap in gene expression patterns is unknown, and it is not yet clear whether functional specification depends on amino acid sequence differences or gene-specific expression patterns among the A subunits.

PP2A activity has been implicated in biological events ranging from regulation of cell cycle transitions, cell division and MAPK signaling cascades, to regulation of metabolic enzyme activities (reviewed in Smith and Walker, 1996; Millward *et al.*, 1999; Virshup 2000). In line with this apparent substrate promiscuity, PP2A activity shows relatively little specificity *in vitro*, and it is thought that activity *in vivo* is targeted and regulated through protein-protein interactions and post-translational modifications. For instance, the identification of several kinases that bind to PP2A suggests that some PP2A functions may be directed and facilitated by co-localization of antagonistic kinase and phosphatase enzymes within specific signaling complexes (Millward *et al.*, 1999). Existing evidence from yeast, *Drosophila* and mouse systems indicates that PP2A function is essential in eukaryotes (Ronne *et al.*, 1991; Wassarman *et al.*, 1996; Götz *et al.*, 1998). In interpreting the *rcn1* phenotype, it is important to remember that the *rcn* mutation is likely to affect only a subset of PP2A functions overall, but that these functions may nonetheless affect multiple mechanisms or pathways.

Reduced PP2A activity in *rcn1* mutant seedlings

In organ elongation assays, roots and etiolated hypocotyls of *rcn1* seedlings are 20–30% shorter than those of wild-type plants; hypocotyls of light-grown *rcn1* seedlings are normal (Deruère *et al.*, 1999). The *rcn1* defects in root and hypocotyl elongation, hypocotyl hook formation, root waving and root curling can be mimicked by treating wild-type seedlings with the phosphatase inhibitor cantharidin. This indicates that these defects are caused by decreased PP2A activity *in vivo*. Enzymatic assays *in vitro* confirm that extracts from mutant seedlings contain 45–50% less PP2A activity than is present in extracts from wild-type seedlings. Thus mutant seedlings show reduced PP2A activity *in vivo* and *in vitro*, indicating that the RCN1 A subunit normally performs a positive regulatory role (see Figure 2). Decreased PP2A activity in *rcn1* seedlings is not due to C subunit degradation, as C subunit protein is present at normal levels in mutant seedlings. In seedling roots, RCN1 mRNA is expressed in root tips, lateral root primordia, and cells of the pericycle and stele, as shown by both *in situ* hybridization and an *RCN1-GUS* fusion construct (Deruère *et al.*, 1999). A similar pattern of *RCN1-GUS* expression is also observed in the roots of adult plants, suggesting that the role of *RCN1* in root growth is not limited to seedling stages (J. Song and A. DeLong, unpublished results). *RCN1* is also expressed in the basal portion of etiolated hypocotyls, but hypocotyl expression of *RCN1* is greatly reduced in light-grown seedlings (Deruère *et al.*, 1999). As noted above, the *rcn1* mutation reduces hypocotyl elongation only in etiolated seedlings, while root elongation is defective both in light- and dark-grown mutant seedlings. Thus the *RCN1* gene is expressed in specific cell types in roots and in dark-grown hypocotyls, and loss of the *RCN1* A subunit decreases seedling PP2A activity required for elongation in those same organs (Deruère *et al.*, 1999).

Auxin transport defects in *rcn1* roots

The growth phenotypes of *rcn1* mutant seedlings suggest alterations in auxin transport. Mutant seedlings exhibit increased sensitivity to NPA in an auxin accumulation assay with hypocotyl tissue. However, this assay measures accumulation rather than polar transport, and furthermore, relevant *rcn1* phenotypes are manifested in the root (Garbers *et al.*, 1996). Specific assays for auxin transport in *Arabidopsis* roots (Reed *et al.*, 1998b; Rashotte *et al.*, 2000) now allow for more direct analysis of *rcn1* auxin transport phenotypes.

With recently developed methods, two auxin transport streams can be measured in *Arabidopsis* roots (Figure 3). Acropetal transport (from the root/shoot junction towards the root tip) is required for lateral root growth (Reed *et al.*, 1998b). Basipetal transport (from the root tip towards the root/shoot junction) is required for gravity response (Rashotte *et al.*, 2000), and may be required for early cell divisions dur-

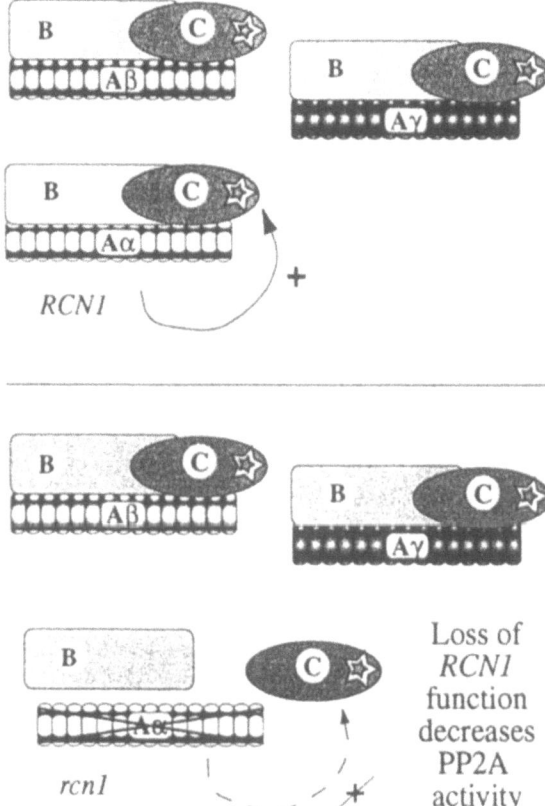

Figure 2. RCN1 positively regulates PP2A activity. The *RCN1* gene encodes a regulatory A subunit (shown as Aα) of the heterotrimeric PP2A enzyme, and in wild-type seedlings this protein positively regulates PP2A activity (arrow, upper panel). *In vivo* and *in vitro* assays show that PP2A activity is decreased in *rcn1* mutant seedlings, which lack the RCN1 A subunit (lower panel). The A subunit is required for interaction of the catalytic (C) and B regulatory subunits. Stars represent the active sites on the catalytic subunits.

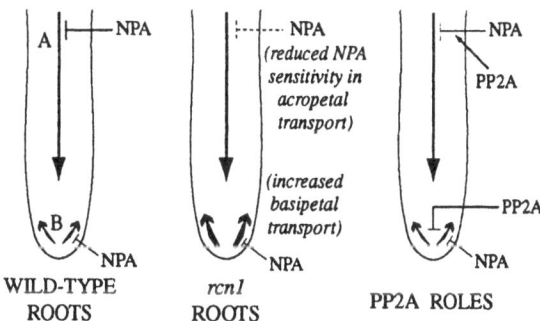

Figure 3. Roles of PP2A in regulating polar auxin transport in *Arabidopsis* roots. Acropetal (A) and basipetal (B) auxin transport can be measured in *Arabidopsis* roots (left; Reed *et al.*, 1998b; Rashotte *et al.*, 1999). Auxin transport assays reveal increased basipetal transport and reduced NPA sensitivity in acropetal transport in *rcn1* seedlings and in seedlings treated with the phosphatase inhibitor cantharidin (middle; Rashotte *et al.*, 2001). Normal PP2A activity therefore is involved in negative regulation of basipetal transport, and is required for NPA sensitivity of acropetal transport (right).

ing lateral root initiation (Casimiro *et al.*, 2001). By analogy with *Phaseolus* and *Vicia* roots, acropetal transport is expected to occur in the central cylinder, while basipetal transport should occur primarily in the epidermal and cortical cell layers near the root tip (Mitchell and Davies, 1975; Tsurumi and Ohwaki, 1978; Rashotte *et al.*, 2000). Strikingly, the root tip, pericycle and central cylinder also show the most abundant *RCN1* expression in seedlings (Deruère *et al.*, 1999). Both transport streams are sensitive to NPA; 40–50% of IAA movement is blocked by application of NPA either locally (i.e. with the labeled IAA) or globally (i.e. in growth media). The *eir1-1* mutant, an *AGR1/EIR1/PIN2* loss-of-function mutant (Luschnig *et al.*, 1998), exhibits a similar reduction in basipetal transport and no sensitivity to NPA, but shows normal levels of acropetal transport (Rashotte *et al.*, 2000), supporting the hypothesis that this efflux carrier is involved in auxin transport in the root tip (Chen *et al.*, 1998; Luschnig *et al.*, 1998; Müller *et al.*, 1998). It is not yet known whether the residual basipetal IAA movement is due to diffusion or to other carriers with lower NPA sensitivity. The *AUX1* gene is also expressed in root tips and *aux1* mutant seedlings show reduced auxin uptake, but quantitative assays of basipetal transport in *aux1* seedlings have not yet been published (Bennett *et al.*, 1996; Marchant *et al.*, 1999).

Recent work indicates that auxin transport activities are altered in roots of *rcn1* seedlings. Basipetal transport is increased to over 150% of the normal wild-type level in tips of *rcn1* roots, but shows normal sensitivity to inhibition by NPA. In contrast, acropetal transport levels are normal, but the NPA sensitivity of acropetal transport is abrogated in *rcn1* roots. Growth of wild-type seedlings in the presence of the phosphatase inhibitor cantharidin produces a phenocopy of *rcn1* in both basipetal and acropetal transport assays (Rashotte *et al.*, 2001). Thus auxin transport alterations in *rcn1* seedlings arise from loss of PP2A activity, and clearly indicate a role for this enzyme in regulation of both transport streams in seedling roots (Figure 3). However, the distinct effects of *rcn1* on basipetal vs. acropetal transport suggest differential regulation of the two transport streams. PP2A normally acts as a negative regulator of the basipetal transport stream. In the acropetal transport stream, PP2A activity is required for normal NPA sensitivity, although 'basal' activity of this stream is unaffected (Rashotte *et al.*, 2001). In each case it is reasonable to assume that activity of one or more protein kinase(s)

must counterbalance the effects of PP2A. Further experiments will be required to determine whether PP2A directly affects transport in roots, and to identify the targets of kinase and phosphatase regulation in auxin transport.

MAPK pathway signaling

Signals for numerous growth and developmental responses in eukaryotes are communicated through sequential protein phosphorylation within mitogen-activated protein kinase (MAPK) pathways (reviewed by Meskiene and Hirt, 2000). Several expression and localization studies now correlate MAPK pathway kinases to mitogenic processes, and recently MAPK activity was shown to be rapidly activated by auxin and correlated to auxin responses in roots (Mockaitis and Howell, 2000), as described below.

The first element of the pathway core, the MAPK kinase kinase (MAPKKK), may be one from among several structurally diverse groups (Jouannic *et al.*, 1999) and its activity may be regulated by phosphorylation as well as a variety of intermolecular interactions potentially relevant to current data in auxin signaling, as discussed below. Phosphorylation by MAPKKK activates the associated MAPK kinase (MAPKK), which in turn activates its target MAPK through phosphorylation on Tyr and Thr residues. The heterogeneous substrates of MAPKs share semi-conserved sequence motifs around Ser or Thr phosphoacceptor sites, and include cytoplasmic proteins and other proteins associated with nuclear, membrane and cytoskeletal structures. Young tissues exhibit higher transcript levels of some MAPK pathway kinases than do mature organs (Shibata *et al.*, 1995; Decroocq-Ferrant *et al.*, 1995; Morris *et al.*, 1997) and tissue specificities of protein abundance in some kinase isoforms are becoming apparent (Ichimura *et al.*, 1998; Bögre *et al.*, 1999; Mockaitis, unpublished results).

Intracellular compartmentalization and proteins interacting with MAPK pathway kinases, such as scaffolding proteins and adaptor proteins, can be critical for both directing the specificity and maximizing the amplitude of signal propagation and attenuation (see Pouysségur, 2000). Appropriate cellular responses to MAPK cascade action depend in part on kinetic control of the core kinases. Several classes of protein phosphatases, including PP2A, protein tyrosine phosphatases (PTPs) and dual specificity phosphatases are known to inactivate MAPKs in various systems and are sometimes responsible for kinetic regulation of a given pathway.

Since some protein kinase activities are highly sensitive to touch and minor stresses that are unavoidable in experimental manipulations, it has been difficult to observe kinase responses to exogenously applied hormones in plant systems (see Tena and Renaudin, 1998; Bögre *et al.*, 1996). In the first report describing MAPKs from *Arabidopsis*, Mizoguchi *et al.* (1994) observed a recombinant *Arabidopsis* MAPK becomes activated *in vitro* when incubated with lysates of tobacco suspension culture treated with 2,4-D. However, stimuli other than 2,4-D were not tested in this study, leaving open the possibility that activation had been induced by an abiotic stress such as cytoplasmic acidification. It was later demonstrated in the same system that simple acids mimic the 2,4-D effect on MAPK activation and that mitogenic responses are not linked to this signaling (Tena and Renaudin, 1998).

MAPK activation in response to auxin

The possibility that auxin signals are transduced through positive MAPK pathway signaling, in a manner partially analogous to mitogenic signals in non-plant systems, was presented by Mockaitis and Howell (2000) in a study with *Arabidopsis* roots. MAPK activity is dramatically and transiently stimulated within 5 min upon application of biologically relevant concentrations of auxin to young roots. Auxin-activated kinase activity is immunoprecipitated with antibodies to members of mammalian ERKs, but not with antibodies to divergent kinase subfamilies, placing the activity among the MAPK isoforms of *Arabidopsis*. Further data to support this assignment come from correlations of the increased activity with immunoprecipitable phosphotyrosine, a characteristic unique to MAPKs among known protein kinases in plants, and substrate selectivity tested in *in vitro* assays.

As a positive link of MAPK pathway activity to auxin signaling, it is first notable that in the above study rapid kinase activation is observed in an organ that is very sensitive to auxin with respect to elongation and tropic growth, cell division, and lateral root initiation. Second, the response correlates, temporally and in dose dependence, with the transcriptional activation of a reporter directed by elements of an auxin-induced immediate early gene promoter (Oono *et al.*, 1998). Importantly, this study tested the response in an auxin-insensitive mutant background. Plants carrying the *axr4* lesion are impaired in gravitropism, lateral root development, auxin-induced inhibition of root elongation (Hobbie and Estelle, 1995) and the above transcriptional activation response (Oono *et al.*, 1998). Auxin-inducible MAPK activation is also deficient in *axr4* seedlings (Mockaitis and Howell, 2000). Intriguingly, *axr4* roots exhibit a wild-type level of early osmotic stress-induced MAPK activation, providing further evidence for the strong auxin specificity of AXR4 function.

Hallmarks of MAPK pathway regulation (e.g., post-translational activation of kinases, cellular compartmentahzation and interactions that enhance signal propagation) enable the rapidity in cellular responses that might be expected to occur in cells sensing changes in auxin levels. Time courses of up to 1 h show MAPK activation to be transient, reaching a maximum of ca. 10–15-fold induction after 5 min of IAA treatment in roots (Mockaitis and Howell, 2000). Among the auxin-influenced processes that might be mediated by MAPK signaling, therefore, are primary responses such as immediate early gene transcription (reviewed by Hagen and Guilfoyle, this issue), modulation of ion channel activity (reviewed by Becker and Hedrich, this issue), reorientation of cytoskeletal structures (Shibaoka and Nagai, 1994) and production of second messengers (reviewed by Scherer, this issue).

Potential MAPK pathway involvement in auxin-influenced cell division

MAPK pathway involvement in the signaling of plant growth-promoting hormones is suggested by the expression of the pathway core kinases during mitotic processes; however, hormone influences have not been demonstrated directly. Links to the cell cycle control apparatus, especially entry into the cell cycle and in mitotic transitions, are some of the intriguing possibilities to explain roles for plant hormone signaling in cell proliferation.

Several reports correlate elevated mRNA levels of MAPK pathway kinases in synchronized suspension-cultured cells with specific stages of cell cycle progression (Jonak *et al.*, 1993; Banno *et al.*, 1993; Jouannic *et al.*, 2001). Studies of MMK3 protein and activation levels in cycling alfalfa cells (Bögre *et al.*, 1999), and of Ntf6, the putative tobacco orthologue of MMK3 (Calderini *et al.*, 1998), link this MAPK with cell division. Discrete synchronization of cells in different stages revealed that MMK3 is expressed prior to mitosis and becomes activated after metaphase, during the time window of microtubule polymerization and phragmoplast formation. Its activation appears to depend on prior entry into mitosis and on the presence of intact microtubules. By direct immunodetection, endogenous alfalfa MMK3 and tobacco Ntf6 localize to the midplane of dividing cells, implying a role in the formation of division structures. Recently proteins recognized by anti-Ntf6 antibodies were observed in the nuclei of cells undergoing heat-shock-induced microspore embryogenesis in *Brassica*, tomato and pepper (Testillano *et al.*, 2000). Similar detection is observed in vacuolate microspores and in dividing somatic cells of these plants (Prestamo *et al.*, 1999). Implications of the above results with regard to cell cycle regulation have been reviewed recently by Bögre *et al.* (2000).

Data from studies of a MAPKKK family member in tobacco currently offer a more detailed picture of MAPK pathway involvement in cell division processes. Expression of *NPK1* in leaf discs treated with both auxin and cytokinin precedes DNA replication (Banno *et al.*, 1993), and in plants appears specific to meristems, lateral root primordia, and developing organs (Nakashima *et al.*, 1998). Recently, real-time expression studies using GFP-tagged NPK1 revealed an association of the kinase with lateral growth of the cell plate and the completion of cytokinesis (Nishihama *et al.*, 2001). Whereas active NPK1 localizes to the equatorial zone of the phragmoplast throughout cell plate expansion, cell plates labeled with a kinase-inactive mutant of NPK1 do not expand fully. Over-expression of the inactive mutant kinase produces multinucleate cells with incomplete cell plates in both cultured cells and stably transformed *Arabidopsis* (Nishihama *et al.*, 2001). These data clearly link MAPKKK function and phragmoplast expansion. In addition, it is interesting to note that plants contain MAPK pathway components similar to those in yeast that act in mitotic exit and in control of cytokinesis (Jouannic *et al.*, 2001). Continued study of these and other potential signaling inputs regulating cell division programs are expected to reveal roles of kinase cascades in influencing cell proliferation as well as plant development.

The dependence upon (or dispensability of) MAPK pathway signaling for auxin-mediated responses at several other levels relevant to plant development needs to be examined. It is not known whether the same activities, or any shared pathway components, mediate the diverse roles of auxin signaling in the plant. The tobacco MAPK Ntf4 is synthesized under the developmental controls of pollen maturation, then activated once the rehydration process essential for germination is initiated (Wilson *et al.*, 1997). Since an immediate consequence of pollen hydration is cell expansion and cytoskeletal rearrangement during swelling of the pollen grain, auxin could play a role in this developmental progression by regulating Ntf4 function. It will be interesting to determine whether or not MAPK pathways are involved in auxin-mediated cell expansion (Jones *et al.*, 1998), receptor transport/relocalization (see Napier *et al.*, 2002), modulation of auxin transport processes (reviewed by Friml and Palme, this issue), polarized growth in root hairs or pollen tubes (Kost *et al.*, 1999), or receptor-mediated developmental signaling in meristems (Brand *et al.*, 2000).

MAPK pathway networks in auxin-influenced growth and development

Prior to the demonstration that auxin specifically induces rapid MAPK activation in roots (Mockaitis and Howell, 2000), the molecular evidence for MAPK pathway involvement in auxin signaling existed solely in data indicating auxin responses are negatively regulated by MAPK pathway kinases induced by non-growth-promoting factors (Kovtun *et al.*, 1998, 2000). Although the function of tobacco NPK1 is linked to promotion of cell division (above), the latter studies implicate its action in oxidative stress signaling and in the inhibition of auxin-inducible transcription.

When tobacco *NPK1* or one of its three putative *Arabidopsis* orthologues (the ANPs; Nishihama *et al.*, 1997) is transfected into a protoplast transient expression system, auxin-inducible GFP expression driven by the soybean *GH3* promoter is inhibited (Kovtun *et al.*, 1998, 2000). The effect of NPK1/ANP on promoter activity depends upon NPK1/ANP kinase activity, and constitutively active forms, produced by deletion of the regulatory domain, are more effective than the wild-type kinases in inhibiting the reporter response. Correlations made between the activation of recombinant MAPKs (stress-activated isoforms 3 and 6), heightened ANP activity, and H_2O_2-inducible promoter activation (Kovtun *et al.*, 2000) suggest that kinase cascade(s) including ANP1-like activity mediate oxidative stress signals and simultaneously establish an anti-auxin-response program in these cells.

Cross-talk among signaling pathways is also apparent in intact plant experiments pharmacologically targeting MAPKKs (Mockaitis and Howell, 2000). PD098059, a compound inhibitory to specific mammalian MAPKK activation and now known to inhibit MAPK-mediated pathogen and oxidative-burst signaling in plants (Romeis et al., 1999; Grant et al., 2000), blocks auxin-dependent activation of the BA3:GUS reporter (Oono et al., 1998) in roots. This result indicates a role for a MAPK cascade in this well-characterized transcriptional response to auxin (Mockaitis and Howell, 2000). However, a dramatic enhancement of auxin-stimulated MAPK activation in these roots suggests that the drug targets kinase cascade(s) other than the auxin-responsive pathway. Neither the putative MAPKK targets for PD098059 in plants nor the MAPKK activities affected by NPK1/ANP transient over-expression have been identified.

One notable aspect of certain well-studied MAP-KKKs in mammalian and yeast systems is their ability to serve as molecular junctions among multiple pathways responding to different signals. The yeast MAPKKK Ste11 plays a signaling role in the mating pathway in response to pheromones, in metabolic responses induced by osmotic stress, and in the promotion of invasive growth in starvation conditions. Given the data discussed above, one candidate junction component functioning among plant pathways is likely to be one or more MAPKKK(s) of the NPK1 class. It is interesting to note that ANP1 exists in two forms in plant tissues (Nishihama et al., 1997), one encoding the full-length predicted protein, and the other a splice variant lacking most of the region outside of the kinase catalytic domain. The expression of splice variants suggests that plants may in some cases regulate signaling partially through production of alternative isoforms of a single MAPK pathway component. Furthermore, non-catalytic intermolecular interactions may influence the direction of pathway signaling. In the case of the mammalian MAPKKK Raf, expression level, localization, and relative proportions of several developmentally regulated isoforms are all important factors directing its signaling role in multiple pathways (reviewed by Hagemann and Rapp, 1999; Jordan et al., 2000). Just as in deletion mutants used for production of constitutively active ANPs and NPK1 (above), the naturally truncated form of ANP1 shows increased kinase activity

and functions in an unregulated manner to partially complement the yeast mutation in Ste11 (Nishihama et al., 1997). It is worth noting that deletion of the regulatory domain may cause some loss of kinase specificity in vivo, especially at high expression levels.

In transgenic plants, active NPK1 uncoupled from its normal regulation confers a stress-protective phenotype (Kovtun et al., 2000). One primary means of NPK1 activation in vivo appears to be its interaction with NACK1, a kinesin-like protein co-localized to the phragmoplast (Nishihama and Machida, 2000). The possibility remains that auxin or other growth regulators influence NPK1-mediated processes in a mitosis-promoting interaction such as NACK1 binding, and that constitutively active NPK1 reveals a separate function of NPK1 activity in protecting the cell against oxidative damage. Downstream effects of signaling through NPK1 might bifurcate to modify mitotic structures for continued cell cycle progression or protective arrest, depending on additional cellular signals.

Results from the studies described above are consistent with the idea that one or more MAPK pathway(s) signaling stress agents exerts a negative influence upon the activity of another MAPK pathway that positively transduces auxin signals. Possibilities for signaling cross-talk among MAPK pathways that each transduce separate signals are diagrammed in Figure 4. An activated stress pathway might interfere upstream to inhibit auxin activation of MAPK or might block the productivity of the MAPK activation by preventing its proper localization or substrate interactions. Alternatively, or additionally, an activated stress pathway might block MAPK-mediated auxin responses downstream of an auxin-activated MAPK, through an effect on transcriptional regulators. Here it is pertinent to point out that transcription directed by composite elements in auxin-inducible promoters can be dramatically inhibited when Aux/IAA proteins are over-expressed (Ulmasov et al., 1997b). Alterations in network signaling by the inhibition of selected MAPK pathway(s), and/or the over-expression of active NPK1/ANPs in transfected protoplasts, might cause transient overproduction of Aux/IAA transcriptional inhibitors, possibly due to hyperactivated MAPK activity. Preliminary evidence that constitutively active ANP1 transiently expressed in protoplasts by different methods can indeed lead to activation of auxin-inducible promoters suggests that this response may be partially titratable with this

298

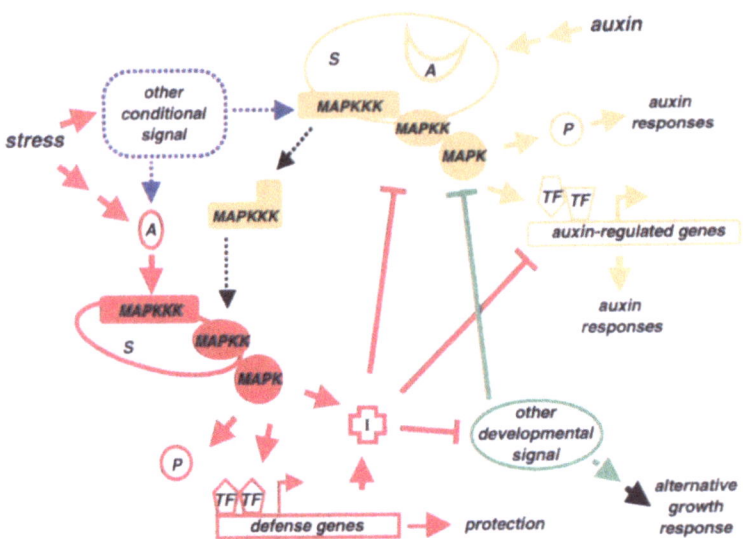

Figure 4. A model for cross-talk among MAPK signaling cascades. A hypothetical auxin-activated MAPK pathway is shown in yellow. Other signaling pathways, such as a stress-activated MAPK pathway (red components) or a developmental signaling pathway (green components), may modulate activity or output of the auxin pathway. The three kinases of each pathway core are shown as as solid shapes. **S** indicates structures such as scaffolding proteins or cytoskeletal complexes that partially determine specificity of signaling in each complex, **A**, activators upstream of MAPKKKs, **TF**, transcription factors influenced by the pathways, and **P**, other substrate proteins phosphorylated by MAPKs. **I** indicates a hypothetical inhibitor protein. A stress-activated pathway might induce production or modification of an inhibitor that blocks interactions within the auxin-stimulated cascade core or blocks the immediate pathway output. Stress signaling may likewise inhibit the activity of another signaling pathway impinging upon auxin response, such as that of another hormone or developmental signal (green). Components upstream of both stress and auxin pathways might interact conditionally, depending on other environmental factors (blue box). In response to this interaction, modification of a junction component, such as a MAPKKK of the auxin pathway, would provide a direct link between activities of the two pathway core components. Stimulating the activity of this junction kinase towards an alternative substrate in the stress-activated pathway would antagonize auxin-activated responses.

kinase activity and may be dependent on cell type (L. Bögre, personal communication).

Signaling components involved in responses to other stimuli such as light and developmental cues such as other hormones are likely to impact MAPK-mediated auxin signaling responses as well. As suggested in Figure 4, an inhibitory pathway may operate to suppress auxin-induced genes in the absence of auxin or in auxin-insensitive cells. Recent reports on the characterization of plants mutant in MAPK pathway kinase genes provide important new insights into the subtle but stringent control some MAPK pathways can exert over cellular response programs. In plants bearing insertion mutations in genes encoding the MAPKKK EDR1 (Frye *et al.*, 2000) and the MAPK phosphatase MKP1 (Ulm *et al.*, 2001), as well as in tobacco plants overexpressing constitutively active NPK1 (Kovtun *et al.*, 2000), visible phenotypes are observed only under stress-inducing conditions. Other evidence shows that components of plant MAPK pathways, each of which may perform specialized functions in response to external stimuli,

serve critical homeostatic roles in the absence of these stimuli. The potential for multiple modes of regulation for some MAPK pathway components is suggested by two recent reports on the *Arabidopsis* MAPK MPK4. While MPK4 is activated *in vivo* upon pathogen encounter, suggesting an active role in defense (Ichimura *et al.*, 2000), heightened systemic acquired resistance in *mkp4* null plants shows a constitutive role for this MAPK in suppressing plant defense in the absence of pathogens (Petersen *et al.*, 2000).

Regardless of the origin of signaling pathway crosstalk in auxin-induced responses, the realization that kinase cascades are certainly not simply linear, parallel paths linking individual receptor action to individual cellular target should inform all experimentation monitoring outcomes of auxin signaling in plants.

Conclusions

Niches for protein kinase and phosphatase action are being identified as detailed mechanisms underlying auxin responses and auxin transport are eluci-dated. Approaches ranging from the pharmacological and physiological to the genetic and molecular have yielded a complex picture in which a number of ele-ments may span the junctions between auxin biology and other signaling pathways. Future research will flesh out this picture, distinguishing direct connections from indirect effects, and identifying targets and con-trol points at which protein phosphorylation adjusts the input and output of the auxin signaling system.

Three areas currently offer particular promise for identifying phosphorylation control components. First, the recent progress in identification and analysis of auxin transport and auxin response proteins pro-vide a platform that may allow identification of key regulators and targets through interaction cloning and through new mutant screens. For instance, a screen for second-site mutations that stabilize Aux/IAA pro-teins might identify a kinase required to phospho-rylate these proteins and initiate their entry into the ubiquitin-mediated proteolysis pathway. Second, the growing availability of functional genomics resources for isolating mutations in genes of interest offers an approach to gene families in which functional re-dundancy and/or lethality have posed obstacles to a forward genetic approach. Because systems for gen-eration of insertion, deletion and point mutations have been developed, gene function may be defined using information from both knock-outs and more subtle lesions that affect specific outputs. Finally, ad-vances in proteomics and the increasing feasibility of spectroscopy-based protein identification provide invaluable tools for researchers investigating protein kinase and phosphatase activities and their substrates.

Acknowledgements

Work in the laboratory of A.D.L. has been sup-ported by grants from the National Science Foun-dation (IBN9986017 and the NIH/NCRR (1 P20 RR15578.01). K.M. is supported by the Centre Na-tional de la Recherche Scientifique. We thank R. Of-fringa and L. Bögre for permission to cite unpublished data, and G. Muday for helpful comments on the manuscript.

References

Ariño, J., Perez-Calderon, E., Cunillera, N., Camps, M., Posas, F. and Ferrer, A. 1993. Protein phosphatases in higher plants: multiplicity of type 2A phosphatases in *A. thaliana*. Plant Mol. Biol. 21: 475–485.

Banno, H., Hirano, K., Nakamura, T., Irie, K., Nomoto, S., Matsumoto, K. and Machida, Y. 1993. NPK1, a tobacco gene that encodes a protein with a domain homologous to yeast BCK1, STE11, and Byr2 protein kinases. Mol. Cell. Biol. 13: 4745–4752.

Becker, D. and Hedrich, D. 2002. Channelling auxin action: mod-ulation of ion transport by indole-3-acetic acid. Plant Mol. Biol. 49: 349–356.

Bennett, M.J., Marchant, A., Green, H.G., May, S.T., Ward, S.P., Millner, P.A., Walker, A.R., Schulz, B. and Feldmann, K.A. 1996. *Arabidopsis AUX1* gene: a permease-like regulator of root gravitropism. Science 273: 948–950.

Bennett, S.R.M., Alvarez, J., Bossinger, G. and Smyth, D.R. 1995. Morphogenesis in *pinoid* mutants of *Arabidopsis thaliana*. Plant J. 8: 505–520.

Berleth, T. and Jürgens, G. 1993. The role of the *monopteros* gene in organising the basal body region of the *Arabidopsis* embryo. Development 118: 575–587.

Bernasconi, P. 1996. Effect of synthetic and natural protein tyrosine kinase inhibitors on auxin efflux in zucchini (*Cucurbita pepo*) hypocotyls. Physiol. Plant. 96: 205–210.

Bögre, L., Calderini, O., Binarova, P., Mattauch, M., Till, S., Kiegerl, S., Jonak, C., Pollaschek, C., Barker, P., Huskisson, N. S., Hirt, H. and Heberle-Bors, E. 1999. A MAP kinase is activated late in plant mitosis and becomes localized to the plane of cell division. Plant Cell 11: 101–114.

Bögre, L., Ligterink, W., Heberle-Bors, E. and Hirt, H. 1996. Mechanosensors in plants. Nature 383: 489–490.

Bögre, L., Meskiene, I., Heberle-Bors, E. and Hirt, H. 2000. Stress-ing the role of MAP kinases in mitogenic stimulation. Plant Mol. Biol. 43: 705–718.

Brand, U., Fletcher, J.C., Hobe, M., Meyerowitz, E.M. and Si-mon, R. 2000. Dependence of stem cell fate in *Arabidopsis* on a feedback loop regulated by CLV3 activity. Science 289: 617–619.

Briggs, W.R., Beck, C.F., Cashmore, A.R., Christie, J.M., Hughes, J., Jarillo, J.A., Kagawa, T., Kanegae, H., Liscum, E., Nagatani, A., Okada, K., Salomon, M., Rudiger, W., Sakai, T., Takano, M., Wada, M. and Watson, J.C. 2001. The phototropin family of photoreceptors. Plant Cell 13: 993–998.

Brown, D., Rashotte, A., Murphy, A., Tague, B., Peer , W., Taiz, L. and Muday, G. 2001. Flavonoids act as negative regulators of auxin transport *in vivo* in *A. thaliana*. Plant Physiol., 126: 524–535.

Calderini, O., Bögre, L., Vicente, O., Binarova, P., Heberle-Bors, E. and Wilson, C. 1998. A cell cycle regulated MAP kinase with a possible role in cytokinesis in tobacco cells. J. Cell Sci. 111: 3091–3100.

Casamayor, A., Perez-Callejon, E., Pujol, G., Ariño, J. and Ferrer, A. 1994. Molecular characterization of the fourth isoform of the catalytic subunit of protein phosphatase 2A from *Arabidopsis thaliana*. Plant Mol. Biol. 26: 523–528.

Casimiro, I., Marchant, A., Bhalerao, R.P., Beeckman, T., Dhooge, S., Swarup, R., Graham, N., Inzé, D., Sandberg, G., Casero, P.J. and Bennett, M. 2001. Auxin transport promotes *Arabidopsis* lateral root initiation. Plant Cell 13: 843–852.

Chang, S. and Kaufman, P. 2000. Effects of staurosporine, okadaic acid and sodium fluoride on protein phosphorylation in gravire-sponding oat shoot pulvini. Plant Physiol. Biochem. 38: 315–323.

Chen, R., Hilson, P., Sedbrook, J., Rosen, E., Caspar, T. and Masson, P.H. 1998. The *Arabidopsis thaliana AGRAVITROPIC*

1 gene encodes a component of the polar-auxin-transport efflux carrier. Proc. Natl. Acad. Sci. USA 95: 15112–15117.

Christensen, S.K., Dagenais, N., Chory, J. and Weigel, D. 2000. Regulation of auxin response by the protein kinase PINOID. Cell 100: 469–478.

Christie, J.M., Reymond, P., Powell, G.K., Bernasconi, P., Raibekas, A.A., Liscum, E. and Briggs, W.B. 1998. *Arabidopsis* NPH1: a flavoprotein with the properties of a photoreceptor for phototropism. Science 282: 1698–1701.

Colón-Carmona, A., Chen, D.L., Yeh, K.-C. and Abel, S. 2000. Aux/IAA proteins are phosphorylated by phytochrome *in vitro*. Plant Physiol. 124: 1728–1738.

Corum, J.W., Hartung, A.J., Stamey, R.T. and Rundle, S.J. 1996. Characterization of DNA sequences encoding a novel isoform of the 55 kDa B regulatory subunit of the type 2A protein serine/threonine phosphatase of *Arabidopsis thaliana*. Plant Mol. Biol. 31: 419–427.

Darwin, C. 1880. The Power of Movements in Plants. John Murray, London.

Decroocq-Ferrant, V., Decroocq, S., van Went, J., Schmidt, E. and Kreis, M. 1995. A homologue of the MAP-ERK family of protein kinase genes is expressed in vegetative and in female reproductive organs of *Petunia hybrida*. Plant Mol. Biol. 27: 339–350.

Delbarre, A., Muller, P. and Guern, J. 1998. Short-lived and phosphorylated proteins contribute to carrier-mediated efflux, but not to influx, of auxin in suspension-cultured tobacco cells. Plant Physiol. 116: 833–844.

Deruère, J., Jackson, K., Garbers, C., Söll, D. and DeLong, A. 1999. The *RCN1*-encoded A subunit of protein phosphatase 2A increases phosphatase activity *in vivo*. Plant J. 20: 389–399.

Dharmasiri, S. and Estelle, M. 2002. The role of regulated protein degradation in auxin response. Plant Mol. Biol. 49: 401–408.

Frye, C.A., Tang, D. and Innes, R.W. 2001. Negative regulation of defense responses in plants by a conserved MAPKK kinase. Proc. Natl. Acad. Sci. USA 98: 373–378.

Gälweiler, L., Guan, C., Müller, A., Wisman, E., Mendgen, K., Yephremov, A. and Palme, K. 1998. Regulation of polar auxin transport by AtPIN1 in *Arabidopsis* vascular tissue. Science 282: 2226–2230.

Garbers, C., DeLong, A., Deruère, J., Bernasconi, P. and Söll, D. 1996. A mutation in protein phosphatase 2A regulatory subunit A affects auxin transport in *Arabidopsis*. EMBO J 15: 2115–2124.

Götz J., Probst A., Ehler E., Hemmings B. and Kues W. 1998. Delayed embryonic lethality in mice lacking protein phosphatase 2A catalytic subunit Cα. Proc. Natl. Acad. Sci. USA 95: 12370–12375.

Grant, J.J., Yun, B.W. and Loake, G.J. 2000. Oxidative burst and cognate redox signalling reported by luciferase imaging: identification of a signal network that functions independently of ethylene, SA and Me-JA but is dependent on MAPKK activity. Plant J. 24: 569–582.

Gray, W.M. and Estelle, M. 2000. Function of the ubiquitin-proteasome pathway in auxin response. Trends Biochem. Sci. 25: 133–138.

Groves, M.R., Hanlon, N., Turowski, P., Hemmings, B.A. and Barford, D. 1999. The structure of the protein phosphatase 2A PR65/A subunit reveals the conformation of its 15 tandemly repeated HEAT motifs. Cell 96: 99–110.

Hagemann, C. and Rapp, U.R. 1999. Isotype-specific functions of Raf kinases. Exp. Cell Res. 253: 34–46.

Hagen, G. and Guilfoyle, T. 2002. Auxin-responsive gene expression: genes, promoters and regulatory factors. Plant Mol. Biol. 49: 373–385.

Hanks, S.K., Quinn, A.M., and Hunter, T. 1988. The protein kinase family: conserved features and deduced phylogeny of the catalytic domains. Science 241: 42–52.

Hardie, D.G. 1999 Plant protein serine/threonine kinases: classification and functions. Annu. Rev. Plant Physiol. Plant Mol. Biol. 50: 97–131.

Hardtke, C.S. and Berleth, T. 1998. The *Arabidopsis* gene *MONOPTEROS* encodes a transcription factor mediating embryo axis formation and vascular development. EMBO J. 17: 1405–1411.

Harper, R.M., Stowe-Evans, E.L., Luesse, D.R., Muto, H., Tatematsu, K., Watahiki, M. K., Yamamoto, K. and Liscum, E. 2000. The *NPH4* locus encodes the auxin response factor ARF7, a conditional regulator of differential growth in aerial *Arabidopsis* tissue. Plant Cell 12: 757–770.

Haynes, J.G., Hartung, A.J., Hendershot, J.D.I., Passingham, R.S. and Rundle, S.J. 1999. Molecular characterization of the B′ regulatory subunit gene family of *Arabidopsis* protein phosphatase 2A. Eur. J. Biochem. 260: 127–137.

Hobbie, L. and Estelle, M. 1995. The *axr4* auxin-resistant mutants of *Arabidopsis thaliana* define a gene important for root gravitropism and lateral root initiation. Plant J. 7: 211–220.

Hsieh, H.-L., Okamoto, H., Wang, M., Ang, L.-H., Matsui, M., Goodman, H. and Deng, X. W. 2000. *FIN219*, an auxin-regulated gene, defines a link between phytochrome A and the downstream regulator COP1 in light control of *Arabidopsis* development. Genes Dev. 14: 1958–1970.

Huala, E., Oeller, P.W., Liscum, E., Han, I.-S., Larsen, E. and Briggs, W. 1997. *Arabidopsis* NPH1: a protein kinase with a putative redox-sensing domain. Science 278: 2120–2123.

Ichimura, K., Mizoguchi, T., Hayashida, N., Seki, M. and Shinozaki, K. 1998. Molecular cloning and characterization of three cDNAs encoding putative mitogen-activated protein kinase kinases (MAPKKs) in *Arabidopsis thaliana*. DNA Res. 5: 341–348.

Ichimura, K., Mizoguchi, T., Yoshida, R., Yuasa, T. and Shinozaki, K. 2000. Various abiotic stresses rapidly activate *Arabidopsis* MAP kinases ATMPK4 and ATMPK6. Plant J. 24: 655–665.

Jacobs, M. and Rubery, P.H. 1988. Naturally occurring auxin transport regulators. Science 241: 346–349.

Johnson, P.R. and Ecker, J.R. 1998. The ethylene gas signal transduction pathway: a molecular perspective. Annu. Rev. Genet. 32: 227–254.

Jonak, C., Pay, A., Bögre, L., Hirt, H. and Heberle-Bors, E. 1993. The plant homologue of MAP kinase is expressed in a cell cycle-dependent and organ-specific manner. Plant J. 3: 611–617.

Jones, A.M., Im, K.H., Savka, M.A., Wu, M.J., DeWitt, N.G., Shillito, R. and Binns, A.N. 1998. Auxin-dependent cell expansion mediated by overexpressed auxin-binding protein 1. Science 282: 1114–1117.

Jordan, J.D., Landau, E.M. and Iyengar, R. 2000. Signaling networks: the origins of cellular multitasking. Cell 103: 193–200.

Jouannic, S., Hamal, A., Leprince, A.S., Tregear, J.W., Kreis, M. and Henry, Y. 1999. Plant MAP kinase kinase kinases structure, classification and evolution. Gene 233: 1–11.

Jouannic, S., Champion, A., Segui-Simarro, J.-M., Salimova, E., Picaud, A., Tregear, J., Testillano, P., Risueno, M.-C., Simanis, V., Kreis, M. and Henry, Y. 2001. The protein kinases AtMAP3Kϵ1 and BnMAP3Kϵ1 are functional homologues of *S. pombe* cdc7p and may be involved in cell division. Plant J., 26: 637–649.

Kim, B.C., Soh, M.S., Hong, S.H., Furuya, M. and Nam, H.G. 1998. Photomorphogenic development of the *Arabidopsis shy2-1D* mutation and its interaction with phytochromes in darkness. Plant J. 15: 61–68.

Kim, B.C., Soh, M.S., Kang, B.J., Furuya, M. and Nam, H.G. 1996. Two dominant photomor-phogenic mutations of *A. thaliana* identified as suppressor mutations of *hy2*. Plant J. 9: 441–456.

Kost, B., Lemichez, E., Spielhofer, P., Hong, Y., Tolias, K., Carpenter, C. and Chua, N.-H. 1999. Rac homologues and compartmentalized phosphatidylinositol 4,5-bisphosphate act in a common pathway to regulate polar pollen tube growth. J. Cell Biol. 145: 317–330.

Kovtun, Y., Chiu, W.L., Zeng, W. and Sheen, J. 1998. Suppression of auxin signal transduction by a MAPK cascade in higher plants. Nature 395: 716–720.

Kovtun, Y., Chiu, W.L., Tena, G. and Sheen, J. 2000. Functional analysis of oxidative stress-activated mitogen-activated protein kinase cascade in plants. Proc. Natl. Acad. Sci. USA 97: 2940–2945.

Latorre, K.A., Harris, D.M. and Rundle, S.J. 1997. Differential expression of three *Arabidopsis* genes encoding the B′ regulatory subunit of protein phosphatase 2A. Eur. J. Biochem. 245: 156–163.

Lee, J.S., Chang, W.-K. and Evans, M.L. 1990. Effects of ethylene on the kinetics of curvature and auxin redistribution in gravistimulated roots of *Zea mays*. Plant Physiol. 94: 1770–1775.

Leyser, H.M.O., Lincoln, C.A., Timpte, C., Lammer, D., Turner, J. and Estelle, M. 1993. *Arabidopsis* auxin-resistance gene *AXR1* encodes a protein related to ubiquitin-activating enzyme E1. Nature 364: 161–164.

Li, Y., Hagen, G. and Guilfoyle, T. 1991. An auxin-responsive promoter is differentially induced by auxin gradients during tropisms. Plant Cell 3: 1167–1175.

Lincoln, C., Britton, J.H. and Estelle, M. 1990. Growth and development of the *axr1* mutants of *Arabidopsis*. Plant Cell 2: 1071–1080.

Liscum, E. and Briggs, W.R. 1995. Mutations in the *NPH1* locus of *Arabidopsis* disrupt the perception of phototropic stimuli. Plant Cell 7: 473–485.

Liscum, E. and Reed, J.W. 2002. Genetics of Aux/IAA and ARF action in plant growth and development. Plant Mol. Biol. 49: 387–400.

Lomax, T.L., Muday, G.K. and Rubery, P. 1995. Auxin transport. In: P.J. Davies (Ed.) Plant Hormones: Physiology, Biochemistry, and Molecular Biology, Kluwer Academic Publishers, Dordrecht, Netherlands, pp. 509–530.

Lu, Y.T. and Feldman, L.J. 1997. Light-regulated root gravitropism: a role for, and characterization of, a calcium/calmodulin-dependent protein kinase homolog. Planta 203 (Suppl.): S91–S97.

Luschnig, C., Gaxiola, R.A., Grisafi, P. and Fink, G.R. 1998. EIR1, a root-specific protein involved in auxin transport, is required for gravitropism in *A. thaliana*. Genes Dev. 12: 2175–2187.

MacKintosh, C. and MacKintosh, R.W. 1994. Inhibitors of protein kinases and phosphatases. Trends Biochem Sci. 19: 444–448.

Marchant, A., Kargul, J., May, S.T., Muller, P., Delbarre, A., Perrot-Rechenmann, C. and Bennett, M.J. 1999. AUX1 regulates root gravitropism in *Arabidopsis* by facilitating auxin uptake within root apical tissues. EMBO J. 18: 2066–2073.

Mayer-Jaekel, R.E. and Hemmings, B.A. 1994. Protein phosphatase 2A, a 'ménage à trois'. Trends Cell Biol.4: 287–291.

Meskiene, I. and Hirt, H. 2000. MAP kinase pathways: molecular plug-and-play chips for the cell. Plant Mol Biol. 42: 791–806.

Millward, T.A., Zolnierowicz, S. and Hemmings, B.A. 1999. Regulation of protein kinase cascades by protein phosphatase 2A. Trends Biol. Sci. 24: 186–191.

Mitchell, E.K. and Davies, P.J. 1975. Evidence for three different systems of movement of indoleacetic acid in intact roots of *Phaseolus coccineus*. Physiol. Plant. 33: 290–294.

Mizoguchi, T., Gotoh, Y., Nishida, E., Yamaguchi-Shinozaki, K., Hayashida, N., Iwasaki, T., Kamada, H. and Shinozaki, K. 1994. Characterization of two cDNAs that encode MAP kinase homologues in *A. thaliana* and analysis of the possible role of auxin in activating such kinase activities in cultured cells. Plant J. 5: 111–122.

Mockaitis, K. and Howell, S.H. 2000. Auxin induces mitogenic activated protein kinase (MAPK) activation in roots of *Arabidopsis* seedlings. Plant J. 24: 785–796.

Morelli, G. and Ruberti, I. 2000. Shade avoidance responses. Driving auxin along lateral routes. Plant Physiol. 122: 621–626.

Morris, D.A. and Robinson, J.S. 1998. Targeting of auxin carriers to the plasma membrane: differential effects of brefeldin A on the traffic of auxin uptake and efflux carriers. Planta 205: 606–612.

Morris, D.A., Rubery, P.H., Jarman, J. and Sabater, M. 1991. Effects of inhibitors of protein synthesis on transmembrane auxin transport in *Cucurbita pepo* L. hypocotyl segments. J. Exp. Bot. 42: 773–783.

Morris, P.C., Guerrier, D., Leung, J. and Giraudat, J. 1997. Cloning and characterization of MEK1, an *Arabidopsis* gene encoding a homologue of MAP kinase kinase. Plant Mol. Biol. 35: 1057–1064.

Motchoulski, A. and Liscum, E. 1999. *Arabidopsis* NPH3: A NPH1 photoreceptor-interacting protein essential for phototropism. Science 286: 961–964.

Muday, G. 2000. Maintenance of asymmetric cellular localization of an auxin transport protein through interaction with the actin cytoskeleton. J. Plant Growth Regul., 19: 385–396.

Müller, A., Guan, C., Gälweiler, L., Tanzler, P., Huijser, P., Marchant, A., Parry, G., Bennett, M., Wisman, E. and Palme, K. 1998. *AtPIN2* defines a locus of *Arabidopsis* for root gravitropism control. EMBO J. 17: 6903–6911.

Murphy, A., Peer, W. and Taiz, L. 2000. Regulation of auxin transport by aminopeptidases and endogenous flavonoids. Planta 211: 315–324.

Nagpal, P., Walker, L.M., Young. J.C., Sonawala, A., Timpte, C., Estelle, M. and Reed, J.W. 2000. *AXR2* encodes a member of the Aux/IAA protein family. Plant Physiol. 123: 563–573.

Nakashima, M., Hirano, K., Nakashima, S., Banno, H., Nishihama, R. and Machida, Y. 1998. The expression pattern of the gene for NPK1 protein kinase related to mitogen-activated protein kinase kinase (MAPKKK) in a tobacco plant: correlation with cell proliferation. Plant Cell Physiol. 39: 690–700.

Napier, R.M., David, K.M. and Perrot-Rechenmann, C. 2002. A short history of auxin-binding proteins. Plant Mol. Biol. 49: 339–348.

Nishihama, R., Banno, H., Kawahara, E., Irie, K. and Machida, Y. 1997. Possible involvement of differential splicing in regulation of the activity of *Arabidopsis* ANP1 that is related to mitogen activated protein kinase kinase kinases (MAPKKKs). Plant J. 12: 39–48.

Nishihama, R., Ishikawa, M., Araki, S., Soyano, T., Asada, T. and Machida, Y. 2001. The NPK1 mitogen-activated protein kinase kinase kinase is a regulator of cell-plate formation in plant cytokinesis. Genes Dev. 15: 352–363.

Nishihama, R. and Machida, Y. 2000. The MAP kinase cascade that includes MAPKKK-related protein kinase NPK1 controls a mitotic process in plant cells. Res. Probl. Cell Differ. 27: 119–130.

Okada, K., Ueda, J., Komaki, M.K., Bell, C.J. and Shimura, Y. 1991. Requirement of the auxin polar transport system in early stages of *Arabidopsis* floral bud formation. Plant Cell 3: 677–684.

Oono, Y., Chen, Q.G., Overvoorde, P.J., Kohler, C. and Theologis, A. 1998. *age* Mutants of *Arabidopsis* exhibit altered auxin-regulated gene expression. Plant Cell 10: 1649–1662.

Parry, G., Delbarre, A., Marchant, A., Swarup, R., Napier, R., Perrot-Rechenmann, C. and Bennett, M.J. 2001. Novel auxin transport inhibitors phenocopy the auxin influx carrier mutation *aux1*. Plant J. 25: 399–406.

Petersen, M., Brodersen, P., Naested, H., Andreasson, E., Lindhart, U., Johansen, B., Nielsen, H.B., Lacy, M., Austin, M.J., Parker, J. E., Sharma, S. B., Klessig, D.F., Martienssen, R., Mattsson, O., Jensen, A.B. and Mundy, J. 2000. *Arabidopsis* MAP kinase 4 negatively regulates systemic acquired resistance. Cell 103: 1111–1120.

Pouyssegur, J. 2000. Signal transduction. An arresting start for MAPK. Science 290: 1515–1518.

Prestamo, G., Testillano, P.S., Vicente, O., Gonzalez-Melendi, P., Coronado, M.J., Wilson, C., Heberle-Bors, E. and Risueno, M.C. 1999. Ultrastructural distribution of a MAP kinase and transcripts in quiescent and cycling plant cells and pollen grains. J Cell Sci. 112: 1065–1076.

Price, N.E. and Mumby, M.C. 2000. Effects of regulatory subunits on the kinetics of protein phosphatase 2A. Biochemistry 39: 11312–11318.

Przemeck, G.K.H., Mattsson, J., Hardtke, C.S., Sung, Z.R. and Berleth, T. 1996. Studies on the role of the *Arabidopsis* gene *MONOPTEROS* in vascular development and plant cell axialization. Planta 200: 229–237.

Rahman, A., Ahamed, A., Amakawa, T., Goto, N. and Tsurumi, S. 2001. Chromosaponin I specifically interacts with AUX1 protein in regulating the gravitropic response of *Arabidopsis* roots. Plant Physiol. 125: 990–1000.

Rashotte, A., Brady, S., Reed, R., Ante, S. and Muday, G. 2000. Basipetal auxin transport is required for gravitropism in roots of *Arabidopsis*. Plant Physiol. 122: 481–490.

Rashotte, A.M., DeLong A. and Muday, G.K 2001. Genetic and chemical reductions in protein phosphatase activity alter auxin transport, gravity response and lateral root growth. Plant Cell, 13: 1683–1697.

Reed, J.W., Elumalai, R.P. and Chory, J. 1998a. Supressors of an *Arabidopsis thaliana phyB* mutation identity genes control light signaling and hypocotyl elongation. Genetics 148: 1295–1310.

Reed, R.C., Brady, S.R. and Muday, G.K. 1998b. Inhibition of auxin movement from the shoot into the root inhibits lateral root development in *Arabidopsis*. Plant Physiol. 118: 1369–1378.

Romeis, T., Piedras, P., Zhang, S., Klessig, D.F., Hirt, H. and Jones, J.D. 1999. Rapid Avr9- and Cf-9-dependent activation of MAP kinases in tobacco cell cultures and leaves: convergence of resistance gene, elicitor, wound, and salicylate responses. Plant Cell 11: 273–287.

Ronne, H., Carlberg, M., Hu, G.Z. and Nehlin, J.O. 1991. Protein phosphatase 2A in *Saccharomyces cerevisiae*: effects on cell growth and bud morphogenesis. Mol. Cell. Biol. 11: 4876–4884.

Rubery, P.H. 1990. Phytotropins: receptors and endogenous ligands. Symp. Soc. Exp. Biol. 44: 119–146.

Ruediger, R., Hentz, M., Fait, J., Mumby, M. and Walter, G. 1994. Molecular model of the A subunit of protein phosphatase 2A: interaction with other subunits and tumor antigens. J. Virol. 68: 123–129.

Rundle, S.J., Hartung, A.J., Corum, J.W. and O'Neill, M. 1995. Characterization of a cDNA encoding the 55 kDa B regulatory subunit of *Arabidopsis* protein phosphatase 2A. Plant Mol. Biol. 28: 257–266.

Sabatini, S., Beis, D., Wolkenfelt, H., Murfett, J., Guilfoyle, T., Malamy, J., Benfey, P., Leyser, O., Bechtold, N., Weisbeek, P. and Scheres, B. 1999. An auxin-dependent distal organizer of pattern and polarity in the *Arabidopsis* root. Cell 99: 463–472.

Sakai, T., Kagawa, T., Kasahara, M., Swartz, T.E., Christie, J.M., Briggs, W.R., Wada, M. and Okada, K. 2001. *Arabidopsis nph1* and *npl1*: blue light receptors that mediate both phototropism and chloroplast relocation. Proc. Natl. Acad. Sci. USA, in press.

Sakamoto, K. and Nagatani, A. 1996. Nuclear localization activity of phytochrome B. Plant J. 10: 859–868.

Scherer, G.F.E. 2002. Secondary messengers and phospholipase A_2 in auxin signal transduction Plant. Mol. Biol. 49: 357–372.

Sessions, R.A. and Zambryski, P.C. 1995. *Arabidopsis* gynoecium structure in the wild type and *ettin* mutants. Development 121: 1519–1532.

Sessions, A., Nemhauser, J.L., McColl, A., Roe, J.L., Feldmann, K.A. and Zambryski, P.C. 1997. ETTIN patterns the *Arabidopsis* floral meristem and reproductive organs. Development 124: 4481–4491.

Shibata, W., Banno, H., Ito, Y., Hirano, K., Irie, K., Usami, S., Machida, C. and Machida, Y. 1995. A tobacco protein kinase, NPK2, has a domain homologous to a domain found in activators of mitogen-activated protein kinases (MAPKKs). Mol. Gen. Genet. 246: 401–410.

Shibaoka, H. and Nagai, R. 1994. The plant cytoskeleton. Curr. Opin. Cell Biol. 6: 10–15.

Shu Y., Yang H., Hallberg E. and Hallberg R. 1997. Molecular genetic analysis of Rts1p, a B' regulatory subunit of *Saccharomyces cerevisiae* protein phosphatase 2A. Mol. Cell. Biol. 17: 3242–3245.

Sieberer, T., Seifert, G.J., Hauser, M.-T., Grisafi, P., Fink, G.R. and Luschnig, C. 2000. Post-transcriptional control of the *Arabidopsis* efflux carrier EIR1 requires AXR1. Curr. Biol. 10: 1595–1598.

Smith, R.D. and Walker, J.C. 1996. Plant protein phosphatases. Annu. Rev. Plant Physiol. Plant Mol. Biol. 47: 101–125.

Stamey, R.T. and Rundle, S.J. 1996. Characterization of a novel isoform of a type 2A serine/threonine protein phosphatase from *Arabidopsis thaliana*. Plant Physiol. 110: 335.

Suttle, J.C. 1988. Effect of ethylene treatment on polar IAA transport, net IAA uptake and specific binding of n-1-naphthylphthalamic acid in tissues and microsomes isolated from etiolated pea epicotyls. Plant Physiol. 88: 795–799.

Tena, G. and Renaudin, J.P. 1998. Cytosolic acidification but not auxin at physiological concentration is an activator MAP kinases in tobacco cells. Plant J 16: 173–182.

Testillano, P.S., Coronado, M.J., Segui, J.M., Domenech, J., Gonzalez-Melendi, P., Raska, I. and Risueno, M.C. 2000. Defined nuclear changes accompany the reprogramming of the microspore to embryogenesis. J. Struct. Biol. 129: 223–232.

Thakore, C.U., Livengood, A.J., Hendershot, J.D.I., Corum, J.W., Latorre, K.A. and Rundle, S.J. 1999. Characterization of the promoter region and expression pattern of three *Arabidopsis* protein phosphatase type 2A subunit genes. Plant Sci. 147: 165–176.

Thomason, P. and Kay, R. 2000. Eukaryotic signal transduction via histidine-aspartate phosphorelay. J. Cell Sci. 113: 3141–3150.

Tian, Q. and Reed, J.W. 1999. Control of auxin-regulated root development by the *Arabidopsis thaliana SHY2/IAA3* gene. Development 126: 711–721.

Timpte, C., Wilson, A.K. and Estelle, M. 1994. The *axr2-1* mutation of *Arabidopsis thaliana* is a gain-of-function mutation that disrupts an early step in auxin response. Genetics 138: 1239–1249.

Timpte, C., Lincoln, C., Pickett, F.B., Turner, J. and Estelle, M. 1995. The *AXR1* and *AUX* genes of *Arabidopsis* function in separate auxin-response pathways. Plant J. 8: 561–569.

Tsurumi, S. and Ohwaki, Y. 1978. Transport of ^{14}C-labeled indoleacetic acid in *Vicia* root segments. Plant Cell Physiol. 19: 1195–1206.

Ulm, R., Revenkova, E., di Sansebastiano, G.P., Bechtold, N. and Paszkowski, J. 2001. Mitogen-activated protein kinase phosphatase is required for genotoxic stress relief in *Arabidopsis*. Genes Dev. 15: 699–709.

Ulmasov, T., Hagen G. and Guilfoyle, T.J. 1997a. ARF1, a transcription factor that binds to auxin response elements. Science 276: 1865–1868.

Ulmasov, T., Murfett, J., Hagen, G. and Guilfoyle, T.J. 1997b. Aux/IAA proteins repress expression of reporter genes containing natural and highly active synthetic auxin response elements. Plant Cell 9: 1963–1971.

Utsuno, K., Shikanai, T., Yamada, Y. and Hashimoto, T. 1998. *AGR*, an Agravitropic locus of *A. thaliana*, encodes a novel membrane-protein family member. Plant Cell Physiol. 39: 1111–1118.

Virshup, D.M. 2000. Protein phosphatase 2A: a panoply of enzymes. Curr. Opin. Cell Biol. 12: 180–185.

Wassarman, D.A., Solomon, N.M., Chang, H.C., Karim, F.D., Therrien, M. and Rubin, G.M. 1996. Protein phosphatase 2A positively and negatively regulates Ras1-mediated photoreceptor development in *Drosophila*. Genes Dev. 10: 272–278.

Wilson, C., Voronin, V., Touraev, A., Vicente, O. and Heberle-Bors, E. 1997. A developmentally regulated MAP kinase activated by hydration in tobacco pollen. Plant Cell 9: 2093–2100.

Plant Molecular Biology **49**: 305–317, 2002.
Perrot-Rechenmann and Hagen (Eds.), Auxin Molecular Biology.
© 2002 *Kluwer Academic Publishers.*

305

Complex physiological and molecular processes underlying root gravitropism

Rujin Chen, Changhui Guan, Kanokporn Boonsirichai and Patrick H. Masson*
*Laboratory of Genetics, University of Wisconsin-Madison, 445 Henry Mall, Madison, WI 53706, USA (*author for correspondence; e-mail phmasson@facstaff.wisc.edu)*

Received 8 May 2001; accepted in revised form 11 July 2001

Key words: gravitropism, gravity sensing, gravity signal transduction, roots, statolith

Abstract

Gravitropism allows plant organs to guide their growth in relation to the gravity vector. For most roots, this response to gravity allows downward growth into soil where water and nutrients are available for plant growth and development. The primary site for gravity sensing in roots includes the root cap and appears to involve the sedimentation of amyloplasts within the columella cells. This process triggers a signal transduction pathway that promotes both an acidification of the wall around the columella cells, an alkalinization of the columella cytoplasm, and the development of a lateral polarity across the root cap that allows for the establishment of a lateral auxin gradient. This gradient is then transmitted to the elongation zones where it triggers a differential cellular elongation on opposite flanks of the central elongation zone, responsible for part of the gravitropic curvature. Recent findings also suggest the involvement of a secondary site/mechanism of gravity sensing for gravitropism in roots, and the possibility that the early phases of graviresponse, which involve differential elongation on opposite flanks of the distal elongation zone, might be independent of this auxin gradient. This review discusses our current understanding of the molecular and physiological mechanisms underlying these various phases of the gravitropic response in roots.

Abbreviations: CEZ, central elongation zone; DEZ, distal elongation zone; ER, endoplasmic reticulum; EZ, elongation zone; IAA, indole-3-acetic acid; InsP$_3$, inositol 1,4,5-trisphosphate; NAA, 1-naphthaleneacetic acid; NPA, naphthylphthalamic acid; 2,4-D, 2,4-dichlorophenoxyacetic acid

Introduction

Plasticity in growth behavior allows plants to survive dramatic changes in their environments, despite sessility. Early in their life cycles, seeds are often dropped off on the ground in a random orientation. Upon germination, roots, as well as shoots of young seedlings may be oriented upward, laterally or downward. Consequently, these organs will have to reorient themselves in order to assume a position that properly suits their functions, i.e. roots orienting downward into the soil for anchorage as well as water and nutrient uptake, and shoots orienting upward into the sunlight for photosynthesis.

Throughout their life cycles, plants use environmental parameters to guide their organs' growth so that they can adapt to and take full advantage of the changing environment. For instance, shoots can direct their growth in relation to the vectors of light (phototropism) or gravity (gravitropism). These tropic responses allow shoots to resume vertical upward growth after prostration by high winds or heavy rains. In agriculture, this important process allows salvaging of significant quantities of crop products.

Roots guide their growth in relation to gravity, light, gradients of temperature (thermotropism), humidity (hydrotropism), ions, chemicals (chemotropism), and oxygen (oxytropism). Quite amazingly, the information provided by these environmental parameters is integrated by specific and interconnected signal transduction pathways, and interpreted into specific signals that dictate the pattern of organ growth (Chen et al., 1999).

Hence, gravitropism guides the growth of plant organs throughout the life cycle of a plant, directing their growth at a specific angle for the gravity vector (Firn and Digby, 1997). Typically, it dictates upward and downward growth for shoots and roots, respectively. It has recently received a great deal of attention, and the molecular mechanisms underlying its functioning are being elucidated. This review summarizes some of the recent advances in our understanding of gravitropism in higher plants, using roots as a model.

The root cap signals the root elongation zones of incorrect orientation within the gravity field

A combination of cell division at the apical meristem and cell expansion in the elongation zones allows roots to grow. New cells generated by divisions of meristematic initial cells first expand in all 3 dimensions, and then transit into an anisotropic elongation phase. In the meantime, cells differentiate into specific types, including epidermis, cortex, endodermis, pericycle, and vasculature (Figure 1A, B).

The transition between quasi-isotropic and anisotropic growth is progressive, even though the two regions defined above received distinct names (distal and central elongation zones, respectively). In fact, the distal elongation zone (DEZ) was arbitrarily defined as the region between the meristem and the point within the elongation zone at which the rate of elongation reaches 0.3 of the peak rate (Ishikawa and Evans, 1993), whereas the central elongation zone (CEZ) was defined as centered on the region of maximal growth rate (Ishikawa and Evans, 1993; Fasano et al., 2001). Interestingly, CEZ and DEZ cells displayed substantially different physiological properties, including different abilities to respond to exogenous auxin or osmotic shock (Ishikawa and Evans, 1995; Baluska et al., 1996a; Fasano et al., 2001).

The curvature response to root reorientation within the gravity field (gravistimulation) involves a complex differential growth on opposite flanks of the elongation zones. First, cells at the bottom side of both DEZ and CEZ stop elongating, while cells at the topside of the DEZ increase their rates of elongation. Consequently, a curvature is initiated at the DEZ (Figure 1C, D). Then, the curvature is progressively transmitted basipetally through the CEZ into the mature zone. When the cells involved in this curvature response reach the mature zone, the curvature is fixed in place, and the root tip has resumed vertical downward growth (Barlow and Rathfelder, 1985; Zieschang and Sievers, 1991; Ishikawa and Evans, 1993).

Even though the curvature response to gravistimulation occurs entirely in the DEZ and CEZ, stimulus perception occurs in the cap, a cellular structure that covers the root apical meristem (Figure 1A, B). Hence, a physiological signal has to be generated in the root cap upon gravistimulation, and transmitted to the elongation zones where it regulates the curvature response. As discussed below, this signal may correspond to an auxin gradient generated across the gravistimulated root cap, possibly along with electrical and/or other chemical signals (Evans and Ishikawa, 1997).

Starch or not starch? A question without a complete answer

The first question that needs to be answered when attempting to explain gravitropism is the identity of the gravity receptor. How can a plant organ sense gravity? Physically speaking, gravity is a directional force that exerts on any object of defined mass, displacing or distorting them. Hence, it is reasonable to assume that perception of gravity would involve an object of specified mass (susceptor). Displacement or distortion of that object would then be perceived by a receptor, which should transduce the corresponding physical information into a physiological signal (Sack, 1991).

Early centrifugation experiments indicated that gravity sensing occurs in a region of the root tip that includes the root cap. Later studies suggested that it might involve the sedimentation of starch-containing plastids (amyloplasts) in highly specialized cells at the center of the cap, named columella cells (reviewed in Poff and Martin, 1989).

The columella cells are different from other types of cells present in a plant (Figure 2). The cellular structures within these cells are highly polarized, such that the nucleus is located on the top or in the middle, and the cytoplasm is divided into two microscopically distinct zones. A central zone is characterized by the presence of a dynamic network of actin filaments that permits sedimentation of the starch-containing dense amyloplasts along defined paths (Driss-Ecole et al., 2000; Collings et al., 2001; Yoder et al., 2001). A peripheral zone, on the other hand, contains most

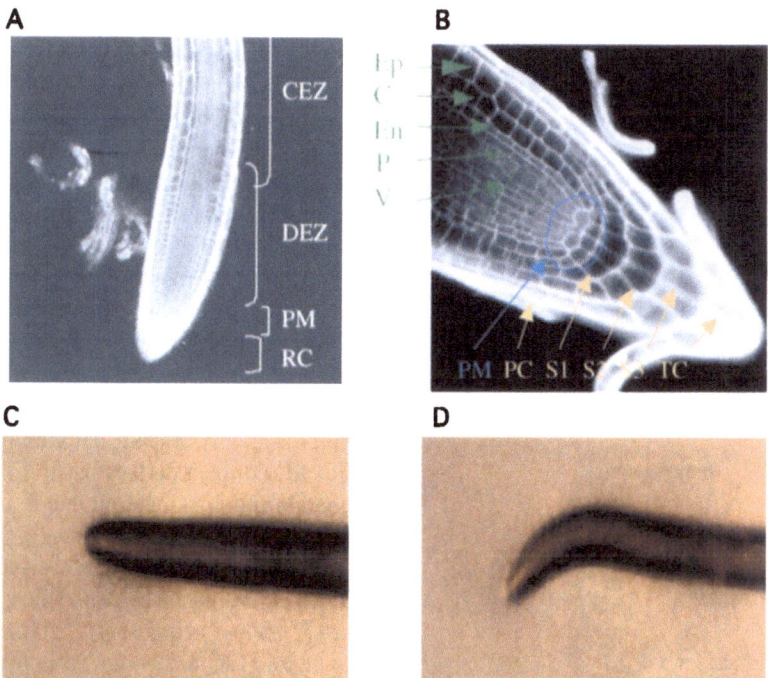

Figure 1. Root-curvature response to gravistimulation involves differential cellular elongation on opposite sides of the elongation zones. A, B. Confocal images of root tips of propidium iodide-stained 5-day old *Arabidopsis* seedlings showing the different regions involved in root growth (A), including the root cap (RC), the promeristem (PM), the distal elongation zone (DEZ) and the central elongation zone (CEZ). B. Close-up of a root tip including the DEZ, the PM and the RC, The different cell types that constitute a root are indicated: epidermis (Ep), cortex (Co), endodermis (En), pericycle (P), provasculature (V), promeristem (PM), peripheral root cap cells (PC), S1, S2 and S3 layers of columella cells, and root-cap tip cells (TC). C, D. Graviresponse of a 3-day old *Arabidopsis thaliana* root at time 0 (C), or 3 h after gravistimulation (D). One root width equals ca. 160 μm. Experimental procedures were as described in Cnops *et al.* (2000) (A), and in Rosen *et al.* (1999) (B, C), respectively.

of the endoplasmic reticulum (ER), other organelles, and thin transverse microfilaments and cortical microtubules. Interestingly, most of the peripheral ER is tubular. However, a small fraction of this ER forms nodal structures which are not found in other cell types, and are distributed asymmetrically within the most peripheral cells of the columella. These interesting structures are composed of a central rod to which up to 7 lateral rough ER cisternae are attached (Zheng and Staehelin, 2001). Their function may well be to protect putative focal attachment points of the cytoskeleton to the plasma membrane and cell wall from potential damage by the sedimenting amyloplasts (Zheng and Staehelin, 2001).

Because the root-cap columella cells contain sedimenting amyloplasts (also named statoliths), Haberlandt and Nemec independently proposed that the sedimentation of amyloplasts is the gravity-sensing mechanism in higher plants (starch-statolith hypothesis; Figure 2B, C) (Haberlandt, 1900; Nemec, 1900). Numerous studies have since strengthened this corre-

lation (reviewed in Sack, 1997). Furthermore, Fukaki and collaborators have demonstrated that two *Arabidopsis* mutants lacking a fully differentiated endodermal layer in both the hypocotyl and the root (*scr* and *shr*) were not capable of hypocotyl gravitropism, but developed significant gravi-responses in roots (Fukaki *et al.*, 1998). Because statolith-containing cells are only located in the endodermis of stems and hypocotyls, but not of roots, this result supported the starch-statolith hypothesis for susception in stems and hypocotyls.

A creative approach was developed to test the starch-statolith model (Kuznetsov and Hasenstein, 1997). Because starch is diamagnetic, it can be displaced in a predictable manner when exposed to a high-gradient magnetic field. Kuznetsov and Hasenstein placed a vertically grown root tip within a high-gradient magnetic field, and observed both a lateral displacement of amyloplasts within the columella cells and the consequent development of a root-tip curvature in the same direction as the displacement of

308

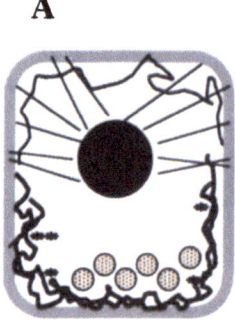

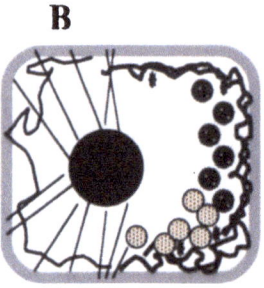

Figure 2. Amyloplast sedimentation within the root-cap columella cells constitutes the primary gravity-sensing mechanism in roots. A. Schematic representation of a root-cap columella cell, showing the nucleus (black circle) connected to the plasma membrane and cell wall by cytoskeleton elements (black bars), peripheral ER made of tubular (black convoluted lines) and nodal ER elements (asterisks), and amyloplasts sedimenting to the bottom of the cell (patterned circles). B. Schematic representation of the same cell as in B, after the corresponding root has been reoriented by 90° within the gravity field (gravistimulation). Amyloplasts sediment from their original positions (marked by dotted circles) into their new positions (patterned circles) at the new physical bottom of the cell.

amyloplasts. Importantly, columella plastids were not displaced in starch-deficient mutants, and the corresponding curvature was not observed.

When coleoptiles and hypocotyls were exposed to the same magnetic gradient, a curvature in the direction opposite to that of the starch-containing plastid displacement was observed. This result was consistent with an involvement of amyloplast sedimentation in gravity sensing because the directions of shoot and root gravitropism are also opposite. However, the *lazy-2* mutant of tomato is an exception. In this case, mutant shoots developed a curvature response in the direction of amyloplast sedimentation when exposed to light. Interestingly, this mutant was previously shown to display positive shoot gravitropism upon exposure to light. Hence, this experimental result supported a role for amyloplast sedimentation in gravity sensing (Hasenstein and Kuznetsov, 1999).

Root-tip ablation experiments demonstrated that the root cap is necessary for root gravitropism (Sack, 1991; Blancaflor *et al.*, 1998). They also led to the identification of root-cap cells that contribute to most of the gravity-sensing capability of roots. Indeed, laser-based cell-specific micro-ablation of subsets of columella cells in *Arabidopsis thaliana* root caps indicated that the first two layers of columella cells contributed to most of the gravity sensing potential in *Arabidopsis* roots (Blancaflor *et al.*, 1998). Interestingly, these cells contained the largest and most sedimentation-competent amyloplasts.

Hence, amyloplast sedimentation appears important for gravity sensing. This ability of columella plastids to sediment derives partly from the fact that they contain starch, a material that is 1.4 times denser than the cytoplasm, and from their weak connections with a loosely organized and highly dynamic microfilament network within these cells (Sack, 1997; Driss-Ecole *et al.*, 2000; Collings *et al.*, 2001; Yoder *et al.*, 2001). In theory, starch would be expected to play an important role in gravity sensing, considering its high density. In reality, starchless and starch-deficient mutants were shown to be only partially defective in root and hypocotyl gravitropism. They displayed intermediate levels of graviresponsiveness that correlated with their plastids' ability to sediment (Kiss *et al.*, 1996; MacCleery and Kiss, 1999).

Even though amyloplast sedimentation appears to be the primary mechanism for gravity sensing in higher plants, an increasing body of indirect evidence suggests that alternative mechanisms may also contribute to graviperception (Staves, 1997; Staves *et al.*, 1997; Blancaflor *et al.*, 1998; Fasano *et al.*, 2001). For instance, decapped roots still showed some evidence of graviresponsiveness (Blancaflor *et al.*, 1998). Also, starchless mutants were capable of some graviresponsiveness, even though plastids did not appear to sediment upon gravistimulation (Kiss *et al.*, 1996; MacCleery and Kiss, 1999). Furthermore, exposing rice roots to solutions of increasing densities while maintaining the osmolarity constant did not alter the rate of amyloplast sedimentation or the overall rate of root growth. Yet, it still resulted in decreased rates of gravitropic curvature (Staves *et al.*, 1997).

Other lines of evidence are also compatible with the existence of an alternative gravity sensing mechanism that involves either the root-cap statocytes, or cells of the meristem or DEZ. For instance, changes in proton secretion were observed on the topside of graviresponding roots, within 20 min of gravistimulation (Behrens *et al.*, 1982), and were associated with acidification of the wall in both the DEZ and the CEZ (Fasano *et al.*, 2001). Even though the root cap was necessary for development of these electrical responses, it appears that an alternative mechanism possibly, not involving amyloplast sedimentation, might be responsible for their occurrence in DEZ cells. Indeed, acidification of the cell wall on the topside of the DEZ was similar in wild-type and starch-deficient mutant roots, while wall acidification at the topside of the CEZ was altered in starch deficient mutants compared to wild type (Fasano *et al.*, 2001).

Changes in the membrane potential of cortical cells at the DEZ of mung bean roots also occurred within 30 s of gravistimulation. The kinetics of these changes were faster than expected if they would derive from a chemical signal originated in the cap (Ishikawa and Evans, 1990). Hence, these changes in membrane potential could be the consequence of gravity sensing by DEZ cells. Alternatively, they could be activated by an electrical signal that originated in the root-cap statocytes upon gravistimulation.

Hence, an alternative or secondary mechanism for gravity sensing may exist in roots. Unfortunately, its molecular identity and overall contribution to gravitropism remain undefined. One might speculate that it could involve the entire cell protoplast as a gravity susceptor (Staves, 1997). Indeed, plant protoplasts are embedded in a rigid cell wall, and are connected to it by transmembrane cytoskeletal linkages to regulate membrane tension at specific sites. The sensing machinery could involve an integration system that allows subtraction of the total force exerted by the protoplast on opposing sides of the cell, thereby discriminating the gravity pressure from other background pressures (e.g. osmotic) directed outwardly on all walls. Coincidentally, large internodal cells of *Chara* appear to utilize this whole-cell model of gravity sensing to modulate the direction of cytoplasmic streaming (Staves, 1997).

Although most of the evidence available points to the primary role of root-cap amyloplasts in gravity susception, the corresponding gravireceptors remain elusive. More than 10 years ago, Andreas Sievers and collaborators proposed that amyloplast sedimentation promotes tensions in the plasma membrane by pulling on the membrane-associated actin microfilaments. Membrane stretching would promote the opening of mechanosensitive Ca^{2+}-selective channels and trigger a signal transduction cascade to result in a gravitropic response (Sievers *et al.*, 1989, 1991).

Evidence for amyloplast connections to the actin network comes from cell-biological, pharmacological and physiological experiments. Firstly, both the central and peripheral regions of the columella cytoplasm contain a discrete and dynamic network of actin microfilaments, with increased density around the amyloplasts and nucleus (Hensel, 1985; Driss-Ecole *et al.*, 2000; Collings *et al.*, 2001; Yoder *et al.*, 2001; Zheng and Staehelin, 2001). Secondly, the disruption of root-cap microfilaments by treatments with cytochalasin B resulted in altered distribution of plastids and nuclei (Hensel, 1985; Lorenzy and Perbal, 1990), and increased sedimentation of amyloplasts within the statocytes (Sievers *et al.*, 1989). The treatment of cress roots with cytochalasin D, antagonized drug that disrupts actin microfilaments, also reduced the decrease in intracellular membrane potential of root statocytes generated by gravistimulation (Sievers *et al.*, 1995). Thirdly, sedimented amyloplasts accomplish saltatory movements (Sack *et al.*, 1986). Furthermore, they were displaced toward the proximal pole of statocytes when seedlings were exposed to reduced gravitational field during parabolic flights of rockets, suggesting that they might be suspended on actin microfilaments (Volkmann *et al.*, 1991; Buchen *et al.*, 1993).

It is interesting to note that similar molecules (microfilaments and mechanosensitive channels as receptors) were also proposed as essential players in gravity reception in situations where the whole-cell model of gravity sensing appears valid (Staves, 1997). Accordingly, in *Chara* internodal cells, the gravity receptors appear to be specialized stretch-activated calcium channels located on the upper and lower sides of the cell (Staves, 1997).

How is the information derived from gravity reception interpreted and converted into a curvature response in the DEZ and CEZ?

Because the gravity receptor functioning in the columella cells remains uncharacterized, any model attempting to explain the transduction of gravity signals will be speculative, at best. Several potential secondary messengers have, however, been identified, and may play important roles in this pathway. These include cytosolic Ca^{2+}, pH and InsP$_3$, as discussed below. The protein encoded by *ARG1* may also function in root and hypocotyl gravitropism (Sedbrook *et al.*, 1999).

310

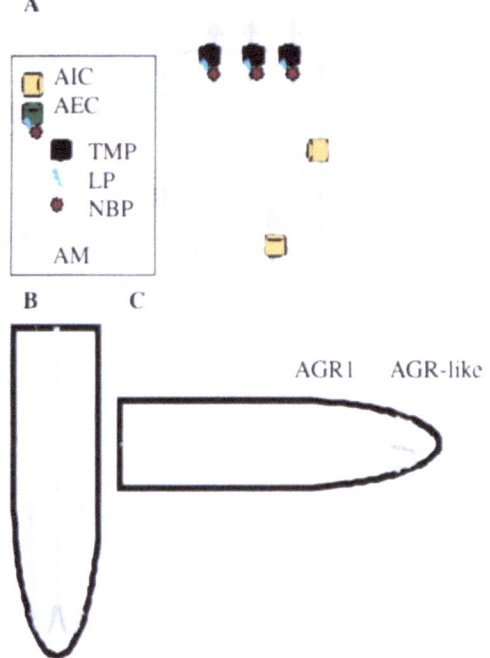

Figure 3. A model of auxin transport in *Arabidopsis* root tips. A. Polar auxin transport occurs from cell to cell within specified cell files. Individual cells carry transmembrane transporters that mediate auxin uptake (auxin influx carrier, AIC) and efflux (auxin efflux carrier, AEC). The auxin efflux carrier is believed to be composed of 3 elements: a transmembrane transporter (TMP), a regulatory NPA binding protein (NBP), and a putative linker component (LP). It is located on the basal side of the cells, thereby defining the polarity of transport. Auxin is taken up by free diffusion through the membranes and/or by transport through an influx carrier (encoded by *AUX1* in *Arabidopsis* root tips). It is then exported out of the cell by specific efflux carriers. Arrows indicate the direction of auxin movement (AM). B. IAA is primarily synthesized in young shoot tissues and transported through the vasculature into the root tip. There, it is redistributed to more lateral tissues through a transport system that may include members of the *AGR1* gene family (AGR-like) in *Arabidopsis thaliana*. From there, peripheral auxin is transported back to the elongation zones through a retrograde polar transport system that involves the *AUX1* and *AGR1* genes of *Arabidopsis thaliana*. Dotted and light gray arrows indicate where less auxin is transported. C. Upon gravistimulation, auxin redistribution to more lateral tissues of the root tip would become asymmetric, resulting in the formation of a lateral auxin gradient, responsible for part of the curvature.

Several physiological and pharmacological experiments have suggested a role for Ca^{2+} in gravity signal transduction. Pharmacological agents that are believed to interfere with the activity of Ca^{2+} channels, calmodulin or Ca^{2+}-ATPases were shown to affect gravitropism (Biro *et al.*, 1982; Björkman and Leopold, 1987a; Stinemetz *et al.*, 1987; Sievers and Busch, 1992; Busch and Sievers, 1993; Sinclair *et al.*, 1996). Also, the columella cells of the root cap con-

tain unusually high levels of Ca^{2+} and calmodulin or calmodulin activity compared to other cells in the plant (Chandra *et al.*, 1982; Allan and Trewavas, 1985; Dauwalder *et al.*, 1986; Busch and Sievers, 1993; Sinclair *et al.*, 1996). Unfortunately, multiple attempts at demonstrating the existence of cytosolic Ca^{2+} fluxes in response to gravistimulation have been largely unsuccessful (Legue *et al.*, 1997; Sedbrook and Masson, unpublished data). It should be noted, however, that even small changes in cytosolic Ca^{2+} levels might have a significant signal-transducing effect in cells that contain high levels of calmodulin (Sinclair and Trewavas, 1997).

Fast gravity-induced changes in cytosolic pH have been observed within the columella cells of the root cap, and in the root-cap apoplast. These responses included an acidification of the apoplast from pH 5.5 to 4.5 within 2 min of gravistimulation, and an alkalinization of the cytoplasm within the columella cells shown to contribute to most of graviperception (Legue *et al.*, 1997). Both apoplastic and cytosolic pH changes occurred before initiation of a graviresponse, and the increase in cytosolic pH was necessary for gravitropism (Scott and Allen, 1999; Fasano *et al.*, 2001). The corresponding pH changes might be related to the rapid changes in membrane potential of *Lepidium* columella cells upon gravistimulation (Behrens *et al.*, 1985), and to the gravity-induced changes in ionic fluxes observed around the root cap (Björkmann and Leopold, 1987b; Björkman and Cleland, 1991; Zieschang *et al.*, 1993).

There is disagreement in the literature as to whether or not the cytosolic pH changes that occur in the root-cap columella cells upon gravistimulation differ between the lower and upper halves of the cap (Scott and Allen, 1999; Fasano *et al.*, 2001). Scott *et al.* reported different changes in cytosolic pH between columella cells on the upper and lower halves of gravistimulated root caps, within the same layer (Scott and Allen, 1999). If these results are correct, the data may support a role for pH changes in establishing the lateral transport polarity described below (Scott and Allen, 1999). However, Fasano *et al.* were not able to reproduce these results (Fasano *et al.*, 2001). Their data suggested that the gravity-induced changes in cytosolic pH within the columella cells do not provide vectorial information to the gravity signal transduction pathway. Instead, these pH changes may generate conditions around the gravisensing root-cap cells that are favorable to the development of a physiological signal that may include an auxin gradient (see below). According to this model, the gravity signal transduction pathway would trigger both the pH changes described above and the development of a lateral polarity responsible for lateral auxin redistribution, as discussed below.

Inositol 1,4,5-trisphosphate (InsP$_3$) may also be involved in gravity signal transduction as a second messenger, at least in cereal pulvini. InsP$_3$ levels and phosphatidylinositol-4-phosphate-5-kinase activity were found to fluctuate in both oat and corn pulvini (Perera et al., 1999, 2001). InsP$_3$ levels increased 3-fold in both upper and lower halves of gravistimulated oat pulvini within 15 s of gravistimulation, and fluctuated over the first few minutes. Then, InsP$_3$ levels increased 3-fold in the lower half over the upper between 10 and 30 min of gravistimulation. Phospholipase C activity was essential for both long-term InsP$_3$ increases and full gravitropic response, but not for short-term changes in InsP$_3$. It was proposed that short-term changes in InsP$_3$ levels might constitute a general stress signal, while the long-term increase on the lower side could contribute to the generation of a biochemical asymmetry between lower and upper halves. This asymmetry would precede the differential growth, which initiates within 30 minutes of gravistimulation (Perera et al., 2001).

Even though InsP$_3$ appears to function in various signal-transduction pathways in part by modulating Ca^{2+} homeostasis within the cell, it remains possible that activation of the phosphatidylinositol pathway mediates its positive effect on plant gravitropism by other means. Considering the general effect of gravistimulation on anisotropic cell expansion, it is interesting to speculate that the gravity-induced changes in levels of phosphoinositide-pathway intermediates and/or products might regulate the vesicular trafficking of membranes or transmembrane receptors within the responding cells, or cytoskeletal restructuring (Stevenson et al., 2000). The regulation of membrane biogenesis and vesicular trafficking could establish the biochemical asymmetry responsible for lateral auxin transport (Steinmann et al., 1999; Stevenson et al., 2000). More work is needed to test this interesting possibility.

Surprisingly, despite the accumulating evidence supporting an involvement of pH, InsP$_3$ and possibly Ca^{2+} in gravity signal transduction, few mutants have thus far been identified that affect this phase of gravitropism. The rhg and arg1 mutations may be the exceptions. These mutations appeared to affect root and hypocotyl gravitropism without affecting

phototropism or starch synthesis in the root cap. They also did not affect root-growth responses to phytohormones. Taken together, these data strongly suggested that both genes are involved in early phases of gravity sensing or signal transduction (Fukaki et al., 1997; Sedbrook et al., 1999).

The ARG1 gene encodes a dnaJ-like protein carrying a J domain at the N-terminus and a putative coiled coil domain toward the C-terminus. While the J domain of several dnaJ-like proteins has been implicated in interactions with HSP70, the putative coiled coil domain of ARG1 shares homology with coiled coils found in a number of proteins that are known to interact with the cytoskeleton (Sedbrook et al., 1999).

ARG1 belongs to a small gene family that also includes ARL1 and ARL2. ARG1 and ARL1 appear to be expressed in all tissues of the plant, whereas ARL2 expression level is below the limits of detection by northern blot analysis of total RNA. Interestingly, mutations in ARL2 result in phenotypes similar to those associated with arg1 (Sedbrook et al., 1999; Rosen, Guan, Boonsirichai and Masson, unpublished data). Hence, both ARG1 and ARL2 are involved in gravity signal transduction in roots and hypocotyls. Although the mechanisms by which ARG1 and ARL2 function in this process have not been elucidated yet, it appears possible that both proteins facilitate the formation of, or are involved in, a gravity signal transduction complex associated with the cytoskeleton (Sedbrook et al., 1999). The identification of proteins that interact biochemically or genetically with ARG1 and/or ARL2 should provide important information on other components of the gravity signal transduction pathway (Boonsirichai, Guan, Sedbrook and Masson, unpublished data).

A lateral gradient of auxin may contribute to signaling EZ cells of a gravistimulus perceived in the root cap

The Cholodny-Went theory proposes that gravistimulation promotes the development of a lateral gradient in auxin which, upon transmission to the elongation zones, promotes the differential growth responsible for the curvature response (discussed in Trewavas, 1992). In fact, several lines of evidence support the involvement of auxin and its transport in root gravitropism. Firstly, auxin-transport and auxin-response mutants show defects in root gravitropism (reviewed in Lomax, 1997; Dolan, 1998). Secondly, inhibitors of auxin

312

transport affect gravitropism at concentrations which are not sufficient to interfere with root growth (Muday and Haworth, 1994; Rashotte et al., 2000). Thirdly, root gravistimulation appears to promote the differential activation of auxin-response genes at opposite flanks of the root (Luschnig et al., 1998; Sabatini et al., 1999).

These three observations can be explained by the Cholodny-Went theory, or by a model postulating that gravistimulation promotes differential tissue sensitivity to auxin on opposite flanks of the root. Alternatively, auxin could simply constitute a permissive signal for gravitropism, without carrying directional information for curvature development. Interestingly, auxin appears to be redistributed laterally across gravistimulated root tips, and this lateral redistribution appears to depend upon active metabolism and Ca^{2+} (Young et al., 1990; Evans, 1991). Hence, auxin redistribution appears to be one of the targets for the gravity signal transduction pathway that occurs in the root cap. However, how this gravity transduction pathway regulates lateral auxin transport remains unclear. Recent work may provide tools to elucidate this process.

In plants, auxin transport occurs between adjacent cells within cell files (Figure 3). Apoplastic auxin enters cells by passive diffusion through the plasma membrane, as well as by transport through specific auxin-import carriers. It can then be transported out of cells by an efflux carrier which appears to be composed of a transmembrane carrier protein, a naphthylphthalamic acid (NPA)-binding regulatory moiety, and a putative linker module (Figure 3A). Transport polarity within specific cell files is mediated by the polar distribution of the auxin efflux carrier complex (Lomax et al., 1998; Chen et al., 1999, for reviews).

Molecular genetic studies of root gravitropism in Arabidopsis thaliana allowed the identification of genes encoding the auxin influx carrier (AUX1) and the transmembrane component of the efflux carrier (AGR1, also named EIR1, PIN2 or WAV6). Mutations in aux1 resulted in altered root gravitropism and increased root-growth resistance to auxin and ethylene. The root gravitropic defect was rescued by adding intermediate concentrations of 1-naphthaleneacetic acid (1-NAA) to the growth medium (an auxin known to penetrate freely through the membranes), but not by adding 2,4-dichlorophenoxyacetic acid (2,4-D) or indole-3-acetic acid (IAA) (two auxins known to require the auxin influx carrier to enter the cell). In addition, auxin-transport assays revealed a deficiency in auxin uptake by aux1 mutant roots. Furthermore,

the AUX1 protein shared homologies with tryptophan transporter proteins. Taken together, these data strongly suggested that AUX1 encodes a component of the auxin influx carrier in roots (Bennett et al., 1996; Yamamoto and Yamamoto, 1998; Marchant et al., 1999).

Mutations in the AGR1 gene resulted in altered root gravitropism, increased root-growth resistance to ethylene and inhibitors of polar auxin transport, and increased sensitivity to high concentrations of NAA (Bell and Maher, 1990; Chen et al., 1998; Luschnig et al., 1998; Müller et al., 1998a, b; Utsuno et al., 1998). Furthermore, agr1 mutant roots were less able to export radiolabeled IAA than wild type, and the basipetal transport of auxin was reduced (Chen et al., 1998; Rashotte et al., 2000). Molecular cloning revealed that AGR1 encodes a transmembrane protein with homologies to bacterial transporters (Chen et al., 1998; Luschnig et al., 1998; Müller et al., 1998a, b; Utsuno et al., 1998). When expressed in yeast, this protein conferred increased resistance to toxic IAA derivatives (5-fluoroindole and 5-fluoroindoleacetic acid) (Chen et al., 1998; Luschnig et al., 1998), and increased the export of pre-loaded ^{3}H-IAA from the cells (Chen et al., 1998). Hence, AGR1 appears to encode a component of the auxin efflux carrier in roots (Chen et al., 1998; Luschnig et al., 1998; Müller et al., 1998a,b; Utsuno et al., 1998). Immuno-localization studies demonstrated that the protein is localized in the basal membrane of root-tip EZ epidermal and cortical cells, as well as in the outermost lateral membrane of cortical cells, providing a molecular basis to transport polarity in cluster cells of roots (Müller et al., 1998a, b).

Hence, the AUX1 and AGR1 genes of Arabidopsis appear to be components of the machinery that mediates the transport of auxin from the root tip to the responding EZ cells (Figure 3). However, these genes are not expressed in the root cap (Chen et al., 1998; Müller et al., 1998a, b; Marchant et al., 1999). This raises two possibilities. These proteins may be needed for proper transmission of the gravity-induced lateral auxin gradient from the cap to the EZ (Figure 3B). Alternatively, AGR1 could contribute to the formation, maintenance or regulation of the lateral auxin gradient in this region of the root (Müller et al., 1998a, b). The presence of some AGR1 protein in the outer periclinal membrane of DEZ cortical cells is compatible with such a model. In this scenario, electrical signals transmitted from the root cap, or an inherent ability of

[64]

DEZ cells to sense gravity (as discussed below), could control AGR1 activity.

Regardless of the mode of AGR1 action in gravitropism, it appears likely that its activity is modulated by post-transcriptional processes, including protein turnover. Indeed, Sieberer *et al.* have recently shown that auxin decreases the stability of AGR1/EIR1-GUS protein fusions, suggesting that auxin transport is regulated through a feedback regulatory loop that affects protein stability in response to auxin (Sieberer *et al.*, 2000). This result is in agreement with earlier pharmacological experiments on suspension-cultured tobacco cells that suggested that some components of the efflux system were being turned over at the plasma membrane with a half-life less than 10 min (Delbarre *et al.*, 1998).

That lateral auxin redistribution occurs in the root tip in the absence of AGR1 seems substantiated by the fact that an auxin-response reporter gene that fuses the β-glucuronidase open reading frame to a minimal core promoter carrying multiple copies of the auxin-response element (*DR5::GUS*) is activated on the bottom side of gravistimulated root tips, laterally to the quiescent center and first columella layers, in *agr1* mutant roots (Sabatini *et al.*, 1999). However, the mechanisms that mediate lateral auxin transport in the root tip have not been characterized yet. Interestingly, both *AUX1* and *AGR1* belong to gene families. It is possible that other members of each gene family mediate the transport functions required for the proper establishment of this gradient in the cap. In fact, one member of the *AGR1* gene family has already been shown to express in the quiescent center and columella cells of the root cap. The corresponding protein appears to predominantly localize evenly at the plasma membranes of these cells in vertically grown roots (Chen and Masson, manuscript in preparation). An interesting possibility currently under investigation is that this protein, and/or other AGR1-like proteins that may also be expressed in the root cap, is/are differentially localized in the lateral membranes of statocytes in response to gravistimulation. This model would be compatible with previous reports on the regulation of auxin efflux in other systems, which showed that some components of the efflux complex may be turned over at the plasma membrane with a half-life less than 10 min (Delbarre *et al.*, 1998), and that vesicle trafficking may establish cell polarity in developing *Arabidopsis* embryos, resulting in polar auxin transport (Steinmann *et al.*, 1999). Alternatively, the gravity signal transduction pathway could modulate

the transport activity of these proteins by phosphorylation. This alternative model is not exclusive with the first one, and is compatible with observations made in cultured tobacco cells, indicating that protein phosphorylation is essential to sustain the activity of auxin efflux carriers (Delbarre *et al.*, 1998).

The curvature response to gravistimulation involves differential cellular elongation on opposite flanks of the EZs

As already emphasized in this review, initiation of the curvature response to gravistimulation involves the differential elongation of cells on opposite flanks of the DEZ. Cells at the topside increase their rates of elongation, while those at the bottom side display decreased elongation rates. Several studies have suggested that this DEZ-mediated phase of gravitropism is independent of any auxin gradient. Firstly, the DEZ still drives a vigorous gravitropic curvature when the tip of a root is gravistimulated in the presence of high concentrations of auxin, sufficient to completely inhibit growth (Ishikawa and Evans, 1993). Furthermore, the cell walls on the upper side of gravistimulated DEZs acidifies within 10 minutes of gravistimulation in both wild type and starch-deficient mutants (Fasano *et al.*, 2001). This response was much faster than one would have expected if it was triggered by an auxin signal generated in the root-cap statocytes (Fasano *et al.*, 2001). Hence, DEZ cells may receive some other signals from the root cap upon gravistimulation, possibly apoplastic Ca^{2+} or other electrical or chemical signals (Baluska *et al.*, 1996a; Björkman and Cleland, 1991; Lee *et al.*, 1983; Stevenson *et al.*, 2000), or they are capable of sensing gravity on their own.

On the other hand, the second phase of gravitropism, which involves the propagation of the curvature through the CEZ, appears to be auxin-gradient dependent. Therefore, in order for a gravitropic response to proceed until completion, proper auxin redistribution, transport and response by CEZ cells are needed. The processes involved in cellular responses to auxin have also been extensively studied. Important breakthroughs in our understanding of these processes will be discussed in several reviews within this issue of *Plant Molecular Biology*. We will simply mention that auxin acts on cellular elongation in roots by regulating the activity of plasma-membrane proton pumps responsible for the acidification of the cell wall which,

314

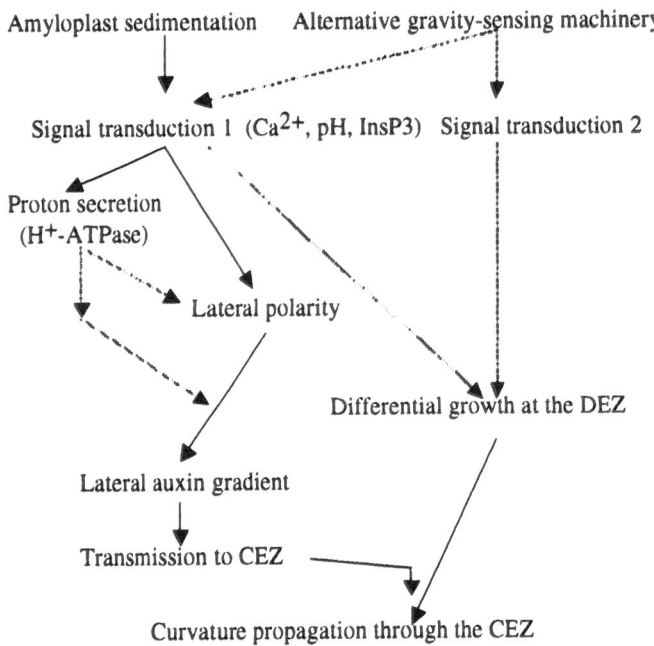

Figure 4. Model summarizing the different phases of root gravitropism, as discussed in this review. Plain arrows indicate putative regulatory steps in the pathway. Dotted arrows and question marks indicate unresolved alternative ways of regulating specific steps in the pathway, or unresolved alternative activities of specific components in the pathway.

in turn, regulates cell-wall extensibility. Auxin also acts on intracellular processes which result in the activation of auxin-specific signal transduction pathways involving protein degradation and regulation of gene expression (del Pozo and Estelle, 1999). Several genes involved in these pathways have been characterized, and shown to be essential for root gravitropism (del Pozo and Estelle, 1999).

Conclusions

Recent advances in our understanding of the molecular mechanisms involved in gravitropism have derived from the appropriate utilization of integrated analytical approaches involving molecular genetics, physiology, biochemistry, and cell biology. All data currently available point to the fascinating molecular complexity that underlies this rather simple growth behavior. A model attempting to explain the main phases of root gravitropism can be set forth at this time, based on the discussion included in this review (Figure 4).

The sedimentation of amyloplasts within the columella cells of the root cap appears to be the primary mechanism allowing roots to perceive the gravity vector. However, evidence is accumulating toward the existence of another sensing machinery that may con-

tribute to gravitropism. Several molecules have been implicated in the transduction of gravity signals in gravi-sensing cells, including Ca^{2+}, pH and InsP3. We can speculate that the gravity signal transduction pathway(s) regulates the activity of auxin efflux carriers on opposite lateral membranes of the statocytes, possibly through differential phosphorylation, protein turnover and/or targeting. Consequently, auxin is redistributed laterally and accumulates at the physical bottom of the root tip. Acidification of the apoplast surrounding the statocytes, also triggered by the gravity signal transduction pathway(s), may facilitate this process. The corresponding auxin gradient is then transmitted to the elongation zones where it regulates cellular elongation. Additionally, the gravity signal transduction pathway(s) may also generate another signal, possibly electrical, which regulates the differential cell elongation that is responsible for the initial phase of the gravi-curvature at the DEZ.

Unfortunately, many steps in this process remain poorly characterized, and many questions remain unsolved, as discussed in this review. However, the basic tools required to answer some of these fundamental questions are becoming available. These include the development of *in vivo* markers, allowing the tracing of potential signal transducers (Ca^{2+}, pH, auxin, InsP$_3$) and of cytoskeletal elements in real time and space, and the identification and characterization of new molecules involved in the regulation of auxin transport and membrane channels. Forward and reverse genetics, genomics and proteomics provide new ways to identify additional components in the pathway. These strategies, coupled to more classical approaches (physiology, biochemistry, cell biology), should also allow us to better understand the order of the components of this complex perception and transduction machinery. At the same time, an integrated approach will also provide the tools required to define how roots can integrate the information coming from a multitude of environmental parameters, in addition to gravity, and use that global information to decide on a more appropriate growth behavior in compliance with environmental constraints. We can finally foresee a time when we can not only better understand some of these fascinating plant growth behaviors, but also use this new information to engineer plants that better respond to extreme environments.

Acknowledgements

This review was made possible by grants from NASA (NAG2-1189 and NAG2-1336), NSF (MCB-9905675), and HATCH funds (WIS04310). We thank Kai ('Billy') Hung, Jessica Will, Li-Sen Young and Christen Yuen for critically reading this manuscript. This is manuscript number 0000 of the Laboratory of Genetics.

References

Allan, E. and Trewavas, A. 1985. Quantitative changes in calmodulin and NAD kinase during early cell development in the root apex of *Pisum sativum*. Planta 165: 493–501.

Baluska, F., Hauskrecht, M., Barlow, P. and Sievers, A. 1996a. Gravitropism of the primary root of maize: a complex pattern of differential cellular growth in the cortex independently of the microtubular cytoskeleton. Planta 198: 310–318.

Baluska, F., Volkmann, D. and Barlow, P. 1996b. Specialized zones of development in roots: view from the cellular level. Plant Physiol. 112: 3–4.

Barlow, P.W. and Rathfelder, E.L. 1985. Distribution and redistribution of extension growth along vertical and horizontal gravireacting maize roots. Planta 165: 134–141.

Behrens, H., Weisenseel, M. and Sievers, A. 1982. Rapid changes in the pattern of electric current around the root tip of *Lepidium sativum* L. following gravistimulation. Plant Physiol. 70: 1079–1083.

Behrens, H., Gradmann, D. and Sievers, A. 1985. Membrane-potential responses following gravistimulation in roots of *Lepidium sativum* L. Planta 163: 463–472.

Bell, C.J. and Maher, E.P. 1990. Mutants of *Arabidopsis thaliana* with abnormal gravitropic response. Mol. Gen. Genet. 220: 289–293.

Bennett, M., Marchant, A., Green, H., May, S., Ward, S., Millner, P., Walker, A., Schulz, B. and Feldmann, K. 1996. *Arabidopsis AUX1* gene: a permease-like regulator of root gravitropism. Science 273: 948–950.

Biro, R., Hale, C., Wiegang, O. and Roux, S. 1982. Effect of chlorpromazine on gravitropism in *Avena* coleoptiles. Ann. Bot. 50: 737–742.

Björkman, T. and Cleland, R. 1991. The role of extracellular free Ca^{2+} gradients in gravitropic signalling in maize roots. Planta 185: 379–384.

Björkman, T. and Leopold, A. 1987a. Effect of inhibitors of auxin transport and of calmodulin on a gravisensing-dependent current in maize roots. Plant Physiol. 84: 847–850.

Björkmann T. and Leopold, A. 1987b. An electric current associated with gravisensing in maize roots. Plant Physiol 83: 841–846.

Blancaflor, E., Fasano, J. and Gilroy, S. 1998. Mapping the functional roles of cap cells in the response of *Arabidopsis* primary roots to gravity. Plant Physiol. 116: 213–222.

Büchen, B., Braun, M., Hejnowicz, Z. and Sievers, A. 1993. Statoliths pull on microfilaments. Experiments under microgravity. Protoplasma 172: 38–42.

Busch, M. and Sievers, A. 1993. Membrane traffic from the endoplasmic reticulum to the Golgi apparatus is disturbed by an inhibitor of the Ca^{2+} ATPase in the ER. Protoplasma 177: 23–31.

Chandra, S., Chabot, J., Morrison, G. and Leopold, A. 1982. Localization of Ca^{2+} in amyloplasts of root cap cells using ion microscopy. Science 216: 1221–1223.

Chen, R., Hilson, P., Sedbrook, J., Rosen, E., Caspar, T. and Masson, P. 1998. The *Arabidopsis thaliana AGRAVITROPIC 1* gene encodes a component of the polar-auxin-transport efflux carrier. Proc. Natl. Acad. Sci. USA 95: 15112–15117.

Chen, R., Rosen, E. and Masson, P. 1999. Update: Gravitropism in higher plants. Plant Physiol. 120: 343–350.

Cnops, G., Wang, X., Linstead, P., Van Montagu, M., Van Lijsebettens, M. and Dolan, L. 2000. TORNADO1 and TORNADO2 are required for the specification of radial and circumferential pattern in the *Arabidopsis* root. Development 127: 3385–3394.

Collings, D., Zsuppan, G., Allen, N. and Blancaflor, E. 2001. Demonstration of prominent actin filaments in the root columella. Planta 212: 392–403.

Dauwalder, M., Roux, S. and Hardison, L. 1986. Distribution of calmodulin in pea seedlings: immunochemical localization in plumules and root apices. Planta 168: 461–470.

Delbarre, A., Muller, P. and Guern, J. 1998. Short-lived and phosphorylated proteins contribute to carrier-mediated efflux, but not to influx, of auxin in suspension-cultured tobacco cells. Plant Physiol 116: 833–844.

del Pozo, J. and Estelle, M. 1999. Function of the ubiquitin-proteosome pathway in auxin response. Trends Plant Sci. 4: 107–112.

Dolan, L. 1998. Pointing roots in the right direction: The role of auxin transport in response to gravity. Genes Dev. 12: 2091–2095.

Driss-Ecole, D., Vassy, J., Rembur, J., Guivarc'h, A., Prouteau, M., Dewitte, W. and Perbal, G. 2000. Immunolocalization of actin in root statocytes of *Lens culinaris* L. J. Exp. Bot. 51: 521–528.

Evans, M. 1991. Gravitropism: Interaction of sensitivity modulation and effector redistribution. Plant Physiol. 95: 1–5.

316

Evans, M. and Ishikawa, H. 1997. Cellular specificity of the gravitropic motor response in roots. Planta 203: S115–S122.

Fasano, J., Swanson, S., Blancaflor, E., Dowd, P., Kao, T. and Gilroy, S. 2001. Changes in root cap pH are required for the gravity response of the *Arabidopsis* root. Plant Cell 13: 907–921.

Firn, R.D. and Digby, J. 1997. Solving the puzzle of gravitropism: has a lost piece been found? Planta 203: S159–S163.

Fukaki, H., Fujisawa, H. and Tasaka, M. 1997. The *RHG* gene is involved in root and hypocotyl gravitropism in *Arabidopsis thaliana*. Plant Cell Physiol. 38: 804–810.

Fukaki, H., Wysocka-Diller, J., Kato, T., Fujisawa, H., Benfey, P. and Tasaka, M. 1998. Genetic evidence that the endodermis is essential for shoot gravitropism in *Arabidopsis thaliana*. Plant J. 14: 425–430.

Haberlandt, G. 1900. Über die Perzeption des geotropischen Reizes. Ber. Dt. Bot. Ges. 18: 261–272.

Hasenstein, K.H. and Kuznetsov, Ol.A. 1999. The response of *lazy-2* tomato seedlings to curvature-inducing magnetic gradients is modulated by light. Planta 208: 59–65.

Hensel, W. 1985. Cytochalasin B affects the structural polarity of statocytes from cress roots (*Lepidium sativum* L.). Protoplasma 129: 178–187.

Ishikawa, H. and Evans, M. 1990. Gravity-induced changes in intracellular potentials on elongating cortical cells of mung bean roots. Plant Cell Physiol. 31: 457–462.

Ishikawa, H. and Evans, M. 1993. The role of the distal elongation zone in the response of maize roots to auxin and gravity. Plant Physiol 102: 1203–1210.

Ishikawa, H. and Evans, M. 1995. Specialized zones of development in roots. Plant Physiol. 109: 725–727.

Kiss, J., Wright, J. and Caspar, T. 1996. Gravitropism in roots of intermediate-starch mutants of *Arabidopsis*. Physiol. Plant. 97: 237–244.

Kuznetsov, O. and Hasenstein, K. 1996. Intracellular magnetophoresis of amyloplasts and induction of root curvature. Planta 198: 87–94.

Kuznetsov, O. and Hasenstein, K. 1997. Magnetophoretic induction of curvature in coleoptiles and hypocotyls. J. Exp. Bot. 48: 1951–1957.

Lee, J., Mulkey, T. and Evans, M. 1983. Gravity induced polar transport of calcium across root tips of maize. Plant Physiol. 73: 874–876.

Legue, V., Blancaflor, E., Wymer, C., Perbal, G., Fantin, D. and Gilroy, S. 1997. Cytoplasmic free Ca^{2+} in *Arabidopsis* roots changes in response to touch but not gravity. Plant Physiol. 114: 789–800.

Lomax, T. 1997. Molecular genetic analysis of plant gravitropism. Gravit. Space Biol Bull. 10: 75–82.

Lomax, T., Muday, G. and Rubery, P. 1998. Auxin transport. In: P.J. Davies (Ed.) Plant Hormones: Physiology, Biochemistry, and Molecular Biology, Kluwer Academic Publishers, Dordrecht, Netherlands, pp. 509–530.

Lorenzy, G. and Perbal, G. 1990. Actin filaments responsible for the location of the nucleus in the lentil statocyte are sensitive to gravity. Biol. Cell 68: 259–263.

Luschnig, C., Gaxiola, R., Grisafi, P. and Fink, G. 1998. EIR1, a root-specific protein involved in auxin transport, is required for gravitropism in *Arabidopsis thaliana*. Genes Dev. 12: 2175–2187.

MacCleery, S. and Kiss, J. 1999. Plastid sedimentation kinetics in roots of wild-type and starch-deficient mutants of *Arabidopsis*. Plant Physiol. 120: 183–192.

Marchant, A., Kargul, J., May, S., Muller, P., Delbarre, A., Perrot-Rechenmann, C. and Bennett, M. 1999. AUX1 regulates root gravitropism in *Arabidopsis* by facilitating auxin uptake within root apical tissues. EMBO J. 18: 2066–2073.

Muday, G. and Haworth, P. 1994. Tomato root growth, gravitropism, and lateral root development: correlation with auxin transport. Plant Physiol. Biochem. 32: 193–203.

Müller, A., Guan, C., Galweiler, L., Tanzler, P., Huijser, P., Marchant, A., Parry, G., Bennett, M., Wisman, E. and Palme, K. 1998. AtPIN2 defines a locus of *Arabidopsis* for root gravitropism control. EMBO J. 17: 6903–6911.

Nemec, B. 1900. Über die Art der Wahrnehmung des Schwerkraftreizes bei den Pflanzen. Ber. Dt. Bot. Ges. 18: 241–245.

Perera, I., Heilmann, I. and Boss, W. 1999. Transient and sustained increases in inositol-1,4,5-trisphosphate precede the differential growth response in gravistimulated maize pulvini. Proc. Natl. Acad. Sci. USA 96: 5838–5843.

Perera, I., Heilmann, I., Chang, S., Boss, W. and Kaufman, P. 2001. A role for inositol 1,4,5-trisphosphate in gravitropic signaling and the retention of cold-perceived gravistimulation of oat shoot pulvini. Plant Physiol. 125: 1499–1507.

Poff, K. and Martin, H. 1989. Site of graviperception in roots: a re-examination. Physiol. Plant. 76: 451–455.

Rashotte, A., Brady, S., Reed, R., Ante, S. and Muday, G. 2000. Basipetal auxin transport is required for gravitropism in roots of *Arabidopsis*. Plant Physiol. 122: 481–490.

Rosen, E., Chen, R. and Masson, P.H. 1999. Gravitropism: a complex response to a simple stimulus? Trends Plant Sci. 4: 407–412.

Sabatini, S., Beis, D., Wolkenfelt, H., Murfett, J., Guilfoyle, T., Malamy, J., Benfey, P., Leyser, O., Bechtold, N., Weisbeek, P. and Scheres, B. 1999. An auxin-dependent distal organizer of pattern and polarity in the *Arabidopsis* root. Cell 99: 463–472.

Sack, F. 1991. Plant gravity sensing. Int. Rev. Cytol. 127: 193–252.

Sack, F. 1997. Plastids and gravitropic sensing. Planta 203: S63–S68.

Sack, F., Suyemoto, M. and Leopold, A.C. 1986. Amyloplast sedimentation and organelle saltation in living columella cells. Am. J. Bot. 73: 1692–1698.

Scott, A. and Allen, N. 1999. Changes in cytosolic pH within *Arabidopsis* root columella cells play a key role in the early signaling pathway for root gravitropism. Plant Physiol. 121: 1291–1298.

Sedbrook, J., Chen, R. and Masson, P. 1999. *ARG1* (Altered Response to Gravity) encodes a novel DnaJ-like protein which potentially interacts with the cytoskeleton. Proc. Natl. Acad. Sci. USA 96: 1140–1145.

Sieberer, S., Seifert, G.J., Hauser, M.T., Grisafi, P., Fink, G.R., Luschnig, G. 2000. Post-transcriptional control of the Arabidopsis auxin efflux carrier EIR1 requires AXR1. Curr. Biol. 10: 1595–1598.

Sievers, A. and Busch, M. 1992. An inhibitor of the Ca^{2+}-ATPases in the sarcoplasmic reticula inhibits transduction of the gravity stimulus in cress roots. Planta 188: 619–622.

Sievers, A., Kruse, S., Kuo-Huang, L. and Wendt, M. 1989. Statoliths and microfilaments in plant cells. Planta 179: 275–278.

Sievers, A., Buchen, B., Volkmann, D. and Hejnowicz, Z. 1991. Role of the cytoskeleton in gravity perception. In: C. Lloyd (Ed.) The Cytoskeletal Basis of Plant Growth and Form, Academic Press, London, pp. 169–182.

Sievers, A., Sondag, C., Trebacz, K. and Hejnowicz, Z. 1995. Gravity induced changes in intracellular potentials in statocytes of cress roots. Planta 197: 392–398.

Sinclair, W. and Trewavas, A. 1997. Calcium in gravitropism: a re-examination. Planta 203: S85–S90.

Sinclair, W., Oliver, I., Maher, P. and Trewavas, A. 1996. The role of calmodulin in the gravitropic response of *Arabidopsis thaliana agr-3* mutant. Planta 199: 343–351.

Staves, M. 1997. Cytoplasmic streaming and gravity sensing in *Chara* internodal cells. Planta 203: S79–S84.

Staves, M., Wayne, R. and Leopold, A. 1997. The effect of external medium on the gravitropic curvature of rice (*Oryza sativa*, Poaceae) roots. Am. J. Bot. 84: 1522–1529.

Steinmann, T., Geldner, N., Grebe, M., Mangold, S., Jackson, C., Paris, S., Gaelweiler, L., Palme, K. and Juergens, G. 1999. Coordinated polar localization of auxin efflux carrier PIN1 by GNOM ARF GEF. Science 286: 316–318.

Stevenson, J., Perera, I., Heilmann, I., Persson, S. and Boss, W. 2000. Inositol signaling and plant growth. Trends Plant Sci. 5: 252–258.

Stinemetz, C., Kuzmanoff, K., Evans, M. and Jarret, H. 1987. Correlations between calmodulin activity and gravitropic sensitivity in primary roots of maize. Plant Physiol. 84: 1337–1342.

Trewavas, A. 1992. What remains of the Cholodny-Went theory? Plant Cell Environ. 15: 759–794.

Utsuno, K., Shikanai, T., Yamada, Y. and Hashimoto, T. 1998. *AGR*, an *Agravitropic* locus of *Arabidopsis thaliana*, encodes a novel membrane-protein family member. Plant Cell Physiol. 39: 1111–1118.

Volkmann, D., Buchen, B., Hejnowicz, Z., Tewinkel, M. and Sievers, A. 1991. Oriented movement of statoliths studied in a reduced gravitational field during parabolic flights or rockets. Planta 185: 153–161.

Yamamoto, M. and Yamamoto, K. 1998. Differential effects of 1-naphthaleneacetic acid, indole-3-acetic acid and 2,4-dichlorophenoxyacetic acid on the gravitropic response of roots in an auxin-resistant mutant of *Arabidopsis*, *aux1*. Plant Cell Physiol. 39: 660–664.

Yoder, T., Zheng, H., Todd, P. and Staehelin, L. 2001. Amyloplast sedimentation dynamics in maize columella cells support a new model for the gravity-sensing apparatus of roots. Plant Physiol. 125: 1045–1060.

Young, L., Evans, M. and Hertel, R. 1990. Correlations between gravitropic curvature and auxin movement across gravistimulated roots of *Zea Mays*. Plant Physiol. 92: 792–796.

Zheng, H. and Staehelin, L. 2001. Nodal endoplasmic reticulum, a specialized form of endoplasmic reticulum found in gravity-sensing root tip columella cells. Plant Physiol. 125: 252–265.

Zieschang, H.E. and Sievers, A. 1991. Graviresponse and the location of its initiating cells in roots of *Phleum pratense* L. Planta 184: 468–477.

Zieschang, H.E., Köhler, K. and Sievers, A. 1993. Changing proton concentrations at the surfaces of gravistimulated *Phleum* roots. Planta 190: 546–554.

Plant Molecular Biology **49**: 319–338, 2002.
Perrot-Rechenmann and Hagen (Eds.), Auxin Molecular Biology.
© 2002 *Kluwer Academic Publishers.*

Evolutionary patterns in auxin action

Todd J. Cooke[1],*, DorothyBelle Poli[1], A. Ester Sztein[1] and Jerry D. Cohen[2]
[1]*Department of Cell Biology and Molecular Genetics, University of Maryland, College Park, MD 20742, USA*
(*author for correspondence; e-mail tc23@umail.umd.edu*); [2]*Department of Horticultural Science, University of Minnesota, Saint Paul, MN 55108, USA*

Received 17 July 2001; accepted 16 October 2001

Key words: auxin, charophytes, evolution, green algae, indole-3-acetic acid, liverworts, mosses, vascular plants

Abstract

This review represents the first effort ever to survey the entire literature on auxin (indole-3-acetic acid, IAA) action in all plants, with special emphasis on the green plant lineage, including charophytes (the green alga group closest to the land plants), bryophytes (the most basal land plants), pteridophytes (vascular non-seed plants), and seed plants. What emerges from this survey is the surprising perspective that the physiological mechanisms for regulating IAA levels and many IAA-mediated responses found in seed plants are also present in charophytes and bryophytes, at least in nascent forms. For example, the available evidence suggests that the apical regions of both charophytes and liverworts synthesize IAA via a tryptophan-independent pathway, with IAA levels being regulated via the balance between the rates of IAA biosynthesis and IAA degradation. The apical regions of all the other land plants utilize the same class of biosynthetic pathway, but they have the potential to utilize IAA conjugation and conjugate hydrolysis reactions to achieve more precise spatial and temporal control of IAA levels. The thallus tips of charophytes exhibit saturable IAA influx and efflux carriers, which are apparently not sensitive to polar IAA transport inhibitors. By contrast, two divisions of bryophyte gametophytes and moss sporophytes are reported to carry out polar IAA transport, but these groups exhibit differing sensitivities to those inhibitors. Although the IAA regulation of charophyte development has received almost no research attention, the bryophytes manifest a wide range of developmental responses, including tropisms, apical dominance, and rhizoid initiation, which are subject to IAA regulation that resembles the hormonal control over corresponding responses in seed plants. In pteridophytes, IAA regulates root initiation and vascular tissue differentiation in a manner also very similar to its effects on those processes in seed plants. Thus, it is concluded that the seed plants did not evolve *de novo* mechanisms for mediating IAA responses, but have rather modified pre-existing mechanisms already operating in the early land plants. Finally, this paper discusses the encouraging prospects for investigating the molecular evolution of auxin action.

Abbreviations: 2,4-D, 2,4-dichlorophenoxyacetic acid; IAA, auxin (indole-3-acetic acid); NAA, naphthalene-acetic acid; NPA, *N*-(1-naphthyl)phthalamic acid; PCIB, *p*-chlorophenoxyisobutyric acid; 2,4,6-T,2,4,6-trichlorophenoxyacetic acid; TIBA, 2,3,5-triiodobenzoic acid

Introduction

It has long been appreciated that the hormone auxin (indole-3-acetic acid, IAA) is an important regulator of developmental processes in seed plants. For instance, IAA acts as the intercellular signal coupling environmental stimuli to growth responses in phototropism and gravitropism (Kaufman *et al.*, 1995;

Dolan, 1998; Marchant *et al.*, 1999). For instance, the progression through different stages of angiosperm embryo development appears to require the sequential activation of two different auxin biosynthetic pathways (Michalczuk *et al.*, 1992a, b; Cooke *et al.*, 1993; Ribnicky *et al.*, 2001). A tryptophan-dependent pathway capable of high biosynthetic rates mediates

320

isodiametric growth of young embryos, and then polar auxin transport, in conjunction with a tryptophan-independent pathway with better homeostatic control, helps to establish polarized growth of older embryos (Schiavone and Cooke, 1987; Liu *et al.*, 1993; Steinmann *et al.*, 1999). Localized synthesis and/or polar transport establish IAA gradients that appear to be absolutely critical for positioning leaf primordia on shoot apices (Meicenheimer, 1981; Lyndon, 1998; Reinhardt *et al.*, 2000) and for initiating lateral root primordia on mature roots (Blakely *et al.*, 1988; Reed *et al.*, 1998; Casimiro *et al.*, 2001). Molecular studies of *Arabidopsis* flower development have implicated IAA-responsive genes as being crucial for the positional relationships in flowers (Sessions *et al.*, 1997; Nemhauser *et al.*, 2000). Finally, it is well documented that IAA exercises predominant control over many aspects of vascular tissue development, including the induction of primary vascular tissues (Roberts *et al.*, 1988; Aloni, 1995), the positioning of primary vascular bundles (Sachs, 1991; Berleth *et al.*, 2000), and the activity of vascular cambia (Uggala *et al.*, 1996, 1998).

Although contemporary research on IAA action remains almost completely focused on the seed plants, there is an extensive, but scattered, older literature devoted to IAA action in other green plants. The few reviews concerning this literature are generally restricted to specific groups, such as the algae (Bradley, 1991), liverworts (Maravolo, 1980), and mosses (Christianson, 1999). By contrast, this paper takes a comparative approach in the effort to characterize IAA action in all plants, with special emphasis on the green plant lineage, including green algae, bryophytes, pteridophytes, and seed plants. It is anticipated that this approach should ultimately allow us to evaluate a number of important questions that cannot be addressed by studying a single organism or specific group. Of special interest for understanding the evolution of IAA action are the results obtained from charophycean green algae (also known as charophytes). Because these algae and land plants share many specialized characteristics, such as cell division mechanism, sperm ultrastructure, and photorespiratory enzymes, the charophytes are thought to be the green algal class most closely related to land plants (Graham, 1993; Graham *et al.*, 2000). Thus, it is assumed that extant charophytes have retained the primitive features of IAA action that were also expressed in the earliest land plants.

Given that every paper ever written about IAA action is potentially relevant to the topic of this review, we have adopted the following strategy for organizing this voluminous literature into a coherent manuscript. In the major sections, we summarize the current understanding of different processes involved in IAA action in the seed plants, and then we discuss the available literature on these processes in the algae, bryophytes, and vascular plants emphasizing the pteridophytes. Consequently, one can synthesize an appealing, albeit still incomplete, picture of the evolutionary changes in IAA metabolism and polar IAA transport, which are the principal processes responsible for regulating intracellular IAA concentration. It is also possible to draw tentative conclusions about how IAA's ability to mediate such developmental processes, such as tropisms, correlative interactions, and positional relationships, in non-seed plants evolved to regulate similar processes in seed plants. Lastly, we shall discuss the molecular events that are presumably responsible for this observed evolution of IAA-mediated development throughout the land plant lineage.

Why consider the evolutionary patterns in IAA action?

As a starting point, it is important to ask the question of why a special issue devoted to IAA molecular biology includes a paper on evolutionary patterns in IAA action. In response, we could simply cite Dobzhansky's dictum that nothing makes sense in biology except in the light of evolution, but philosophical statements do not make for compelling scientific arguments. Instead, a more appropriate justification for this paper lies in the potential of evolutionary approaches for solving certain difficult problems being faced by molecular biologists studying IAA action. For example, our current knowledge suggests that hormonal regulation in angiosperm development may frequently involve complex signal transduction networks composed of multiple, interacting, and somewhat redundant pathways (Bennett *et al.*, this issue). By contrast, it appears that the lower plants utilize simpler mechanisms of IAA regulation (Sztein *et al.*, 2000), which implies that these plants may more readily reveal how the molecular regulation of IAA action is causally related to the developmental processes responsible for plant morphology. Secondly, it is becoming increasingly obvious that due to differential rates of sequence divergence, relative sequence similarity among ho-

mologous genes is not necessarily sufficient to sort
out the pattern of molecular evolution within a gene
family (Eisen, 1998). Eisen (1998) has argued quite
convincingly that an understanding of how different
members of any gene family (e.g., the globin genes
encoding oxygen-binding proteins or the homeobox
genes encoding transcription factors regulating animal
development) evolved over time should greatly im-
prove the ability to make functional predictions about
under-characterized members of that family. Such
knowledge should also help to elucidate the roles of
different homologues with overlapping functions acting in the same regulatory network. Other genomic
approaches are also likely to contribute significantly
to our understanding of auxin action (Theologis, this
issue). Thirdly, although no molecular work has been
done on IAA action in non-seed plants, there is bur-
geoning interest in the several model systems for
these plants, including the moss *Physcomitrella patens*
(Reski, 1999; Wood *et al*, 2000) and the fern *Ceratopteris richardii* (Hickok *et al.*, 1995). Thus, it is
rapidly becoming feasible to apply molecular techniques for studying developmental and evolutionary
problems in selected non-seed plants. *Physcomitrella*
is an especially promising system due to its unique
ability among all land plants studied to date to carry
out homologous recombination efficient enough for
gene knock-out and allele replacement studies (Schaefer and Zryd, 1997). Thus, the evolutionary perspectives contained in this review may become useful
to IAA molecular biologists interested in applying
the knowledge obtained from *Arabidopsis* toward the
study of related problems in non-seed plants.

Because the goal of this paper is to place the
knowledge available about auxin action in all plants
into an evolutionary context, it is essential that the
reader have an appreciation for the emerging consensus concerning the universal phylogeny of life
(Baldauf *et al.*, 2000). It appears as if three lineages of photosynthetic eukaryotes were independently able to evolve multicellular growth forms:
the viridiplants (green plants) being composed of
the chlorophytes (green algae) and the charophytes
(stoneworts and their algal relatives plus the land
plants); the rhodophytes (red plants) also known as
red algae; and the phaeophytes (brown plants) also
known as brown algae. The rhodophytes now appear to be closely related as the sister group to the
viridiplants (Baldauf *et al.*, 2000). By contrast, the lineage that gave rise to the viridiplants and rhodophytes
separated as unicellular heterotrophic flagellates in

deep evolutionary time from another lineage that
would ultimately result in the heterokonts consisting
of oomycetes and phaeophytes (see Figure 1 in Baldauf *et al.*, 2000). The rhodophytes and viridiplants
appear to have acquired their chloroplasts via the primary endosymbiosis of a single cyanobacterium or
a group of related cyanobacteria (Delwiche, 1999;
Grzebyk *et al.*, pers. comm.). The phaeophyte lineage originated in more recent evolutionary time via
secondary endosymbiosis in ancestral heterokonts in
which engulfed rhodophytes were transformed into
chloroplasts (Delwiche, 1999; Grzebzk *et al.*, pers.
comm.). It is important in our discussion of the origins of auxin action that the reader be mindful of these
contemporary perspectives on the phylogenetic relationships of multicellular photosynthetic organisms.
Nevertheless, for the sake of clear communication, we
shall employ the traditional organism names that were
used in the primary literature on auxin action, and
thus, the chlorophytes are referred to as green algae,
the rhodophytes as red algae, and the phaeophytes as
brown algae.

Modern phylogenetic methods have also resolved
the evolutionary relationships among the simplest
charophytes, with the order Charales including the
genera *Chara* and *Nitella*, as being the closest living relatives to the land plants (Karol *et al.*, 2001).
Comparable efforts are being made in the effort to
resolve the phylogenetic relationships among the vascular land plants, including the lycophytes, horsetails,
ferns, and seed plants (e.g., Pryer *et al.*, 2001). However, the evolutionary positions among the simplest
land plants called bryophytes (liverworts, hornworts,
mosses) remain as an unresolved and contentious issue. Certain molecular evidence is consistent with the
liverworts being the first-divergent lineage (viz., the
closest living relatives) of the earliest land plants (e.g.,
Qiu *et al.*, 1998; Qiu and Lee, 2000), while other
data fit the hornworts-basal hypothesis (e.g., Nickrent
et al., 2000). In this paper, the sections devoted to the
bryophytes emphasize the literature on liverworts and
mosses for the simple reason that few published papers
investigated auxin action in the hornworts.

Evolutionary patterns in IAA metabolism

Basic characteristics in seed plants

The metabolic processes that regulate endogenous
IAA levels are: biosynthesis, conjugation, and degradation (for reviews, see Normanly, 1997; Bartel, 1997;

Slovin *et al.*, 1999). The research on auxin metabolism in *Arabidopsis* is fully described in Kowalczyk *et al.* (this issue) so we shall emphasize here only those topics needed for evolutionary interpretations. IAA is synthesized via two different classes of biosynthetic pathways. One, tryptophan-mediated IAA biosynthetic pathways utilize tryptophan itself as the source of the indole ring for the IAA molecule. Earlier research had repeatedly demonstrated that excised organs, tissue sections, cultured cells, and cell-free preparations can utilize tryptophan-dependent pathways to carry out IAA biosynthesis (Wildman *et al.*, 1947; Sembdner *et al.*, 1981; Nonhebel *et al.*, 1993). This interpretation has been supported by analytical work on isolated axes of germinating bean seedlings (Bialek *et al.*, 1992; Sztein *et al.*, 2001), embryogenic cells in carrot cultures (Michalczuk *et al.*, 1992b), and excised maize coleoptiles (Koshiba *et al.*, 1995). Two tryptophan-independent pathways divert the indole ring toward IAA biosynthesis before it is catalyzed into tryptophan. Considerable evidence has been accumulating in support of the notion that these pathways are the predominant IAA biosynthetic pathways operating in intact plants. In entire *Lemna* plants (Baldi *et al.*, 1991; Rapparini *et al.*, 1999), post-globular carrot somatic embryos (Michalczuk *et al.*, 1992b), and wild-type *Arabidopsis* seedlings (Normanly *et al.*, 1993), labeled precursors common to both the tryptophan-independent and tryptophan-dependent pathways result in much greater enrichments of the IAA pool than the enrichment observed with labeled tryptophan. In addition, *Zea mays* and *Arabidopsis* mutants impaired in tryptophan synthesis exhibit much higher levels of IAA than do wild-type plants, thereby providing compelling evidence for the predominant activity of the tryptophan-independent pathway in intact plants. Finally, other experiments have led to the interpretations that the tryptophan-mediated pathway is capable of high biosynthetic rates due to the large tryptophan pool and the apparent lack of feedback inhibition, whereas the tryptophan-independent pathway serves as the low-capacity pathway subject to feedback inhibition (Ribnicky *et al.*, 1996, 2001; Sztein *et al.*, 2001). Thus, it appears that the two pathways have the potential to play distinct roles in the regulation of various developmental processes, including plant embryogenesis (Michalczuk *et al.*, 1992a, b; Ribnicky *et al.*, 2001).

IAA is usually active in developmental processes as the free molecule. Nevertheless, in the seed plants, most IAA-based metabolites accumulate as IAA-ester conjugates to inositol, co-enzyme A, sugars, polysaccharides, or glycoproteins and/or as IAA-amide conjugates to amino acids, small peptides, or proteins. IAA conjugates are generally thought to act as short-term intermediates that can be hydrolyzed to release free IAA (Cohen and Bandurski, 1982; Normanly, 1997). The competing processes of conjugate synthesis and conjugate hydrolysis are thus predicted to help further modulate the free IAA levels. Finally, IAA can be degraded via decarboxylative and oxidative pathways. In conclusion, it appears that the metabolic processes of biosynthesis, conjugation, and degradation allow plant cells to maintain precise homeostatic regulation of intracellular IAA levels (Ljung *et al.*, this issue).

Sztein *et al.* (1995, 1999, 2000) performed a comprehensive survey of IAA metabolism in the major divisions of land plants. This survey was undertaken in large part because the data available from the earlier literature were so flawed that they could not be used to devise plausible hypotheses about evolutionary patterns in IAA metabolism (for discussion, see Sztein *et al.*, 1999). Consequently, Sztein *et al.* (1995, 1999, 2000) employed similar growth conditions for all plants, excised comparable thallus or shoot tips, employed common analytical techniques, and used antibiotic treatments or axenic cultures, whenever possible. Standard isotope dilution methods were used in conjunction with GC-MS to measure the steady-state concentrations of free IAA and IAA metabolites; thus, these measurements were virtually independent of sample size, purification method, and preparative losses. Thin-layer chromatography was used to measure the rate of IAA conjugate formation from radiolabeled IAA and to tentatively identify the chemical nature of these conjugates. It is worth mentioning that the requirement for enough material for GC-MS analysis meant that the apical tips were excised from gametophytic structures in charophytes and bryophytes as opposed to from sporophytic structures in vascular plants. The significance of this limitation for evolutionary interpretations will become apparent in the later section on IAA transport. The following sections are derived in part from Sztein *et al.* (2000).

Algae

Previous work on IAA levels in a few green, red and brown algae utilized non-axenic specimens (Jacobs *et al.*, 1985; Bradley, 1991; Evans and Trewavas, 1991; Ashen *et al.*, 1999; Basu *et al.*, pers. comm.). For example, free IAA levels in thalli of the red alga *Prionitis lanceolata* were measured at 2.5 ng per gram fresh weight (Ashen *et al.*, 1999). Similar levels were also quantified in zygotes and fruiting tips of the brown alga *Fucus distichus* (Basu *et al.*, pers. comm.). However, Evans and Trewavas (1991) cautioned that one should remain skeptical about such measurements due to microbial contamination, which is routinely present in algal cultures. No data on the concentrations of IAA conjugates and other IAA metabolites were presented in these papers.

Since charophytes are considered to represent the closest algal relatives of the land plants, it is assumed that charophytes exhibit the primitive condition for the IAA metabolism of land plants. Under steady-state conditions, the thallus tips of the charophyte *Nitella* exhibited lower absolute levels of free IAA but higher ratios of free IAA to total IAA metabolites, as compared to most land plants (Tables 1 and 2). This alga was capable of only negligible rates of IAA conjugate synthesis in the presence of radiolabeled IAA. Instead, it appeared that exogenous IAA was predominantly converted into degradation products and/or other labeled metabolites synthesized from IAA. Therefore, charophytes, at least as exemplified by *Nitella*, must primarily regulate free IAA levels via the balance between the biosynthesis of new IAA molecules and the degradation of existing molecules. The nature of the IAA biosynthetic pathway has not yet been studied in charophytes although the non-tryptophan pathway is typically observed to maintain free IAA levels similar to those measured in *Nitella* tips.

Bryophytes

What makes the bryophytes so interesting from an evolutionary perspective is that each division, namely, liverworts, hornworts, and mosses, exhibits a distinctive IAA metabolism that may underlie the profound structural differences observed among these groups.

In liverworts, GC-MS analysis showed that thallus tips produced low to intermediate levels of free IAA, as well as high ratios of free IAA to total IAA metabolites, which were similar to the levels and ratios observed in charophytes (Tables 1 and 2). Judging from the results of tryptophan-feeding experiments with *Pallavicinia* tips, IAA biosynthesis in liverworts did not involve tryptophan as its primary intermediate. Liverworts can thus be said to utilize a tryptophan-independent biosynthetic pathway. In ^{14}C-IAA labeling experiments, liverworts tended to exhibit slow, but discernable, rates of IAA conjugate formation, with

most species primarily synthesizing amide conjugates. The slow conjugation rates suggest that liverworts utilize a biosynthesis/degradation strategy for regulating free IAA levels, which means that charophytes and liverworts share the putative ancestral strategy for IAA metabolic regulation.

Thallus tips of the hornwort *Phaeoceros* produced high levels of free IAA, but the ratio of free IAA to total IAA metabolites was significantly lower than those observed in charophytes and liverworts (Tables 1 and 2). ^{14}C-IAA labeling studies demonstrated that hornworts can rapidly synthesize amide conjugates so that the regulation of free IAA levels appears to involve the competing reactions of conjugate synthesis vs. conjugate hydrolysis. Hornworts exhibited IAA conjugate levels comparable to those seen in the vascular plants. However, because IAA conjugates are chemically different in these two groups, it appeared that high conjugate levels had evolved independently at least twice in the land plants. No observations on the primary IAA biosynthetic pathway have been made for hornworts.

Vegetative tips of moss gametophores maintained much lower levels of free IAA and total IAA than did hornwort tips (Tables 1 and 2). Nevertheless, the free IAA to total IAA ratios were essentially identical in these two groups. Just like the hornworts, the mosses almost exclusively accumulated amide conjugates. Lastly, given the similar rates of IAA conjugate formation, it is reasonable to conclude that mosses must also use the conjugation/hydrolysis strategy as their principal mechanism for regulating free IAA levels. Earlier work on *Funaria* chloronema cultures and cell-free homogenates demonstrated that these preparations have the potential to use several intermediates in the tryptophan-dependent pathway for inducing IAA-mediated bud formation (Lehnert and Bopp, 1983; Bhatla and Bopp, 1985; Atzorn *et al.*, 1989), and to use exogenous ^{3}H-tryptophan to synthesize IAA (Jayaswal and Johri, 1985). The principal limitation of these approaches was that they do not compare the relative contributions of the tryptophan-independent and tryptophan-dependent pathways to IAA synthesis (for further discussion, see Sztein *et al.*, 2000). By contrast, the tryptophan-feeding method permitted a direct comparison of the relative contributions of the two pathways, and it showed that mosses use the tryptophan-independent pathway, at least in gametophore tips (Sztein *et al.*, 2000).

Table 1. Major characteristics of IAA levels and biosynthetic pathways operating in the vegetative tips of green plants (Sztein et al. 1999, 2000). The data for the angiosperms were taken from Baldi et al. (1991), Wright et al. (1991), Michalczuk et al. (1992a), and Normanly et al. (1993).

Plants	Number of species	Free IAA level		Total IAA metabolite level ng/g FW	Number of species	Predominant IAA biosynthetic pathway
		ng/g FW	%			
Charophytes	1	11	30	38	0	unknown
Liverworts	5	10–20	20–35	35–75	1	tryptophan-independent
Mosses	4	5–10	8–12	45–80	1	tryptophan-independent
Hornworts	1	35	11	328	0	unknown
Pteridophytes	2	25–35	10–20	100–400	1	tryptophan-independent
Angiosperms	4	10–20	5–10	400–700	4	tryptophan-independent

Table 2. Major characteristics of the IAA conjugation and IAA regulation in the vegetative tips of green plants (Sztein et al., 1995, 1999, 2000).

Plants	Number of species	Major conjugates	Conjugation rate	Regulatory strategy
Charophytes	1	unknown	very slow	biosynthesis/degradation
Liverworts	7	amide conjugates	slow	biosynthesis/degradation
Mosses	5	amide conjugates	intermediate to rapid	conjugation/hydrolysis
Hornworts	1	amide conjugates	intermediate to rapid	conjugation/hydrolysis
Pteridophytes	10	IAsp/Glu and/or IAgluc	rapid	conjugation/hydrolysis
Seed plants	7	IAsp/Glu and/or IAgluc	very rapid	conjugation/hydrolysis

Vascular plants

Little attention has been granted to IAA metabolism in the pteridophyte grade, which includes lycophytes, horsetails, and ferns. Shoot tips of the lycophyte *Selaginella* and the fern *Ceratopteris* exhibited high IAA metabolite levels in proportions similar to those measured in hornworts (Tables 1 and 2). Judging from tryptophan-feeding experiments, the major pathway for IAA biosynthesis in *Selaginella* tips is the tryptophan-independent pathway. Pteridophytes can rapidly synthesize IAA conjugates so that these plants appear to maintain free IAA levels by adjusting the balance between conjugation and deconjugation. The pteridophytes are distinguishable from the bryophytes with respect to the ability of the pteridophytes to synthesize several characteristic conjugates, namely, the ester conjugate IAA-glucose and the amide conjugates IAA-asparate and/or IAA-glutamate.

The IAA metabolism of gymnosperms and angiosperms is thoroughly described in several recent reviews (Normanly, 1997; Bartel, 1997; Slovin et al., 1999; Kowalczyk et al., this issue). Pteridophytes and seed plants manifest only quantitative differences in their IAA metabolism (Tables 1 and 2). In general, seed plants produced the highest levels of total IAA metabolites, with exceptional levels of IAA-glucose, IAA-aspartate, and/or IAA-glutamate. Seed plants must also use the equilibrium between conjugation and hydrolysis for controlling free IAA levels.

Summary

The major features of auxin metabolism in green plants can be organized into an evolutionary framework by placing the features listed in Tables 1 and 2 on a simplified land plant cladogram (Figure 1), which represents a common, but not universally accepted land plant phylogeny deduced from recent molecular and morphological work (Kendrick and Crane, 1997; Qiu et al., 1998; Duff and Nickrent, 1999; Nickrent et al., 2000). It appears plausible that the tryptophan-independent pathway is the predominant class of IAA biosynthetic pathway in growing shoot tips in the

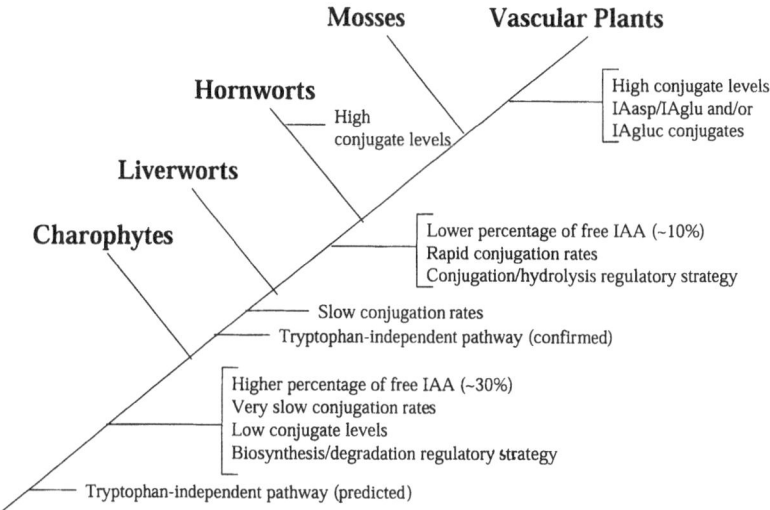

Figure 1. A character map of the major evolutionary events in IAA metabolism of green plants. The characters derived from Tables 1 and 2 are placed on a simplified version of a land plant cladogram in order to illustrate the likely positions of the major evolutionary events in IAA metabolism. Modified from Sztein *et al.* (2000) with permission.

charophytes and land plants. However, a major innovation in the IAA metabolic regulation occurred within the bryophyte grade. Liverworts have retained the putative ancestral strategy for regulating free IAA levels that depends on the biosynthesis of new IAA molecules and the degradation of existing molecules. The nested suite of allied characteristics seen in charophytes and liverworts included a higher percentage of free IAA, slow conjugation rates, and low levels of total IAA metabolites. The other bryophytes and vascular plants evolved the potential to regulate free IAA levels by adjusting the equilibrium between conjugate synthesis and conjugate hydrolysis. It is presumed that this one-step strategy permitted the more precise spatial and temporal regulation of free IAA levels.

Evolutionary patterns in IAA transport

Basic characteristics

IAA transport is characterized by its polarity, direction, distance, and transporting cells (Goldsmith, 1977; Lomax *et al.*, 1995). Most research attention has been devoted to polar IAA transport, which is defined as IAA movement in a specific, often basipetal direction. Polar IAA transport has been extensively studied in seed plants for several reasons: (1) it appears to be causally involved in the polarized growth observed in many plant structures, (2) it can easily be measured in plant explants with a

simple apparatus, and (3) it is sensitive to several inhibitors that act on the IAA-efflux-carrier complex, such as N-(1-naphthyl)phthalamic acid (NPA) and 2,3,5-triiodobenzoic acid (TIBA). In angiosperms, NPA acts as a strong phytotropin in that it reduces the lateral IAA transport associated with tropic responses in addition to its antagonism of polar transport, while the inhibitory effects of TIBA are restricted to polar transport (Lomax *et al.*, 1995). Since NPA acts as competitive inhibitor of the binding of certain flavonoids such as quercitin, which are the presumptive *in vivo* inhibitors of polar IAA transport, it is generally thought that NPA inhibition of any process is a diagnostic indicator of the involvement of polar IAA transport (Jacobs and Rubery, 1988; Brown *et al.*, 2001). Although a thorough discussion of the mode of action of these inhibitors would exceed the scope of this review, recent work (Steinmann *et al.*, 1999) can be interpreted to mean that NPA may function as a general inhibitor of secretory processes as opposed to being a specific inhibitor of auxin transport per se. This review will follow the convention of using these inhibitors as diagnostic indicators of IAA transport, because it is premature to assess the validity of that new perspective.

According to the chemiosmotic model for polar IAA transport, electrochemical H^+ gradients across plasma membranes are the ultimate driving force for polar transport (Rubery and Sheldrake, 1974; Raven, 1975; Goldsmith, 1977). Apoplastic IAA in the cell wall (pH 5) is envisaged to cross the plasma membrane either through passive diffusion of protonated IAA (pK_a 4.7) or via IAA influx carriers acting as proton symporters. Unprotonated IAA in the cytosol (pH 7) is transported back into the cell wall via IAA efflux carriers. The pronounced polarity of IAA transport is proposed to result from the asymmetric localization of IAA carriers at different ends of transporting cells. Recent molecular evidence has lent considerable support to this model (Estelle, 1998; Palme and Galweiler, 1999; Swarup et al., 2000). Bennett et al. (1996) isolated a putative IAA influx carrier known as AUX1, which shares considerable homology with amino acid permeases. Immunolocalization studies suggest that influx carrier proteins are sometimes positioned at the apical ends of transporting cells (Swarup et al., in press). Moreover, the IAA efflux carriers called PIN proteins, which exhibit significant sequence similarity to bacterial proteins responsible for transporting small molecules, are preferentially localized at the basal ends of transporting cells (Muller et al., 1998; Steinmann et al., 1999). Current knowledge of the molecular regulation of polar IAA transport in Arabidopsis is reviewed elsewhere (Galweiler and Palme, this issue).

A few reports on IAA transport can be found in the literature on lower plants. This research utilized conventional agar-block techniques to characterize IAA transport in multicellular structures. The polarity of IAA transport was routinely determined by switching the positions of the donor and receiver blocks and comparing the ratio of basipetal (apex to base) vs. acropetal (base to apex) transport at a selected time. In addition, flux experiments were conducted to quantify the rates of IAA transport across the plasma membranes in vesicle fractions, single cells, and small tissue pieces. In flux experiments where polar IAA transport inhibitors mediated an increase in net IAA accumulation, this result is interpreted to mean that the IAA efflux carriers were active in those systems. Although these experiments disclose the biochemical potential of a multicellular structure to transport IAA in a polar manner, the question of whether the intact structure carries out polar transport depends on whether the transport components are actually localized at specific intracellular sites.

Frankly, the entire literature on lower plants is unfortunately plagued by methodological problems and idiosyncratic units that prevent rigorous quantitative comparisons of IAA transport rates among different structures in different plants. However flawed the data

on transport rates are, polarity ratios calculated from those data are, in general, independent of the methodologies employed. These ratios when considered with inhibitor studies allow us to discern qualitative differences in the abilities of various lower plant structures to carry out IAA transport (Tables 3 and 4).

Algae

In the brown alga *Fucus*, the addition of NPA to the medium mediated a 1.6-fold increase in accumulated IAA in the rhizoids, which can be taken to indicate that the IAA efflux carrier is present in brown algal rhizoids (Basu et al., pers. comm.) (Table 3). Interestingly, NPA caused those rhizoids to undergo precocious branching, which differed from the multiple rhizoids observed in IAA treatments. These results suggested that NPA has developmental effects on *Fucus* rhizoids that cannot be attributed to the direct inhibition of IAA efflux alone.

Insofar as the charophytes are assumed to have retained the basal (or pleisiomorphic) state of the IAA transport system in the land plant lineage, it was hoped that the existing literature would have presented a definitive picture of IAA transport in the charophytes. Dibb-Fuller and Morris (1992) compared IAA transport capabilities of unicellular green alga *Chlorella* vs. thallus tips of the multicellular charophyte *Chara* (Table 3). It was concluded that IAA accumulation in *Chlorella* cells did not depend on specific IAA carriers, but rather appeared to involve the pH-sensitive diffusion of IAA across the plasma membrane. The apical portions of *Chara* thalli did exhibit saturable, i.e., carrier-dependent, IAA fluxes in both directions that were competitively inhibited by unlabeled IAA. IAA efflux was almost totally unaffected by NPA and TIBA, with the exception that the highest TIBA concentration caused a slight inhibition that was attributed to secondary pH effects. Surprisingly, other workers reported that decapitated *Chara* explants with growing rhizoids showed substantial net IAA accumulation in the presence of NPA (Klambt et al., 1992). The most straightforward way to reconcile the results of these two papers is to assign the NPA effect to the possibility that an active NPA-sensitive IAA efflux carrier is localized in the rhizoids that are attached to the thallus explants used in Klambt et al. (1992). However, these authors observed that another NPA effect, namely the inhibition of rhizoid growth, was abolished by the coincidental application of IAA. Because this result suggested that NPA had additional effects unrelated

Table 3. A summary of flux experiments used to characterize transmembrane IAA fluxes and the potential for polar IAA transport in green plants. IAA accumulation increase was calculated as the ratio of the difference in net IAA accumulated in inhibitor vs. control experiments over the net IAA accumulated in control experiments times 100, as reported in the cited references.

Group	Species	Generation	Structure	Inhibitor	IAA accumulation increase (%)	References
Algae	*Fucus distichus*	sporophyte	zygotes	5×10^{-5} M NPA	58	Basu *et al.* (pers. comm.)
	Chorella vulgaris	gametophyte	unicells	3×10^{-6} M NPA	0	Dibb-Fuller and Morris (1992)
Charophytes	*Chara globularis*	gametophyte	thallus segments with rhizoids	1×10^{-4} M NPA	67	Klambt *et al.* (1992)
	Chara vulgaris	gametophyte	thallus	1×10^{-5} M NPA	0	Dibb-Fuller and Morris (1992)
				1×10^{-5} M TIBA	0	Dibb-Fuller and Morris (1992)
Mosses	*Funaria hygrometrica*	gametophyte	protonemal	1×10^{-5} M NPA	32	Geier *et al.* (1990)
			protoplasts	1×10^{-5} M TIBA	29	Geier *et al.* (1990)
			protonemata	1×10^{-5} M TIBA	27	Rose *et al.* (1983)
			rhizoids	1×10^{-5} M TIBA	96	Rose and Bopp (1983)
Angiosperms	*Zea mays*	sporophyte	coleoptile segments	1×10^{-5} M NPA	140	Sussman and Goldsmith (1981)
				1×10^{-5} M TIBA	64	Sussman and Goldsmith (1981)
	Cucurbita pepo	sporophyte	hypocotyl vesicles	1×10^{-5} M NPA	80	Hertel *et al.* (1983)
				1×10^{-5} M TIBA	84	Hertel *et al.* (1983)

to its ability to increase intracellular IAA levels, further research is certainly needed to characterize IAA transport in the charophytes.

Bryophytes

The charophytes are capable of producing large haploid thalli, but zygotes represent the only diploid cells in their life cycles. One major innovation that occurred during bryophyte evolution is the origin of the embryo with its subsequent elaboration into the macroscopic sporophyte body. The bryophyte life cycle is thus said to consist of alternating haploid gametophytic and diploid sporophytic generations. Indeed, the embryo is frequently cited as the quintessential morphological adaptation of land plants, in large part because a multicellular diploid generation is much better suited to produce abundant meiospores for aerial dispersal in terrestrial environments (Graham, 1993; Taylor and Taylor, 1993; Kendrick and Crane, 1997; Niklas, 1997). Given this evolutionary history, it is important in this review to distinguish between the nature of IAA transport in gametophytes vs. sporophytes of different bryophyte groups.

Almost all the published work on IAA transport in bryophytes has focused on the gametophyte generation (Tables 3 and 4). In thallus midribs of the liverwort *Marchantia*, the initial front of labeled IAA

moved at similar rates in the basipetal and acropetal directions (Maravolo, 1976; Gaal *et al.*, 1982). However, the polarity ratio of total transported IAA (i.e., basipetal transport over acropetal transport) exceeded 5.0. Basipetal transport was significantly inhibited by a median ring of TIBA-containing lanolin ring around the thallus explant (Maravolo, 1976), which provided further evidence that liverwort gametophytes can carry out typical polar IAA transport. This interpretation is consistent with other observations that anaerobic conditions and metabolic inhibitors could suppress basipetal transport in liverwort thalli (Gaal *et al.*, 1982). No information concerning IAA transport in liverwort rhizoids has been published in the literature.

In protonemata of the moss *Funaria* grown under low light, high levels of applied IAA appeared to saturate IAA uptake, and TIBA mediated a 27% increase in net IAA accumulation (Rose *et al.*, 1983). These observations demonstrated that this stage of moss gametophyte development produces both influx and efflux carriers associated with polar IAA transport. Identical conclusions were drawn from protoplasts isolated from *Funaria* protonemata grown under similar light conditions (Geier *et al.*, 1990). By contrast, both characteristics of saturable influx and inhibitor sensitivity were not observed in high-light-grown protonemata, which was attributed to the possibility that such pro-

Table 4. A summary of agar-block experiments used to characterize polar IAA transport in green plants. Polarity (B/A) ratio was calculated as the ratio of the basipetal transport over the acropetal transport rates reported in the cited reference. Transport inhibition was calculated as the ratio of the difference in basipetal transport rates in control vs. inhibited structures over the basipetal rate in control structures times 100.

Group	Species	Generation	Structure	Polarity (B/A) ratio	Inhibitor	Transport inhibition (%)	References
Liverworts	*Marchantia polymorpha*	gametophyte	thallus	5.3	10^{-3} M TIBA	62	Maravolo (1976); Gaal *et al.*, 1982
	Pellia epiphylla	sporophyte	seta	0.9–1.1	10^{-5} M NPA	0	Thomas (1980); Poli *et al.* (unpub. obs.)
Hornworts	*Phaeoceros pearsoni*	sporophyte	immature sporangium	0.9	10^{-5} M NPA	0	Poli *et al.* (unpub. obs.)
Mosses	*Funaria hygrometrica*	gametophyte	rhizoid	11.4	10^{-4} M TIBA	54	Rose and Bopp (1983)
	Polytrichum ohioense	sporophyte	seta	9.1	10^{-5} M NPA	11	Poli *et al.* (unpub. obs.)
Pteridophytes	*Selaginella willenovi*	sporophyte	stem	2.1	None applied	–	Wochok and Sussex (1973)
	Osmunda cinnamomea	sporophyte	rachis	190.0	None applied	–	Steeves and Briggs (1960)
	Regnellidium diphyllum	sporophyte	rachis	52.0	None applied	–	Walters and Osborne (1979)
Angiosperms	*Zea mays*	sporophyte	coleoptile	609.3	10^{-5} M NPA	99	Poli *et al.* (unpub. obs.)
	Cucurbita pepo	sporophyte	hypocotyl	20.0	10^{-5} M TIBA	68	Jacobs and Hertel (1978)

tonemata might have higher endogenous IAA levels. In *Funaria* rhizoids, TIBA mediated a pronounced 96% increase in net IAA accumulation, which established that the inhibitor-sensitive IAA efflux carrier is also present in moss rhizoids (Rose and Bopp, 1983). Due to the multicellular nature of moss rhizoids, these authors were able to confirm that moss rhizoids carry out polar IAA transport with a polarity ratio of 11.4.

Because polar IAA transport is an important mechanism for regulating developmental events in the sporophytes of seed plants (for references, see Introduction), our lab decided to examine IAA transport in bryophyte sporophytes in the attempt to gain some insights into the evolutionary origins of IAA transport in land plant sporophytes (Poli *et al.*, unpublished observations). Bryophyte sporophytes are especially intriguing because they undergo polarized growth to form linear tripartite axes (Bold *et al.*, 1987; Crum, 2001). Both liverwort and moss sporophytes develop an apical capsule, a basal foot, and an intermediate seta, although developmental processes responsible for generating these sporophytes are strikingly different in the two groups. The hornwort sporophyte consists of an elongated apical sporangium and a basal foot with an intervening intercalary meristem that divides to generate new sporangial cells throughout sporophytic growth.

In liverworts, elongation rates of IAA-treated setae were more than twice the rates observed in control

setae (Schnepf *et al.*, 1979; Thomas, 1980). In addition, the anti-auxin PCIB caused marked reductions in elongation rates of *Pellia* setae, which suggests very strongly that seta elongation is principally regulated by endogenous IAA under normal conditions. Nevertheless, agar-block studies involving long-term equilibration (Thomas, 1980) and repeated sampling (Poli *et al.*, unpublished observations) demonstrated that IAA must diffuse through developing setae, because these workers found no evidence for either transport polarity or NPA sensitivity (Table 4). In the hornwort *Phaeoceros*, short segments of immature sporangia that were cut from just above their intercalary meristems maintained a slight, non-significant, polarity ratio favoring acropetal transport that was completely resistant to NPA (Table 4). The only structures from bryophyte sporophytes capable of measurable polar IAA transport were young setae of the moss *Polytrichum*, which manifested a polarity ratio favoring basipetal transport of 9.1 (Table 4). Surprisingly, polar IAA transport in this moss was only slightly inhibited by NPA, in marked contrast to its strong inhibition of polar IAA transport in seed plants (Lomax *et al.*, 1995).

Vascular plants

Although there are numerous reports of polar IAA transport in coleoptiles, stems, roots, and other organs of seed plants (for reviews, see Goldsmith, 1977; Lomax *et al.*, 1995), we were able to locate just four reports on IAA transport in pteridophytes (Table 4). Stem segments from the lycophyte *Selaginella* carried out both basipetal and acropetal transport, with a polarity ratio approaching 2.1 (Wochok and Sussex, 1973). Albaum (1938) demonstrated that IAA could move from the apex toward the base in the gametophytes of the fern *Pteris*, but he did not measure the potential for acropetal transport so that it is unknown if IAA transport in this gametophyte was preferentially basipetal. IAA transport was strongly polar in the elongating rachis of the leaves from the ferns *Osmunda* (Steeves and Briggs, 1960) and *Regnellidium* (Walters and Osborne, 1979).

Summary

The known features of polar IAA transport in green plants can be organized into an evolutionary framework by placing the features listed in Tables 3 and 4 on a simplified land plant cladogram (Figure 2), as was done for IAA metabolism in Figure 1. However, for several reasons, Figure 2 should be considered as being provisional. Only a few species represent each lineage in Figure 2; in particular, it includes the observations from two liverworts, one hornwort, and two mosses for the entire bryophyte grade. Furthermore, the limited work available on the Charales means that the origins of transport characteristics must be assigned rather uncertain positions at the base of this cladogram. Keeping these provisions in mind, one can still speculate about broad evolutionary patterns in polar IAA transport, which can, in turn, serve as testable hypotheses for future research. The molecular potential for IAA transport, i.e., IAA influx and efflux carriers, was apparently already present in the charophyte lineage before the divergence of the Charales. The capacity for polar auxin transport in the Charales, at least as judged by the sensitivity to efflux-carrier inhibitors, is apparently restricted to the rhizoids. Thus, there is no present evidence to support the notion that polar IAA transport might be involved in the generation of the branched *Chara* thallus. By contrast, polar IAA transport occurs in all structures of land plant gametophytes tested to date so that it becomes plausible to assert that polar transport may have helped to regulate the development of the gametophytes of the earliest land plants. By contrast, judging from the extant bryophyte lineages, the primitive bryophytes evolved various developmental mechanisms for pro-

ducing axial sporophytes. Only mosses can apparently exploit polar IAA transport as the mechanism for regulating the axial growth of their sporophytes. Although it is appealing to hypothesize that this developmental mechanism evolved in the common ancestor of the moss and vascular plant lineages, the alternative explanation of independent origins in both lineages can not be eliminated by the limited observations available in the literature.

Evolutionary patterns in IAA regulation of plant development

The lower-plant literature contains numerous papers on the developmental effects of IAA; synthetic auxin analogues such as 2,4-dichlorophenoxyacetic acid (2,4-D) and naphthaleneacetic acid (NAA), and so-called anti-auxins such as *p*-chlorophenoxyisobutyric acid (PCIB) and 2,4,6-trichlorophenoxyacetic acid (2,4,6-T) dating back to the 1930s. This literature is typically surveyed in occasional reviews devoted to specific groups, as noted above. In addition, these reviews tend to focus on the hormonal regulation of developmental processes that are unique to that specific group such as the chloronema-to-caulonema transition and subsequent bud formation in mosses (Bopp, 1980; Cove and Ashton, 1984; Bhatla *et al.*, 1996; Christianson, 1999). The present review will instead describe the roles that IAA plays in several developmental processes that occur throughout the land plant lineage. These widespread IAA responses are somewhat arbitrarily separated into IAA responses involving tropisms, correlative interactions, and positional relationships. The following text is thus not intended as an exhaustive review of all IAA responses in non-seed plants, but instead cites selected papers in order to provide a general overview of certain IAA responses.

IAA responses involving tropisms

In the early land plants, novel physiological processes are thought to have evolved as specific adaptations to the localized distributions of essential resources and to the non-buoyant atmosphere found in the terrestrial environment. One topic of considerable interest to this paper is the evolution of tropistic growth responses that can serve to orient multicellular plant structures relative to various environmental stimuli such as light and gravity. Multicellular algae tend not to exhibit

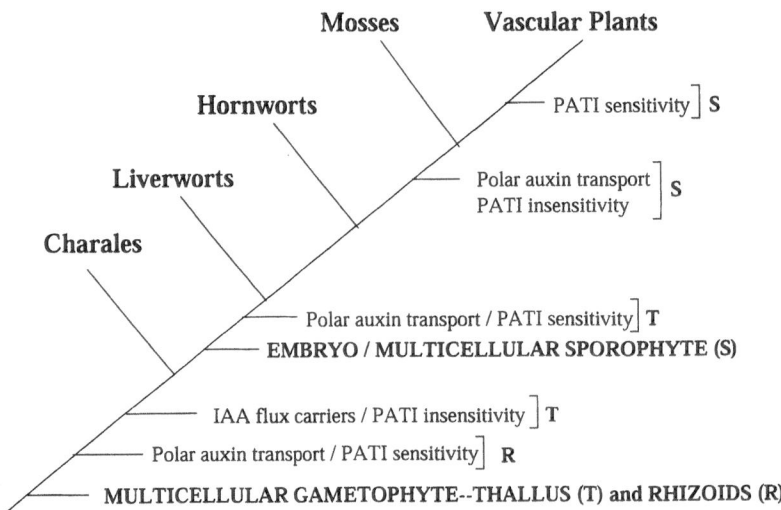

Figure 2. A provisional character map of the major evolutionary events in IAA transport in green plants. The characters derived from Tables 3 and 4 are placed on a simplified version of a land plant cladogram in order to illustrate the provisional positions of the major evolutionary events in IAA transport. The character of polar IAA transport is recorded for those plants that exhibit polar transport in agar-block experiments and/or significant sensitivity to polar transport inhibitors (PATI) in flux experiments. The data on PATI sensitivity is based on the responses to at least one inhibitor. Not enough lower plant structures have been exposed to PATIs to discriminate between NPA and TIBA sensitivities at this time. The letters R, T, and S refer to the first appearance of the bracketed traits in the gametophytic rhizoids, gametophytic thallus, and sporophyte, respectively. For further discussion of this figure, see its description in the text.

tropistic responses but rather they employ buoyancy mechanisms involving air bladders that cause the algae to assume upright orientations in the water column toward the light (Bold and Wynne, 1985). In filamentous charophycean algae and in land plant protonemata, the entire tropic response, including environmental perception, signal transduction, and directional growth, occurs within individual cells (Wada and Kadota, 1989; Sievers *et al.*, 1996; Braun, 1997; Staves, 1997). The evidence reported in those papers indicates that these plants do not use IAA as the intracellular signal for either phototropism or gravitropism.

However, in multicellular plant structures, IAA was apparently recruited early in land plant evolution to serve as the intercellular signal connecting apical cells best positioned for environmental perception to subapical cells most capable of undergoing rapid directional growth. For instance, several workers have investigated the phototropic responses of *Pellia* setae (for review, see Thomas, 1980). Ellis and Thomas (1985) demonstrated that shaded sides of these setae became more acidic prior to the onset of phototropic curvature. This acid efflux and the subsequent curvature were inhibited by both neutral buffers and IAA antagonists. These observations are consistent with the interpretation that seta phototropism is mediated by lateral IAA transport resulting in enhanced proton

efflux and wall loosening on the shaded side. This mechanism appears identical to the mechanism proposed to operate in seed plants (Kaufman *et al.*, 1995), with the intriguing exception that the polar transport inhibitor TIBA blocks seta phototropism in *Pellia* (Ellis and Thomas, 1985) but has no apparent effect on higher plant phototropism (Lomax *et al.*, 1995). In the stems of the lycophyte *Selaginella* (Bilderback, 1984), directional light induced asymmetric IAA accumulation resulting in marked phototropic curvatures. Moreover, the localized application of IAA on either side of *Selaginella* stems mediated tropic curvature independent of the light conditions. TIBA could also block the phototropic responses of these stems. Thus, it can reasonably be concluded that lateral IAA gradients act as the common intercellular signaling mechanism for phototropic responses throughout the land plant lineage; however, the differential sensitivity to transport inhibitors implies that the regulation of these gradients may differ in non-seed vs. seed plants.

IAA responses involving correlative interactions

Among the charophytes, the Charales develop the most complex bodies, which are composed of a shoot-like main axis consisting of alternating single-celled internodes and multicellular nodes bearing lateral secondary branches in a regular pattern. In many species, these secondary branches resemble miniaturized reiterations of the main axis. Thus, given regular morphological patterns of charophyte thalli, it is not unexpected that these algae do not tend to express correlative interactions where one plant part affects the growth of another part.

If one defines apical dominance as the ability of IAA from a growing region to maintain the quiescence of preexisting meristematic regions and/or to inhibit the formation of new meristematic regions, then it follows that this IAA response must also have arisen early in the evolution of land plants. For example, considerable evidence suggests that IAA regulates apical dominance and/or equivalent phenomena in liverworts. Exogenous IAA was observed to suppress the growth of isolated vegetative propagules known as gemmae in several species of liverworts, which led to the hypothesis that IAA transported from the parent thallus blocks the germination of gemmae still attached to that thallus (LaRue and Narayanaswami, 1957; Maravolo and Voth, 1966; Stange, 1971). Davidonis and Munroe (1972) presented surgical evidence that IAA transported from the larger branch of *Marchantia* thalli inhibited the growth of adjacent smaller branches. Several instances of genuine apical dominance were reported for the leafy gametophores of mosses. For example, in *Splachnum*, decapitating gametophore apices resulted in vigorous growth of reactivated lateral buds (MacQuarrie and von Maltzahn, 1959). The application of IAA to the cut apex suppressed bud activation, whereas kinetin overcame the IAA inhibition of bud activation (von Maltzahn, 1959), in a manner very similar to the hormonal regulation of apical dominance observed in seed plants (Tamas, 1995). When lanolin rings containing the polar transport inhibitor TIBA were placed around intact *Plagiomnium* gametophores, bud activation occurred below the rings, as might be expected from reduced IAA transport to those buds (Nyman and Cutter, 1981). However, the simultaneous application of IAA and a cytokinin was required to suppress all microscopic indications of bud activation in this moss. Applied IAA was also seen to block the development of quiescent lateral buds below decapitated shoots in several ferns (Wardlaw, 1946), but its effects were less pronounced in another fern, *Davallia* (Croxdale, 1976). Using ELISA techniques, Pilate *et al.* (1989) attempted to quantify the levels of IAA and several cytokinins in growing, quiescent, and activated buds of the fern *Marsilea*. In general, growing buds exhibited much higher hormone levels than quiescent buds. Interestingly, the relative ability

of lateral buds to become activated upon apical bud removal was directly correlated with the measured levels of inactive cytokinin precursors in those buds. It was hypothesized that these precursors are metabolized into active molecules as an initial step in bud activation (Pilate *et al.*, 1989).

In conclusion, because IAA and cytokinins act as the primary regulators of apical dominance in seed plants (Tames, 1995), the evidence from bryophytes and pteridophytes suggests that this hormonal regulation of apical dominance is widespread in the land plant lineage.

IAA responses involving positional relationships

Each division of land plants can be said to exhibit a characteristic body plan based on such organizational features as axial polarity, embryo structure, meristematic activity, vascular tissue organization, positional relationships among vegetative organs, and positional relationships of reproductive organs on vegetative organs (Bold *et al.*, 1987; Gifford and Foster, 1989). The evidence cited in the Introduction demonstrates that IAA acts as an important regulator of the body plans of seed plants. The limited research available on non-seed plants hints that IAA may also help to regulate the body plans of these plants, at least with respect to the positions of absorptive structures and the differentiation of vascular tissue.

Insofar as IAA mediates increased rhizoid formation on cut sections of *Chara* thalli (Sievers and Schröter, 1971), it appears that this particular IAA response evolved prior to the evolution of land plants. It should therefore be expected that IAA was frequently observed to control the amount and sites of rhizoid initiation in the gametophytes of liverworts (e.g., Kaul *et al.*, 1962; Maravolo and Voth, 1966; Stange, 1977; Kumra and Chopra, 1987), mosses (e.g., Nyman and Cutter, 1981; Chopra and Vashistha, 1990), and pteridophytes (e.g., Haupt, 1957; Kato, 1957; Hickok and Kiriluk, 1984).

The more interesting question is whether or not the IAA response system responsible for rhizoid induction in land plant gametophytes was co-opted in early vascular plant evolution to regulate root initiation. The evidence available from extant pteridophytes is consistent with the hypothesis that endogenous IAA exercises primary control over lateral root initiation in these plants, much like its effects in seed plants. In ferns, IAA mediated the formation of both lateral roots on excised *Pteridium* roots (Partanen and Partanen,

332

1963) and adventitious roots on *Matteuccia* rhizomes (Ma and Steeves, 1992). In *Selaginella*, leafless cylindrical axes called rhizophores emerge at the sites of shoot branching. Although a discussion of the morphological nature of the rhizophore extends far beyond the scope of this review (see Lu and Jernstedt, 1996), the formation of *Selaginella* roots via either direct rhizophore transformation or endogenous root initiation was greatly enhanced by the application of IAA and the IAA analogue indole-3-butyric acid (Williams, 1937; Webster, 1969; Wochok and Sussex, 1975). Furthermore, TIBA induced all lateral meristems to develop as shoots (Wochok and Sussex, 1975). It is expected that recent advances in the molecular regulation of lateral root formation in *Arabidopsis* (e.g., Casimiro *et al.*, 2001) may soon provide the opportunity to address the issue of whether the IAA response genes for root formation were co-opted from the preexisting system for regulating rhizoid formation or from another IAA response system.

Given that IAA is the primary hormone for regulating vascular tissue development in seed plants (for references, see Introduction), it is surprising that this topic has rarely been investigated in non-seed vascular plants, namely the pteridophytes. For instance, Steeves and Briggs (1960) did observe that IAA, when applied in place of excised pinnae, was sufficient to mediate the final stage of xylem maturation in *Osmunda* leaves. Moreover, the limited research available suggests that IAA may also help to regulate the patterning of vascular bundles in pteridophytes, with some intriguing features that require further investigation. Ma and Steeves (1992) studied IAA effects on the primary vascular tissues in *Matteuccia* stems, which are typically arranged as a dictyostele composed of isolated vascular bundles. Suppressing leaf primordia on these apices resulted in the formation of additional vascular tissue so that the vascular tissue was organized as a complete ring or siphonostele in the underlying stem (Ma and Steeves, 1992). The application of IAA-soaked beads to suppressed apices caused the stele to exhibit parenchymatous leaf gaps similar to those formed in the dictyosteles of untreated plants. Thus, while IAA acted to reduce the total amount of vascular differentiation in the suppressed apices of this fern, these experiments could also be interpreted to suggest that IAA was involved in the positioning of the remaining vascular bundles, in a manner reminiscent of its ability to position primary vascular tissues in seed plants (Sachs, 1991; Berleth *et al.*, 2000). Nothing is known about the hormonal regulation of the

simple vascular tissues observed in certain bryophytes (Hébant, 1977; Ligrone *et al.*, 2000).

Discussion

IAA is an ancient signaling molecule

The presence of IAA responses in both the charophyte *Chara* and the brown alga *Fucus* suggests that IAA is an ancient signaling molecule in photosynthetic aquatic organisms. One can imagine that a prototypical IAA effect might be similar to the group effect of *Fucus* rhizoids where IAA appears to act as the pheromone that coordinates the mutual attraction of the rhizoids emerging from adjacent zygotes (Jaffe, 1968). IAA does indeed exhibit several features that could conceivably make it predisposed to serve as a pheromone for communicating among small multicellular aquatic organisms. IAA is a small organic molecule with a pK_a of 4.7, which means that it has the potential for rapid diffusion in aquatic environments and across cell membranes. Its deprotonated form is soluble at low concentrations in aquatic environments with typical pH values. It becomes protonated in acidic cell walls and, thus, it can readily diffuse across adjacent cell membranes. Due to its structural similarity to tryptophan, IAA has some inherent affinity for pre-existing amino acid transporters. It seems quite reasonable to hypothesize that natural selection might have therefore favored the evolution of generalized amino-acid transporters into more specific IAA transporters. Indeed, the IAA influx carrier AUX1 in *Arabidopsis* has apparently evolved from an amino acid permease (Bennett *et al.*, 1996). Finally, IAA can easily be synthesized as a product of amino acid degradation. (Clearly, once IAA was adopted as a signaling molecule, it became advantageous to control its biosynthesis in a more regulated manner, which might have provided the selection pressure favoring the evolution of the tryptophan-independent pathway for synthesizing IAA.)

It is thus plausible to speculate that IAA had originally evolved in photosynthetic aquatic organisms to serve as a pheromone for regulating the growth of nearby members of the same species. What is unexpected, or perhaps even shocking, is that the rhizoids of *Chara* and *Fucus* exhibit similar IAA regulation and inhibitor sensitivities. Such similarity can be attributed to either a common origin or convergent evolution. In the case of a common origin, it must be appreciated that the viridiplantae and heterokont lineages, which ultimately gave rise to charophytes and brown algae, respectively, separated as unicellular flagellates in deep evolutionary time (Baldauf *et al.*, 2000), as described in the Introduction. It seems unlikely that the molecular architecture for an IAA response system for rhizoid regulation would have been functioning in ancestral unicellular flagellates prior to the evolutionary divergence of the viridiplantae and heterokont lineages over a billion years ago. Given the closer phylogenetic relationship between the viridiplants and the red algae (Baldauf *et al.*, 2000), another remote possibility is that the IAA response system was transferred from the red algal lineage into the heterokont lineage during the secondary endosymbiosis that created the brown algal lineage (Delwiche, 1999; Grzebyk *et al.*, pers. comm.). In the case of convergent evolution, it is possible to construct a plausible argument that the chemical properties of IAA mentioned above favored its independent co-optation as a signaling molecule in different lineages. Molecular characterization of the IAA response systems in the brown, red (if present), and green algae will be necessary to distinguish among the alternative explanations of common origin via direct transmission, common origin via secondary endosymbiosis, and convergent evolution.

IAA as an ancient regulator of green plant development

It must first be acknowledged that we have a rather sketchy picture of IAA regulation of developmental processes in several basal groups in the land plant lineage. In particular, the only developmental research published on the charophytes studied the ability of IAA to promote rhizoid initiation in *Chara*; moreover, no research has been done on IAA regulation of hornwort development. Nevertheless, what emerge from the present survey are several surprising perspectives about the evolutionary patterns in auxin action in green plants. One, the entire land plant lineage (from charophytes to angiosperms) exhibits the same fundamental metabolic and transport mechanisms for regulating intracellular IAA levels, with the only significant differences being the increasing regulatory sophistication in IAA metabolism (e.g., conjugation reactions) and IAA transport (e.g., inhibitor sensitivity) observed in certain bryophytes and all vascular plants. Two, several IAA responses, such as tropisms and apical dominance, are seen to play critical roles in the development of bryophytes, pteridophytes, and seed plants. The evidence gathered from the bryophytes

suggests that their tropistic responses involve an asymmetric IAA distribution across the stimulated organs and that their apical dominance results from IAA secretion from the apical meristem. Thus, it appears that both these responses depend on the same regulatory mechanisms in all land plants. Moreover, IAA appears to mediate root initiation and vascular tissue differentiation in the pteridophytes via regulatory mechanisms that are reminiscent of those operating in the seed plants. In summary, it is quite plausible that the seed plants did not evolve *de novo* mechanisms for mediating IAA responses, but have rather modified pre-existing mechanisms already present in the early land plants.

Future prospects for studying the molecular evolution of IAA action

But an important question remains: how does one reconcile the fundamental commonality of IAA action observed throughout the charophyte and land plant lineage with the increasing morphological complexity expressed in this lineage? The tentative answer to that question comes from the new field of evolutionary developmental genetics (Raff, 1996). Its insights into how genetic changes in developmental mechanisms seem to underlie the evolution of animal body plans are truly amazing, at least to most plant biologists gazing in awe over the kingdom boundary.

In essence, the primary working hypothesis in evolutionary developmental genetics is that certain genes responsible for regulating developmental processes in simple organisms experienced repeated duplication and altered transcriptional regulation, with the result that these genes were able to specify more complex body plans. Of course, the paradigmatic example for the genetic regulation of animal body plans involves a conserved group of homeobox (*Hox*) genes known as the *Hox* clusters that encode transcription factors involved in diverse developmental processes (Erwin *et al.*, 1997; Valentine *et al.*, 1999; Knoll and Carroll, 1999; Peterson and Davidson, 2000). It is proposed that a single 'primordial' *Hox* gene present in ancient sponges underwent a series of duplication events that resulted in two *Hox* clusters arising in the basal bilateral animal group in the early Cambrian period. During the Cambrian radiation, rapid diversification and altered regulation of these genes is thought to have greatly contributed to the evolution of the different body plans of bilateral animals.

In plants, initial work on the molecular evolution of developmental control genes is beginning to yield similar interpretations. For instance, the diversity of the MADS-box gene family encoding another group of eukaryotic transcription factors appears to correlate with morphological complexity during plant evolution (Theissen *et al.*, 2000). Two MADS-box genes have been isolated from the moss *Physcomitrella* (Krogan and Ashton, 2000) in contrast to 15 genes from the fern *Ceratopteris* and to even larger numbers from several angiosperm species (Theissen *et al.*, 2000) The ability of *Physcomitrella* to perform homologous recombination should allow these investigators to determine the precise roles of the MADS-box genes in the mosses (Theissen *et al.*, 2001). Of particular importance to angiosperm development are the MADS-box gene subfamilies that act as homeotic selector genes for controlling floral organ identity. On the other hand, although certain other MADS-box genes are seen to regulate various developmental processes in angiosperms, it is not clear at present whether they are involved in the fundamental organization of the angiosperm body plan. Finally, the correlation between gene family size and morphological complexity is also manifested by the actin gene family that encodes one of the major cytoskeletal proteins in plants (Bhattacharya *et al.*, 2000). These authors do, however, note that the actin gene family may be more useful for investigating the mechanisms of gene duplication in green plants as opposed to for elucidating the relationship between gene duplication and morphological complexity.

Aside from the research on the MADS-box genes acting in floral development, current work on the evolutionary significance of plant MADS-box genes is largely focused on the effort to assign developmental functions to these putative regulatory genes. Plant biologists interested in the molecular evolution of IAA action face the opposite problem: we know that IAA responses are critical to the overall organization of the body plans of different land plants, but we must now start to characterize the gene families responsible for regulating those responses throughout the plant lineage. Given the increasing complexity observed in the gene families described above, it seems quite conceivable that gene families involved in IAA action may also be composed of one or a few genes in charophytes, with a progressive increase in the number of homologous genes from the early-divergent bryophytes to the late-divergent angiosperms. Just to cite one of many possible examples of IAA regulatory

genes, the *Arabidopsis* genome contains 16 sequences that share considerable similarity to the AtPIN1 gene encoding the auxin efflux carrier. Although some of these sequences are certainly pseudogenes, a substantial number must be expressed as functional genes because they exhibit mutant phenotypes (Palme and Galweiler, 1999; Galweiler and Palme, this issue). Since there is no evidence to suggest that the charophytes use auxin gradients to establish their body plans, one might expect that the auxin efflux carrier reported to function in *Chara* thalli (Dibb-Fuller and Morris, 1992) is encoded only by a single PIN homologue. It follows from this prediction that the number of PIN gene family members should correlate with increasing IAA transport capability and morphological complexity observed in land plant lineages from charophytes to angiosperms. Similar arguments can be constructed for most other gene families involved in IAA regulation. As a final consideration, we anticipate that the characterization of the molecular evolution of IAA action will also provide considerable insight into the rapid diversification of early land plants during the late Silurian through middle Devonian periods, as is discussed elsewhere (Cooke *et al.*, 2001).

References

Albaum, H.G. 1938. Inhibition due to growth hormones in fern prothallia and sporophytes. Amer. J. Bot. 25: 124–133.

Aloni, R. 1995. The induction of vascular tissues by auxin and cytokinin. In: P.J. Davies (Ed.) Plant Hormones: Physiology, Biochemistry, and Molecular Biology, 2nd ed., Kluwer Academic Publishers, Dordrecht, Netherlands, pp. 531–546.

Ashen, J.B., Cohen, J.D. and Goff, L.J. 1999. GC-SIM-MS detection and quantification of free indole-3-acetic acid in bacterial galls on the marine red alga *Prionitis lanceolata* Harvey. J. Phycol. 35: 493–500.

Atzorn, R., Bopp, M. and Merdes, U. 1989. The physiological role of indole acetic acid in the moss *Funaria hygrometrica* Hedw. II. Mutants of *Funaria hygrometrica* which exhibit enhanced catabolism of indole-3-acetic acid. J. Plant Physiol. 135: 526–530.

Baldauf, S. L., Roger, A. J., Wenk-Siefert, I. and Doolittle, W.F. 2000. A kingdom-level phylogeny of eukaryotes based on combined protein data. Science 290: 972–977.

Baldi, B.G., Maher, B.R., Slovin, J.P. and Cohen, J.D. 1991. Stable isotope labeling, *in vivo*, of D- and L- tryptophan pools in *Lemna gibba* and the low incorporation of label into indole-3-acetic acid. Plant Physiol. 95: 1203–1208.

Bartel, B. 1997. Auxin biosynthesis. Annu. Rev. Plant Physiol. Plant Mol. Biol. 48: 51–66.

Bennett, M.J., Marchant, A., Green, H.G., May, S.T., Ward, S.P., Millner, P.A., Walker, A.R., Schulz, B. and Feldmann, K.A. 1996. *Arabidopsis AUX1* gene: a permease-like regulator of root gravitropism. Science 273: 948–950.

Berleth, T., Mattsson, J., and Hardtke, C.S. 2000. Vascular continuity and auxin signals. Trends Plant Sci. 5: 387–393.

Bhatla, S.C. and Bopp, M. 1985. The hormonal regulation of protonemal development in mosses. III. Auxin-resistant mutants of the moss *Funaria hygrometrica* Hedw. J. Plant Physiol. 120: 233–243.

Bhatla, S.C., Kapoor, S., and Khurana, J.P. 1996. Involvement of calcium in auxin-induced cell differentiation in the protonema of the wild strain and auxin mutants of the moss *Funaria hygrometrica*. J. Plant Physiol. 147: 547–552.

Bhattacharya, D., Aubry, J., Twait, E.C. and Jurk, S. 2000. Actin gene duplication and the evolution of morphological complexity in land plants. J. Phycol. 36: 813–820.

Bialek, K., Michalczuk, L. and Cohen, J.D. 1992. Auxin biosynthesis during seed germination in *Phaseolus vulgaris*. Plant Physiol. 100: 509–517.

Bilderback, D.E. 1984. Phototropism of *Selaginella:* the role of the small dorsal leaves and auxin. Am. J. Bot. 71: 1330–1337.

Blakely, L.M., Blakely, R.M., Colowit, P.M. and Elliott, D.S. 1988. Experimental studies on lateral root formation in radish seedling roots. II. Analysis of dose-response to endogenous auxin. Plant Physiol. 87: 414–419.

Bold, H.C. and Wynne, M. J.1985. Introduction to the Algae. Prentice-Hall, Englewood Cliffs, NJ.

Bold, H.C., Alexopoulos, C.I. and Delevoryas, T. 1987. Morphology of Plants and Fungi, 5th ed. Harper & Row, New York.

Bopp, M. 1980. The hormonal regulation of morphogenesis in mosses. In: F. Skoog (Ed.) Plant Growth Substances 1979, Springer-Verlag, Berlin, pp. 351– 361.

Bradley, P.M. 1991. Plant hormones do have a role in controlling growth and development in algae. J. Phycol. 27: 317–321.

Braun, M. 1997. Gravitropism in tip-growing cells. Planta 203: S11–S19.

Brown, D.E., Rashotte, A.M., Murphy, A.S., Normanly, J., Tague, B.W., Peer, W.A., Taiz, L. and G.K. Muday. 2001. Flavonoids act as negative regulators of auxin transport *in vivo* in *Arabidopsis*. Plant Physiol. 126:524– 535.

Casimiro, I., Marchant, A., Bhalerao, R.P., Beeckman, T., Dhooge, S., Swarup, R., Graham, N., Inzé, D., Sandberg, G., Casero, P.J. and Bennett, M. 2001. Auxin transport promotes *Arabidopsis* lateral root initiation. Plant Cell 13: 843–852.

Chopra, R.N. and Vashistha, B.D. 1990. The effect of auxins and antiauxins on shoot-bud induction and morphology in the moss, *Bryum atrovirens* Will. ex Brid. Aust. J. Bot. 38: 177–184.

Christianson, M.L. 1999. Control of morphogenesis in bryophytes. In: A.J. Shaw and B. Goffinet (Eds.) Bryophyte Biology, Cambridge University Press, Cambridge, UK, pp. 199–224.

Cohen, J.D. and Bandurski, R.S. 1982. Chemistry and physiology of the bound auxins. Annu. Rev. Plant Physiol. 33: 403–430.

Cooke, T.J., Racusen, R.H. and Cohen, J.D. 1993. The role of auxin in plant embryogenesis. Plant Cell 5: 1494–1495.

Cooke, T.J., Poli, DB., Sztein, A.E. and Cohen, J.D. 2001. Evolutionary trends in auxin regulation and their potential role in the rapid diversification of land plants. Evolution of Plant Physiology Symposium Abstracts, Linnean Society, London, p. 12.

Cove, D.J. and Ashton, N.W. 1984. Hormonal regulation of gametophytic development in bryophytes. In: A.F. Dryer and J. F. Duckett (Eds.) The Experimental Biology of Bryophytes, Academic Press, London, pp. 177–201.

Croxdale, J. 1976. Hormones and apical dominance in the fern *Davallia*. J. Exp. Bot. 27: 801–815.

Crum, H. 2001. Stuctural Diversity of Bryophytes. University of Michigan Herbarium, Ann Arbor, MI.

Davidonis, G.H. and Munroe, M.H. 1972. Apical dominance in *Marchantia*: correlative inhibition of neighbor lobe growth. Bot. Gaz. 133: 177–184.

Delwiche, C.F. 1999. Tracing the thread of plastid diversity through the tapestry of life. Am. Nat. 154: S164–S177.

Dibb-Fuller, J.E. and Morris, D.A. 1992. Studies on the evolution of auxin carriers and phytotropin receptors: transmembrane auxin transport in unicellular and multicellular Chlorophyta. Planta 186: 219–226.

Dolan, L. 1998. Pointing roots in the right direction: the role of auxin transport in response to gravity. Genes Dev. 12: 2091–2095.

Duff, R.J. and Nickrent, D.L. 1999. Phylogenetic relationships of land plants using mitochondrial small-subunit rDNA sequences. Am. J. Bot. 86: 372– 386.

Eisen, J.A. 1998. Phylogenomics: improving functional predictions for uncharacterized genes by evolutionary analysis. Genomic Res. 8: 163–167.

Ellis, J.G. and Thomas, R.J. 1985. Phototropism of *Pellia*: evidence for mediation by auxin stimulated acid efflux. J. Plant Physiol. 121: 259– 264.

Erwin, D., Valentine, J. and Jablonski, D.. 1997. The origin of animal body plans. Am. Sci. 85: 126–137.

Estelle, M. 1998. Polar auxin transport: new support for an old model. Plant Cell 10: 1775–1778.

Evans, LV. and Trewavas, A.J. 1991. Is algal development controlled by plant growth substances? J. Phycol. 27: 322–326.

Gaal, D.J., Dufresne, S.J. and Maravolo, N.C. 1982. Transport of ^{14}C-indoleacetic acid in the hepatic *Marchantia polymorpha*. Bryologist 85: 410–418.

Geier, U., Werner, O. and Bopp, M. 1990. Indole-3-acetic acid uptake in isolated protoplasts of the moss *Funaria hygrometrica*. Physiol. Plant. 80: 584–592.

Gifford, E.M. and Foster, A.S. 1989. Morphology and Evolution of Vascular Plants, 3rd ed. Freeman, New York.

Goldsmith, M.H.M. 1977. The polar transport of auxin. Annu. Rev. Plant Physiol. 28: 439–478.

Graham, L.E. 1993. Origin of Land Plants. Wiley, New York.

Graham, L.E., Cook, M.E. and Busse, J.S. 2000. The origin of plants: body plan changes contributing to a major evolutionary radiation. Proc. Natl. Acad. Sci. USA 97: 4535–4540.

Haupt, W. 1957. Die Induktion der Polarität bei der Spore von *Equisetum*. Planta 49: 61–90.

Hébant, C. 1977. The Conducting Tissues of Bryophytes. J. Cramer, Vaduz.

Hertel, R., Lomax, T.L., and Briggs, W.R. 1983. Auxin transport in membrane vesicles from *Cucurbita pepo* L. Planta 157: 193–201.

Hickok, L.G. and Kiriluk, R.M. 1984. Effects of auxins on gametophyte development and sexual differentiation in the fern *Ceratopteris thalictroides* (L.) Brongn. Bot. Gaz. 145: 37–42.

Hickok, L.G., Warne, T.R., and Fribourg, R.S. 1995. The biology of the fern *Ceratopteris* and its uses as a model system. Int. J. Plant Sci. 156: 332–345.

Jacobs, M. and Hertel, R. 1978. Auxin binding to subcellular fractions from *Cucurbita* hypocotyls: *in vitro* evidence for an auxin transport carrier. Planta 142: 1–10.

Jacobs, M. and Rubery P.H. 1988. Naturally occurring auxin transport regulators. Science 241: 346–349.

Jacobs, W.P., Falkenstein, K. and Hamilton, R.H. 1985. Nature and amount of auxin in algae. IAA from extracts of *Caulerpa paspaloides* (Siphonales). Plant Physiol. 78: 844–848.

Jaffe, L.F. 1968. Localization in the developing *Fucus* egg and the general role of localizing currents. Adv. Morphogen. 7: 295–328.

Jayaswal, R.K. and Johri, M.M. 1985. Occurrence and biosynthesis of auxin in protonema of the moss *Funaria hygrometrica*. Phytochemistry 24: 1211–1214.

Karol, K.G., McCourt, R.M., Cimino, M.T., and Delwiche, C.F. 2001. The closest living relatives to the land plants. Science 294: 2351–2353.

Kato, Y. 1957. The effects of colchicine and auxin on rhizoid formation of *Dryopteris erythrosora*. Bot. Mag. 70: 258–263.

Kaufman, P.B., Wu, L., Brock, T.G. and Kim, D. 1995. Hormones and their orientation of growth. In: P.J. Davies (Ed.) Plant Hormones: Physiology, Biochemistry and Molecular Biology, 2nd ed., Kluwer Academic Publishers, Dordrecht, Netherlands, pp. 547–571.

Kaul, K.N., Mitra, G.C. and Tripathi, B.K. 1962. Responses of *Marchantia* in aseptic culture to well-known auxins and antiauxins. Ann. Bot. 26: 447–467.

Kenrick, P. and Crane, P.R. 1997. The Origin and Early Diversification of Land Plants: A Cladistic Study. Smithsonian Institute Press, Washington, D.C.

Klambt, D., Knauth, B. and Dittmann, I. 1992. Auxin dependent growth of rhizoids of *Chara globularis*. Physiol. Plant. 85: 537–540.

Knoll, A.H. and Carroll, S.B. 1999. Early animal evolution: emerging views from comparative biology and geology. Science 284: 2129–2137.

Koshiba, T., Kamiya, Y. and Iino, M. 1995. Biosynthesis of indole-3-acetic acid from l-tryptophan in coleoptile tips of maize (*Zea mays*). Plant Cell Physiol. 36: 1503–1510.

Krogan, N.T. and Ashton, N.W. 2000. Ancestry of plant MADS-box genes revealed by bryophyte (*Physcomitrella patens*) homologues. New Phytol. 147: 505–517.

Kumra, S. and Chopra, R.N. 1987. Callus initiation, its growth and differentiation in the liverwort *Asterella wallichiana* (Lehm. et Lindenb.) Groelle. I. Effect of auxins and cytokinins. J. Hattori Bot. Lab. 63: 237–245.

LaRue, C.D. and Narayanswami, S. 1957. Auxin inhibition in the liverwort *Lunularia*. New Phytol. 56: 61–70.

Lehnert, B. and Bopp, M. 1983. The hormonal regulation of protonema development in mosses. I. Auxin-cytokinin interaction. Z. Pflanzenphysiol. 110: 379–391.

Ligrone, R., Duckett, J.G. and Renzaglia, K.S. 2000. Conducting tissues and phyletic relationships of bryophytes. Phil. Trans. R. Soc. Lond. Ser. B 355: 795–813.

Liu, C.-M., Xu, Z.-H. and Chua, N.-H. 1993. Auxin polar transport is essential for the establishment of bilateral symmetry during early plant embryogenesis. Plant Cell 5: 621–630.

Lomax, T.L., Muday, G.K. and Rubery, P.H. 1995. Auxin transport. In: P. J. Davies (Ed.) Plant Hormones: Physiology, Biochemistry, and Molecular Biology, 2nd ed., Kluwer Academic Publishers, Dordrecht, Netherlands, pp. 509–530.

Lu, P. and Jernstedt, J.A. 1996. Rhizophore and rood development in *Selaginella martensii*: meristem transitions and identity. Int. J. Plant Sci. 157: 180–194.

Lyndon, R.F. 1998. The Shoot Apical Meristem: Its Growth and Development. Cambridge University Press, Cambridge, UK.

Ma, Y. and Steeves, T.A. 1992. Auxin effects on vascular differentiation in Ostrich fern. Ann. Bot. 70: 277–282.

MacQuarrie, G. and von Maltzahn, K. 1959. Correlations affecting regeneration and reactivation in *Splachnum ampullaceum* (L.) Hedw. Can. J. Bot. 37: 121–134.

Maravolo, N.C. 1976. Polarity and localization of auxin movement in the hepatic, *Marchantia polymorpha*. Am. J. Bot. 63: 529–531.

Maravolo, N.C. 1980. Control of development in hepatics. Bull. Torrey Bot. Club 107: 308–324.

Maravolo, N.C. and Voth, P.D. 1966. Morphogenic effects of three growth substances on *Marchantia* gemmalings. Bot. Gaz. 127: 79–86.

Marchant, A., Kargul, J., May, S.T., Muller, P., Delbarre, A., Perrot-Rechenmann, C. and Bennett, M. J. 1999. AUX1 regulates root gravitropism in *Arabidopsis* by facilitating auxin uptake within root apical tissues. EMBO J. 18: 2066–2073.

Meicenheimer, R.D. 1981. Changes in *Epilobium* phyllotaxy induced by N-1-naphthylphthalmic acid and α-4-chlorophenoxyisobutric acid. Am. J. Bot. 68: 1139–1154.

Michalczuk, L., Cooke, T.J. and Cohen, J.D. 1992a. Auxin levels at different stages of carrot somatic embryogenesis. Phytochemistry 31: 1097–1103.

Michalczuk, L., Ribnicky, D.M., Cooke, T.J. and Cohen, J.D. 1992b. Regulation of indole-3-acetic acid biosynthetic pathways in carrot cell cultures. Plant Physiol. 100: 1346–1353.

Muller, A., Guan, C., Galweiler, L., Tanzler, P., Huijser, P., Marchant, A., Parry, G., Bennett, M., Wisman, E. and Palme, K. 1998. AtPIN2 defines a locus of *Arabidopsis* for root gravitropism control. EMBO J. 17: 6903–6911.

Nemhauser, J.L., Feldman, L.J. and Zambryski, P.C. 2000. Auxin and ETTIN in *Arabidopsis* gynoecium morphogenesis. Development 127: 3877–3888.

Nickrent, D.L., Parkinson, C.L., Palmer, J.D. and Dugg, R.J. 2000. Multigene phylogeny of land plants with special reference to bryophytes and the earliest land plants. Mol. Biol. Evol. 17: 1885–1895.

Niklas, K.J. 1997. The Evolutionary Biology of Plants. University of Chicago Press, Chicago.

Nonhebel, H.M., Cooney, T.P. and Simpson, R. 1993. The route, control and compartmentation of auxin synthesis. Aust. J. Plant Physiol. 20: 527–539.

Normanly, J. 1997. Auxin metabolism. Physiol. Plant. 100: 431–442.

Normanly, J., Cohen, J.D. and Fink, G.R. 1993. *Arabidopsis thaliana* auxotrophs reveal a tryptophan-independent biosynthetic pathway for indole-3-acetic acid. Proc. Natl. Acad. Sci. USA 90: 10355–10359.

Nyman, L.P. and Cutter, E.G. 1981. Auxin-cytokinin interaction in the inhibition, release, and morphology of gametophore buds of *Plagiomnium cupidatum* from apical dominance. Can. J. Bot. 59: 750–760.

Palme, K. and Galweiler, L. 1999. PIN-pointing the molecular basis of auxin transport. Curr. Opin. Plant Biol. 2: 375–381.

Partanen, J.N. and Partanen, C.R. 1963. Observations on the culture of roots of the Bracken fern. Can. J. Bot. 41: 1657–1661.

Peterson, K.J. and Davidson, E.H. 2000. Regulatory evolution and the origin of the bilaterians. Proc. Natl. Acad. Sci. USA 97: 4430–4433.

Pilate, G., Sossountzov, L. and Miginiac, E. 1989. Hormone levels and apical dominance in the aquatic fern *Marsilea drummondii* A. Br. Plant Physiol. 90: 907–912.

Pryer, K.M., Schneider, H., Smith, A.R., Cranfill, R., Wolf, P.G., Hunt, J.S. and Sipes, S.D. 2001. Horsetails and ferns are a monoplyletic group and the closest living relatives to seed plants. Nature 409: 618–622.

Qiu, Y.-L., Cho, Y., Cox, J.C. and Palmer, J.D. 1998. The gain of three mitochondrial introns identifies liverworts as the earliest land plants. Nature 394: 671–674.

Qiu, Y.-L. and Lee, J. 2000. Transition to a land flora: a molecular perspective. J. Phycol. 36: 799–802.

Raff, R.A. 1996. The Shape of Life: Genes, Development, and the Evolution of Animal Form. University of Chicago Press, Chicago.

Rapparini, F., Cohen, J.D. and Slovin, J.P. 1999. Indole-3-acetic acid biosynthesis in Lemna gibba studied using stable isotope labeled anthranilate and tryptophan. Plant Growth Regul. 27: 139–144.

Raven, J.A. 1974. Transport of indoleacetic acid in plant cells in relation to pH and electrical potential gradients, and its significance for polar IAA transport. New Phytol. 74: 163–172.

Reed, R.C., Brady, S.R. and Muday, G.K. 1998. Inhibition of auxin movement from the shoot into the root inhibits lateral root development in Arabidopsis. Plant Physiol. 118: 1369–1378.

Reinhardt, D., Mandel, T and Kuhlemeier, C. 2000. Auxin regulates the initiation and radial position of plant later al organs. Plant Cell 12: 507– 518.

Reski, R. 1999. Molecular genetics of Physcomitrella. Planta 208: 301–309.

Ribnicky, D.M., Ilic, N., Cohen, J.D., and Cooke, T.J. 1996. The effects of exogenous auxins on endogenous indole-3-acetic acid metabolism. The implications for carrot somatic embryogenesis. Plant Physiol. 112: 549–558.

Ribnicky, D.M., Cohen, J.D., Hu, W.-S. and T.J. Cooke. 2001. An auxin surge following fertilization in carrots: a general mechanism for regulating plant totipotency. Planta, in press.

Roberts, L.W., Gahan, P.B. and Aloni, R.1988. Vascular Differentiation and Plant Growth Regulators. Springer-Verlag, Berlin.

Rose, S. and Bopp, M. 1983. Uptake and polar transport of indoleacetic acid in moss rhizoids. Physiol. Plant. 58: 57–61.

Rose, S., Rubery, P.H. and Bopp, M. 1983. The mechanism of auxin uptake and accumulation in moss protonemata. Physiol. Plant. 58: 52–56.

Rubery, P.H. and Sheldrake, A.R. 1974. Carrier-mediated auxin transport. Planta 188: 101–121.

Sachs, T. 1991. Cell polarity and tissue patterning in plants. Development (Suppl.) 1: 83–93.

Schaefer, D.G. and Zryd, J.P. 1997. Efficient gene targeting in the moss Physcomitrella patens. Plant J. 11: 1195–1206.

Schiavone, F.M. and Cooke, T.J. 1986. Unusual patterns of somatic embryogenesis in the domesticated carrot: developmental effects of exogenous auxins and auxin transport inhibitors. Cell. Diff. 21: 53–62.

Schnepf, E., Herth, W. and Morre D.J. 1979. Elongation growth of setae of Pellia (Bryophyta): effects of auxin and inhibitors. Z. Pflanzenphysiol. 94: 211–217.

Sembdner, G., Gross, D., Liebisch, W. and Schneider, G. 1981. Biosynthesis and metabolism of plant hormones. In. J. MacMillan (Ed.) Hormonal Regulation of Plant Development. I. Molecular Aspects of Plant Hormones. Encyclopedia of Plant Physiology, Vol. 9, Springer Verlag, Berlin, pp. 281 444.

Sessions, A., Nemhauser, J.L., McColl, A., Roe, J.L., Feldmann, K.A. and Zambryski, P.C. 1997. ETTIN patterns the Arabidopsis floral meristem and reproductive organs. Development 124: 4481–4491.

Sievers, A. and Schröter, K. 1971. Versuch einer Kausalanalyse der geotropischen Reaktionskette im Chara-Rhizoid. Planta 96: 339–353.

Sievers, A., Buchen, B. and Hodick, D. 1996. Gravity sensing in tip-growing cells. Trends Plant Sci. 1: 273–279.

Slovin, J.P., Bandurski, R.S. and Cohen, J.D. 1999. Auxin. In: P.J.J. Hooykaas, M.A. Hall and K.R. Libbenga (Eds.) Biochemistry and Molecular Biology of Plant Hormones, Elsevier Science, Amsterdam, pp. 115–140.

Stange, L. 1971. Effects of morphactins and of auxin on the formation of meristematic centres in Riella helicophylla. Ind. J. Plant Physiol. 14: 44–54.

Stange, L. 1977. Meristem differentiation in Riella helicophylla (Bory et Mont.) Mont. under the influence of auxin and anti-auxin. Planta 135: 289–295.

Staves, M.P. 1997. Cytoplasmic streaming and gravity sensing in Chara internodal cells. Planta 203: S79–S84.

Steeves, T.A. and Briggs, W.R. 1960. Morphogenetic studies on Osmunda cinnamomea L. The auxin relationships of expanding fronds. J. Exp. Bot. 11: 45–67.

Steinmann, T., Geldner, N., Grebe, M., Mangold, S., Jackson, C. L., Paris, S., Galweiler, L., Palme, K. and Jurgens, G. 1999. Coordinated polar localization of auxin efflux carrier PIN1 by GNOM ARF GEF. Science 286: 316– 318.

Sussmann, M.R. and Goldsmith, M.H.M. 1981. The action of specific inhibitors of auxin transport on uptake of auxin and binding of N-1-naphthylphthalamic acid to a membrane site in maize coleoptiles. Planta 152: 13–18.

Swarup, R., Marchant, A. and Bennett, M.J. 2000. Auxin transport: providing a sense of direction during plant development. Biochem. Soc. Trans. 28: 481– 485.

Swarup, R., Friml, J., Marchant, A., Ljung, K., Sandberg, G., Palme, K. and Bennett, M. Localisation of the auxin permease AUX1 in the Arabidopsis root apex reveals two novel functionally distinct hormone transport pathways. Genes Dev., in press.

Sztein, A.E., Cohen, J.D., Slovin, J.P. and Cooke, T.J. 1995. Auxin metabolism in representative land plants. Am. J. Bot. 82: 1514–1521.

Sztein, A.E., Cohen, J.D., de la Fuente, I.G. and Cooke, T.J. 1999 Auxin metabolism in mosses and liverworts. Am. J. Bot. 86: 1544–1555.

Sztein, A.E., Cohen, J.D. and Cooke, T.J. 2000. Evolutionary patterns in the auxin metabolism of green plants. Int. J. Plant Sci. 161: 849–859.

Sztein, A. E., Cohen,J.D. and T. J. Cooke. 2001. Indole-3-acetic acid biosynthesis in isolated axes from germinating been seeds: the effect of wounding on biosynthetic pathway. Plant Growth Regul., in press.

Tamas, I.A. 1995. Apical dominance. In: P. J. Davies (Ed.) Plant Hormones: Physiology, Biochemistry, and Molecular Biology, 2nd ed. Kluwer Academic Publishers, Dordrecht, Netherlands, pp. 572-597.

Taylor, T.N. and Taylor, E.L. 1993. The Biology and Evolution of Fossil Plants. Prentice Hall, Englewood Cliffs, NJ.

Theissen, G., Becker, A., Di Rosa, A., Kanno, A., Kim, J.T., Munster, T., Winter, K.-U. and Saedler, H. 2000. A short history of MADS-box genes in plants. Plant Mol. Biol. 42: 115–149.

Theissen, G., Munster, T and Henschel, K. 2001. Why don't mosses flower? New Phytol. 150: 1–5.

Thomas, R.J. 1980. Cell elongation in hepatics: the seta system. Bull. Torrey Bot. Club 107: 339–345.

Uggla, C., Mellerowicz, E.J. and Sundberg, B. 1998. Indole-3-acetic acid controls cambial growth in Scots pine by positional signaling. Plant Physiol. 117: 113–121.

Uggla, C., Moritz, T., Sandberg, G. and Sundberg, B. 1996. Auxin as a positional signal in pattern formation in plants. Proc. Natl. Acad. Sci USA 93: 9282–9286.

Valentine, J.W., Jablonski, D. and Erwin, D.H. 1999. Fossils, molecules and embryos: new perspectives on the Cambrian explosion. Development 126: 851– 859.

338

von Maltzahn, K.E. 1959. Interaction between kinetin and indoleacetic acid in the control of bud reactivation in *Splachnum ampullaceum* (L.) Hedw. Nature 183: 60–61.

Wada, M. and Kadota, A. 1989. Photomorphogenesis of lower green plants. Annu. Rev. Plant Physiol. Plant Mol. Biol. 40: 169–191.

Walters, J. and Observe, D.J. 1979. Ethylene and auxin-induced cell growth in relation to auxin transport and metabolism and ethylene production in the semi-aquatic plant, *Regnellidium diphyllum*. Planta 146: 309–317.

Wardlaw, C.W. 1946. Experimental and analytical studies of pteridophytes. VIII. Further observations on bud development in *Matteucia struthiopteris*, *Onoclea sensibilis*, and species of *Dryopteris*. Ann. Bot. 10: 117–132.

Webster, T.R. 1969. An investigation of angle-meristem development in excised stem segments of *Selaginella martensii*. Can. J. Bot. 47: 717– 722.

Wildman, S.G., Ferri, M.G. and Bonner, J. 1947. The enzymatic conversion of tryptophan to auxin by spinach leaves. Arch. Biochem. 13: 131–144.

Williams, S. 1937. Correlation phenomena and hormones in *Selaginella*. Nature 139: 966.

Wochok, Z.S. and Sussex, I.M. 1973. Morphogenesis in *Selaginella*: auxin transport in the stem. Plant Physiol. 51: 646–650.

Wochok, Z.S. and Sussex, I.M. 1975. Morphogenesis in *Selaginella*. III. Meristem determination and cell differentiation. Dev. Biol. 47: 376–383.

Wood, A.J., Oliver, M.J. and Cove, D.J. 2000. New frontiers in bryology and lichenology: bryophytes as model systems. Bryologist 103: 128–133.

Wright, A.D., Sampson, M.B., Neuffer, M.G., Michalczuk, L., Slovin, J.P. and Cohen, J.D. 1991. Indole-3-acetic acid biosynthesis in the mutant maize *orange pericarp*, a tryptophan auxotroph. Science 254: 998–1000.

Plant Molecular Biology **49**: 339–348, 2002.
Perrot-Rechenmann and Hagen (Eds.), Auxin Molecular Biology.
© 2002 *Kluwer Academic Publishers.*

A short history of auxin-binding proteins

Richard M. Napier[1], Karine M. David[2] and Catherine Perrot-Rechenmann[2],*
[1]*Horticulture Research International, Wellesbourne, Warwick, CV35 9EF, UK;* [2]*Institut des Sciences du Végétal, CNRS, Avenue de la Terrasse, 91198 Gif sur Yvette Cedex, France (*author for correspondence; e-mail Catherine.rechenmann@isv.cnrs-gif.fr)*

Received 31 August 2001; accepted in revised form 16 October 2001

Key words: auxin-binding, chromotography, photolabelling

Introduction

Plant hormone receptors have proved to be elusive research targets. The successes of describing receptors from animals and bacteria have not yet been matched for plants. Nevertheless, where candidate receptors have been identified, they have been subjected to detailed examination. One such is the protein known as ABP1, an auxin-binding protein first described from maize (*Zea mais* L.).

The first detection of ABP1 was as an auxin-binding activity in crude membrane preparations of etiolated coleoptiles (Hertel *et al.*, 1972). Over the next decade this binding activity was characterized in detail for ligand specificity (Ray *et al.*, 1977), affinity (Ray, 1977a; Batt *et al.*, 1976) and cellular compartmentalization (Ray *et al.*, 1977). In 1985, the binding protein was purified for the first time (Löbler and Klämbt, 1985) leading on to functional studies of ABP1, examination of its cell biology and its structure. This review summarizes the advances made in each of these areas.

Photolabelled proteins

It would be incorrect to suggest that all auxin binding in plants could be ascribed to ABP1. There have been numerous reports of other binding sites for indole-3-acetic acid. Most have been discovered through the use of tritiated azido IAA, a photoactive auxin analogue (Melhado *et al.*, 1982). A number of labelled proteins from a range of plants have been traced and sequenced, almost all turning out to be enzymes (Venis and Napier, 1995).

Maize ABP1 is on the list of photolabelled proteins and we will return to the activities of this below. For the remainder, the interaction with auxin at physiologically relevant concentrations has failed to alter the protein's activity. As such, the binding fails to satisfy one of the key criteria of receptors, namely that ligand binding initiates a biologically relevant response. Although these proteins are not likely to be receptors, this certainly does not mean that they are not relevant to auxin biochemistry or physiology. Indeed, photolabelling has identified conjugate hydrolases, glutathione *S*-transferases and, possibly, components of the auxin transport machinery (Venis and Napier, 1995).

Auxin affinity chromatography

Affinity purification has added a number of new auxin-binding proteins to the list recently. A phenylacetic acid column is often used to purify maize ABP1, but it has also been shown to select for a 44 kDa protein from pea (Reinard *et al.*, 1998). Further characterisation revealed that this protein is an isovaleryl-CoA dehydrogenase, an enzyme necessary for leucine catabolism in mammals and targeted to mitochondria (Reinard *et al.*, 2000).

A number of soluble auxin-binding proteins have been reported by Sakai's group (Sugaya and Sakai, 1996). Their 2,4-D matrix has been shown to purify a glutathione-dependent formaldehyde dehydrogenase. A 2,4-D matrix was also used to purify peach auxin-binding proteins (Ohmiya *et al.*, 1993), later shown to be members of the protein superfamily known as the germins (Dunwell *et al.*, 2000). It is interesting to note that ABP1 is also a member of this superfamily, although outside the germin motif boxes (see below) there is very low homology (less than 5%) between the proteins from peach and ABP1. The same group has identified another 2,4-D binding protein (labelled Pp60) as protein disulphide isomerase (PDI) (Sugaya, *et al.*, 2000). The affinity of the site was K_D 2,4-D = 3.5 × 10^{-5} M (similar to that for the other peach proteins, 4 × 10^{-5} M) which is not very high. Competition with other auxins showed that IAA was a poor ligand. It seems unlikely that protein disulfide isomerase, an ubiquitous endoplasmic reticulum (ER) folding enzyme, is also an auxin receptor. Nevertheless, it clearly does have a binding capacity for auxin, as it does for other hormones, including 3,3^1,5-triiodothyronine in animal ER (Primm and Gilbert, 2001). It is interesting to note that PDI, ABP1, the peach germins and the IAA-amino acid hydrolases (Davies *et al.*, 1999) co-localize in the ER, although the functional significance of this observation remains unclear.

One other affinity matrix has been used to purify a putative auxin receptor, tryptophan linked to sepharose through the primary amine group (Kim *et al.*, 1998). The reported affinity of a protein purified from rice was high, K_D IAA = 1.9 × 10^{-8} M, with four auxin sites. Evidence for receptor activity was presented, although the half-saturating concentration for IAA in the assay was in the micromolar range and no activity was detected below 10^{-7} M IAA. Therefore, the apparent affinity and biological activity data are at variance. In the presence of the purified protein (named ABP$_{57}$) and IAA, proton translocation across plasma membrane was promoted. Tobacco plasma membrane H$^+$-ATPase activities have been linked to auxin in similar assays before (Laporte and Rossignol, 1997), but in this case an increase in proton translocation was seen only when both purified protein and auxin were present together. The protein can also be purified with antibodies specific for bovine serum albumin (Kim *et al.*, 2001). The only auxin to have activity in the assay was IAA and data for tryptophan were not presented, raising questions on the importance of the protein in auxin perception. The protein has yet to be cloned or fully sequenced.

Mutant screens for receptors

Affinity labelling and affinity purification have yielded a long list of auxin-binding proteins, although few remain candidate receptors. The same is true for molecular genetic approaches to receptor isolation. Many screens of mutant populations have been carried out and many auxin-insensitive plants identified. Once again, these huge programmes have been reviewed widely and readers are referred to Hobbie and Estelle (1994), Leyser (1997) and Luschnig and Fink (1999). Again, there are new mutants and cloning reports, but as these do not have any details of auxin binding they will not be covered here. Indeed, it is pertinent to point out that of the many genes cloned through these programmes, all important for auxin action (Dharmasiri and Estelle, Liscum and Reed, DeLong *et al.*, and Friml and Palme in this issue), none has been shown to have a protein product which binds auxin. It is clear that plant hormone receptors generally (with the exceptions of ethylene, cytokinins, and perhaps, brassinosteroids) have proved recalcitrant to molecular genetics so far. Screening activation-tagged lines might change this, but so far no auxin receptors have been derived from direct mutant screens. Forward screens have also failed to pull out ABP1 mutants. However, an ABP1 knock-out mutant has been isolated recently by reverse genetics (Chen *et al.*, 2001). The homozygote is lethal due to defects during early embryogenesis and the implications of this knock-out are discussed below.

ABP1

APB1 is ubiquitous in the vascular plants, including the pteridophytes and bryophytes. A pile-up of full-length translated cDNA sequences is shown in Figure 1, and this includes the sequence of ABP1 from the moss *Ceratodon purpureus*. ABP1 ESTs from a fern and other monocots are also available but are not included here. Very recently, ESTs have become available from *Chlamydomonas*, a lower plant of the green algae. The occurrence of ABP1 homologues in the green algae predates the evolution of both auxin carriers (Dibb-Fuller and Morris, 1992) and IAA catabolic pathways (Sztein *et al.*, 1995). Auxin binding in this homologue has not yet been demonstrated. There are no close homologues to ABP1 outside the plant kingdom, nor in the blue-green alga *Synechocystus*, fungi or yeast. One sequence motif, the germin motif, is partially conserved and this could place ABP1 in a protein superfamily common to all kingdoms (Dunwell *et al.*, 2000) but, apart from this, ABP1 is special to green plants.

The full sequence of ABP1 itself has revealed little about the protein. Sequences from a range of plants (Figure 1) show a high level of residue conservation throughout the mature protein. However, the signal peptide, cleaved co-translationally as the polypeptide enters the ER, is highly variant. Three domains of 13 to 20 amino acids, labelled boxes A (or D16), B and C, are highly conserved amongst all higher plant ABP1s. All ABP1s also contain an ER lumen retention motif (with the possible exception of the *Ceratodon* sequence, discussed below), three cysteines and one conserved N-glycosylation site. Throughout the dicots there is a second conserved N-glycosylation site close to the N-terminus (Massotte *et al.*, 1995). Strawberry ABP1, for example, has an additional third site (Lazarus and Macdonald, 1996).

Secondary structure algorithms suggest the polypeptide folds into β sheets and β turns with only a small stretch of α helix at the C-terminus. The C-terminus carries the KDEL-ER retention motif, the consequences of which are discussed below.

Maize ABP1 expressed in the baculovirus system (Macdonald *et al.*, 1994) has now been crystallised and preliminary X-ray diffraction data collected (Woo *et al.*, 2000). Three crystal types were grown, each showing the basic unit of structure was a homodimer, in agreement with early gel filtration data (Venis, 1977). The resolution of a high-definition protein structure should follow shortly. The protein used for crystallisation has been shown to be indistinguishable from ABP1 purified from maize seedlings, except at the C-terminus where the ER targeting sequence has been modified to KEQL to promote secretion and facilitate purification. Post-translational modifications and binding kinetics of the mutated version remained unchanged from wild type ABP1 and so the crystal structure should be illuminating.

Mass spectrometry analysis of ABP1 purified from maize coleoptiles has recently been reported (Feckler *et al.*, 2001). In the experimental conditions used by the authors to purify the protein and perform mass spectrometry analysis, a disulfide bridge was found between C_2 and C_{61} (in box A, Figure 1) and no intermolecular disulfide bridge involving the remaining C_{155} (in the C-terminal domain) has been revealed. This is at variance with crystallographic data which show that a single disulfide links C_2 with C_{155}, stabilizing the protein both in the presence and absence of bound auxin (Napier and Pickersgill, unpublished). It was recently shown that mutation C158 of tobacco ABP1 (equivalent to C155 in maize) into serine al-

ters the folding of the protein, its capacity to interact with auxin and its activity at the plasma membrane (David *et al.*, 2001). These results demonstrated the importance of this in ABP1 action, but a discrepancy remains at the structural level between experimental results and this needs to be resolved.

An attractive model is that intramolecular disulfide bridge shuffling occurs *in vivo* to control ABP1 folding, targeting and signalling, but it is necessary to test this hypothesis.

Auxin-binding site

The key to the activity of an auxin receptor is its auxin-binding site. Long before ABP1 was purified, models to predict the relationship between the molecular structure of auxins and hormone activity were developed (Thimann, 1963; Farrimond *et al.*, 1978; reviewed by Napier, 2001). These were complemented by mechanistic models for the recognition site (Kaethner, 1977; Katekar, 1979) which became more sophisticated with increasing computational capacity and knowledge (Tomic *et al.*, 1998). In parallel with models based on auxin structures, biochemistry identified amino acids likely to be involved in auxin binding (Venis, 1977; Navé and Benveniste, 1984). Determination of the sequence of ABP1 indicated that the biochemist's data matched a string of residues in the linear peptide. This peptide became known as the D16 box (Venis *et al.*, 1992) or Box A (Brown and Jones, 1994) (Figure 1).

Antibodies raised to the Box A sequence were shown to have auxin-like activity (Venis *et al.*, 1992; Walther *et al.*, 1997). With the antibody able to substitute for auxin in physiological assays, it seemed likely that this part of ABP1 was a major contributor to the auxin site. As new sequences from diverse plants became available, it was also found that this region of ABP1 is fully conserved over 15 residues.

Two other parts of ABP1 are also highly conserved, labelled Box B and C, respectively (Figure 1). The latter overlaps the site at which a photolabelled IAA was found to bind, labelled peptide 11 (Brown and Jones, 1994), this too might be involved in the binding site, but the role of this part of the protein has yet to be fully discovered. A monoclonal antibody named mAb12 has been raised to tobacco ABP1 and found to recognize a discontinuous epitope embracing residues in both boxes A and C (Leblanc *et al.*, 1999a). Crystallography illustrates that β sheets from

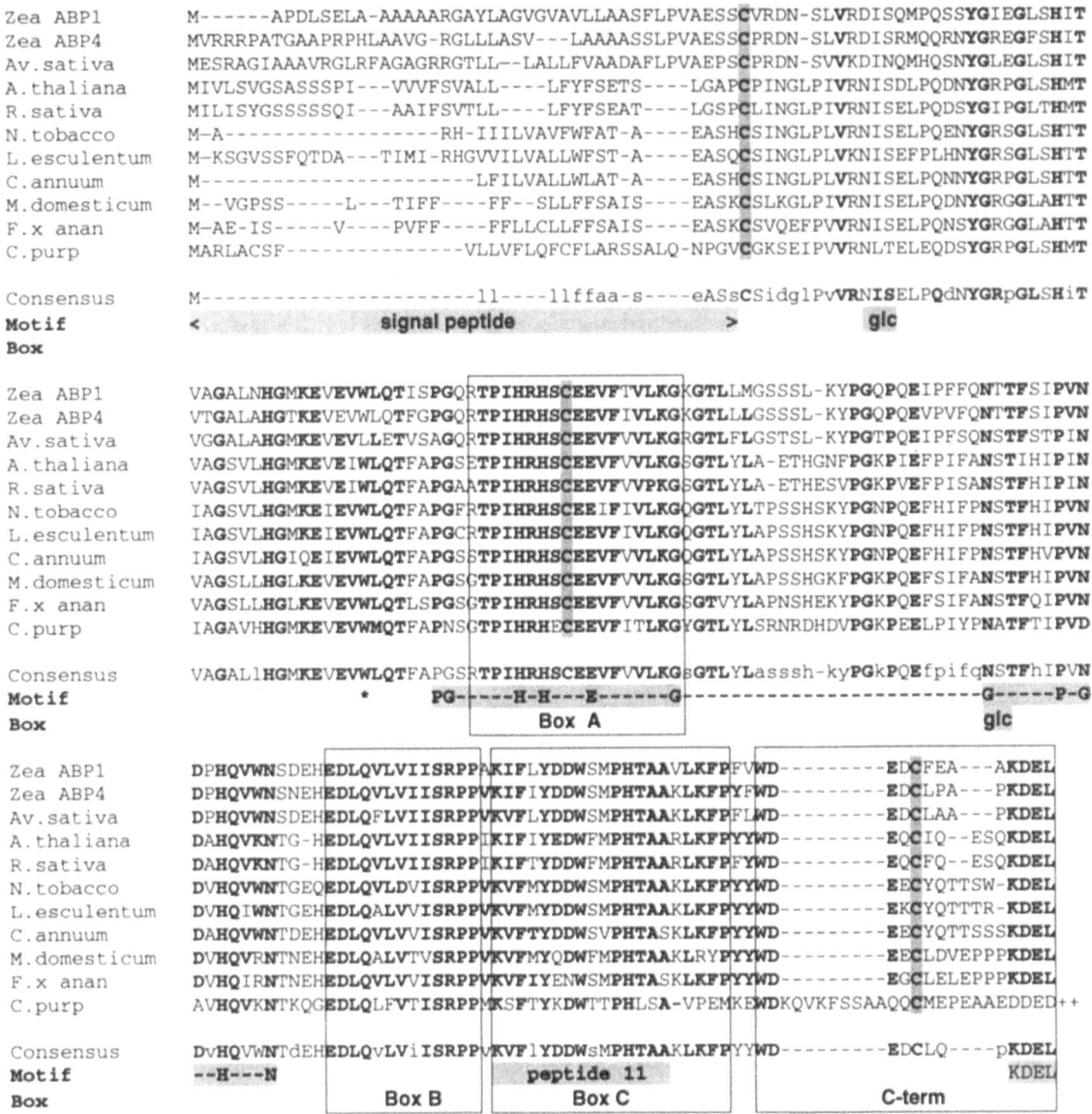

```
Zea ABP1      M------APDLSELA-AAAAARGAYLAGVGVAVLLAASFLPVAESSCVRDN-SLVRDISQMPQSSYGIEGLSHIT
Zea ABP4      MVRRRPATGAAPRPHLAAVG-RGLLLASV---LAAAASSLPVAESSCPRDN-SLVRDISRMQQRNYGREGFSHIT
Av.sativa     MESRAGIAAAVRGLRFAGAGRRGTLL--LALLFVAADAFLPVAEPSCPRDN-SVVKDINQMHQSNYGLEGLSHIT
A.thaliana    MIVLSVGSASSSPI---VVVFSVALL----LFYFSETS----LGAPCPINGLPIVRNISDLPQDNYGRPGLSHMT
R.sativa      MILISYGSSSSSQI---AAIFSVTLL----LFYFSEAT----LGSPCLINGLPIVRNISELPQDSYGIPGLTHMT
N.tobacco     M-A-----------------RH-IIILVAVFWFAT-A----EASHCSINGLPLVRNISELPQENYGRSGLSHTT
L.esculentum  M-KSGVSSFQTDA---TIMI-RHGVVILVALLWFST-A----EASQCSINGLPLVKNISEFPLHNYGRSGLSHTT
C.annuum      M--------------LFILVALLWLAT-A----EASHCSINGLPLVRNISELPQNNYGRPGLSHTT
M.domesticum  M--VGPSS-----L----TIFF----FF--SLLFFSAIS----EASKCSLKGLPIVRNISELPQDNYGRGGLAHTT
F.x anan      M-AE-IS-----V----PVFF----FFLLCLLFFSAIS----EASKCSVQEFPVVRNISELPQNSYGRGGLAHTT
C.purp        MARLACSF-------------VLLVFLQFCFLARSSALQ-NPGVCGKSEIPVVRNLTELEQDSYGRPGLSHMT

Consensus     M-----------------------ll----llffaa-s----eASsCSidglPvVRNISELPQdNYGRpGLSHiT
Motif         <                  signal peptide                 >           glc
Box
```

```
Zea ABP1      VAGALNHGMKEVEVWLQTISPGQRTPIHRHSCEEVFTVLKGKGTLLMGSSSL-KYPGQPQEIPFFQNTTFSIPVN
Zea ABP4      VTGALAHGTKEVEVWLQTFGPGQRTPIHRHSCEEVFIVLKGKGTLLLGSSSL-KYPGQPQEVPVFQNTTFSIPVN
Av.sativa     VGGALAHGMKEVEVLLETVSAGQRTPIHRHSCEEVFVVLKGRGTLFLGSTSL-KYPGTPQEIPFSQNSTFSTPIN
A.thaliana    VAGSVLHGMKEVEIWLQTFAPGSETPIHRHSCEEVFVVLKGSGTLYLA-ETHGNFPGKPIEFPIFANSTIHIPIN
R.sativa      VAGSVLHGMKEVEIWLQTFAPGAATPIHRHSCEEVFVVPKGSGTLYLA-ETHESVPGKPVEFPISANSTFHIPIN
N.tobacco     IAGSVLHGMKEIEVWLQTFAPGFRTPIHRHSCEEIFIVLKGQGTLYLTPSSHSKYPGNPQEFHIFPNSTFHIPVN
L.esculentum  IAGSVLHGMKEIEVWLQTFAPGCRTPIHRHSCEEVFIVLKGQGTLYLAPSSHSKYPGNPQEFHIFPNSTFHIPVN
C.annuum      IAGSVLHGIQEIEVWLQTFAPGSBTPIHRHSCEEVFVVLKGQGTLYLAPSSHSKYPGNPQEFHIFPNSTFHVPVN
M.domesticum  VAGSLLHGLKEVEVWLQTFAPGSGTPIHRHSCEEVFVVLKGSGTLYLAPSSHGKFPGKPQEFSIFANSTFHIPVN
F.x anan      VAGSLLHGLKEVEVWLQTLSPGSGTPIHRHSCEEVFVVLKGSGTVYLAPNSHEKYPGKPQEFSIFANSTFQIPVN
C.purp        IAGAVHHGMKEVEVWMQTFAPNSGTPIHRHECEEVFITLKGYGTLYLSRNRDHDVPGKPEELPIYPNATFTIPVD

Consensus     VAGALlHGMKEVEVWLQTFAPGSRTPIHRHSCEEVFVVLKGSGTLYLasssh-kyPGkPQEfpifqNSTFhIPVN
Motif         *                 PG-----H-H---E------G------------------------G-----P-G
Box                                    Box A                                      glc
```

```
Zea ABP1      DPHQVWNSDEHEDLQVLVIISRPPAKIFLYDDWSMPHTAAVLKFPFVWD--------EDCFEA---AKDEL
Zea ABP4      DPHQVWNSNEHEDLQVLVIISRPPVKIFIYDDWSMPHTAAKLKFPYFWD--------EDCLPA---PKDEL
Av.sativa     DPHQVWNSDEHEDLQFLVIISRPPVKVFLYDDWSMPHTAAKLKFPFLWD--------EDCLAA---PKDEL
A.thaliana    DAHQVKNTG-HEDLQVLVIISRPPIKIFIYEDWFMPHTAARLKFPYYWD--------EQCIQ--ESQKDEL
R.sativa      DAHQVKNTG-HEDLQVLVIISRPPIKIFTYDDWFMPHTAARLKFPFYWD--------EQCFQ--ESQKDEL
N.tobacco     DVHQVWNTGEQEDLQVLDVISRPPVKVFMYDDWSMPHTAAKLKFPYYWD--------EECYQTTSW-KDEL
L.esculentum  DVHQIWNTGEHEDLQALVVISRPPVKVFMYDDWSMPHTAAKLKFPYYWD--------EKCYQTTTR-KDEL
C.annuum      DAHQVWNTDEHEDLQVLVVISRPPVKVFTYDDWSVPHTASKLKFPYYWD--------EECYQTTSSSKDEL
M.domesticum  DVHQVRNTNEHEDLQALVTVSRPPVKVFMYQDWFMPHTAAKLRYPYYWD--------EECLDVEPPPKDEL
F.x anan      DVHQIRNTNEHEDLQVLVVISRPPVKVFIYENWSMPHTASKLKFPYYWD--------EGCLELEPPPKDEL
C.purp        AVHQVKNTKQGEDLQLFVTISRPPMKSFTYKDWTTPHLSA-VPEMKEWDKQVKFSSAAQQCMEPEAAEDDED++

Consensus     DvHQVWNTdEHEDLQvLViISRPPvKVFlYDDWsMPHTAAKLKFPYYWD--------EDCLQ----pKDEL
Motif         --H---N    EDLQvLVi ISRPP              peptide 11                      KDEL
Box              Box B                Box C                        C-term
```

Figure 1. Pile-up of ABP1 sequences. ABP1s are listed against their genus names. Conserved residues are shown in bold type. A consensus sequence is shown at the foot and lower case letters represent residues not conserved. Two boxes of completely conserved residues are outlined, Boxes A and B, and a third box with high identity is labelled Box C. Several sequence motifs are listed; glc represents N-glycosylation sites, KDEL the ER retention motif, the residues listed in the shaded boxes represent the core cupin motif and the peptide reported to carry photo-activated IAA (Brown and Jones, 1994) is labelled peptide 11.

each conserved box fold around the ligand-binding site (Napier and Pickersgill, unpublished), both sets of results being consistent with the photolabelling experiments.

Binding site models

Models of the auxin receptor site have taken a number of forms since that of Kaethner (1977). It should be remembered that all but one of these are based on biological activity measurements and do not necessarily reflect the binding site of ABP1. The exception is the study of Edgerton *et al.* (1994) who used auxin binding data from maize microsomal membranes (Ray,

1977b), later shown to represent the activity of ABP1. The characteristics of this model binding site are very similar to those models based solely on biological data, although specific predictions differ. Essentials include a carboxylic acid binding group, a hydrophobic transitional region and a hydrophobic platform capable of accommodating the π electrons of the aromatic ring system. Details vary, particularly about whether the auxin side chain, the carboxylate group, needs to adopt different confirmations when bound and unbound, unnecessary for the ABP1 model (Edgerton et al., 1994), incorporated in that of (Kaethner, 1977).

By their nature, these binding site models describe the space, hydrophobicity and charge expected around a bound auxin molecule. The first attempt to construct a molecular model of ABP1 itself has just been published (Warwicker, 2001). Crystallographic coordinates of two germins were used as structural framework into which the most homologous stretch of ABP1 sequence was mounted. The model predicts some features with good probability, such as a coordination site for a metal ion within a β barrel structure. Both these predictions have been confirmed by crystallography (Napier and Pickersgill, unpublished). The protein structure of one other germin, oxalate oxidase, has also been published recently and this was shown to have a similar metal site containing manganese. The manganese is likely to contribute to the oxidase and superoxide dismutase activities of oxalate oxidase. A metal ion in ABP1 would form an ideal carboxylic acid coordination group for the binding pocket. However, the metal ion in ABP1 is likely to be Zn^{2+}, not Mn^{2+}, and ABP1 has no oxalate, or IAA oxidase activity (Napier and Marshall, unpublished).

The metal ion in the model is complexed by a cluster of three histidine residues and a glutamic acid. All these are residues of the germin motif, part of which is found in Box A HRHSCEE (Figure 1). The third histidine is contributed by the second part of the germin motif, H106. Further, a tryptophan (W44), ten residues upstream of Box A, was suggested to face into the binding pocket and contribute the π electron acceptor platform for the indole-aromatic ring of auxin. This is a different tryptophan than the one hypothesized to be the hydrophobic platform by Brown and Jones (1994b) (W_{136} in peptide 11, Figure 1), but it is likely that both hypothesis will be corrected once ABP1 crystal data have been analysed (Woo et al., 2000). It is worth noting that Warwicker's tryptophan platform is not conserved in the Arabidopsis sequence.

Warwicker was not able to model either the N- or C-termini of ABP1 due to low sequence homology. However, he did note two tryptophan-aspartic and glutamic acid tripeptides, DDW136 in Box C and WDE153 in the C-terminal domain, and hypothesized that either one of these could occupy the auxin-binding site in the absence of ligand. Displacement of this tripeptide by free auxin would induce a conformation change to initiate signalling. One of these tripeptides, DDW, is adjacent to the residue considered photolabelled by azido-IAA (Brown and Jones, 1994b). This is an attractive model but, again, one that needs to be tested. The acidic residues of the other tripeptide, WDE, were shown by site directed mutagenesis to play a critical role in protein folding and functional activity of ABP1 at the plasma membrane (David et al., 2001).

In addition to the homology with germins, Box A has been shown to share features with some peroxidases (Savitsky et al., 1999). Peroxidases, particularly some plant peroxidase isozymes, share the HxH sequence, for example. In total, five fragments of ABP1 sequence were shown to have some homology, although three of these are overlapping homologies to the same Box A of ABP1 using different fragments of the peroxidases. The protein structure of the peroxidase family has been solved and shown to be principally α-helix different from the β-barrel structure of ABP1α The important observation from the paper is not that ABP1 will have peroxidase activity (it has no haem group) but that plant peroxidases (and not other peroxidases which do not share the homologies) might have a site for auxin binding and oxidation. There is a conserved tryptophan residue in plant peroxidases, its orientation and distance from the HxH motif will be interesting to compare with the coordinates of analogous residues around the auxin-binding site of ABP1. It should be noted, however, that no peroxidase has been found labelled or purified by auxin (see above) and so the affinity and specificity of such a site are likely to be low.

Cell biology

ABP1 has been detected in many tissues of the plant. To those catalogued before in maize (reviewed by Jones, 1994; Napier and Venis, 1995) can be added spores of fern, wheat shoots and roots, barley pre-anthesis flower spikes and developing caryopses, maize tassels and immature embryos, tomato seed,

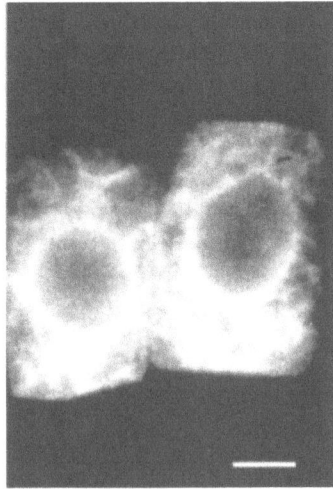

Figure 2. Immunofluorescence image of the distribution of ABP1 in a maize root cell (Henderson *et al.*, 1997).

ovary and flower buds and *Arabidopsis* rosette leaves all according to the available EST databases. Indeed, where ABP1 has been looked for it has been found, suggesting expression throughout the plant even if its expression is weak. The recent ABP1 knockout data (Chen *et al.*, 2001) also suggests a critical role in early embryo development. In maize coleoptiles, the protein is long-lived and expression levels are low, but largely unregulated (Oliver *et al.*, 1995).

All higher-plant ABP1s contain both a signal peptide and an ER retention motif and immunofluorescence, immunogold and biochemical tests have confirmed that most ABP1 is accurately retained (Figure 2). It is unclear whether or not ER retention is a feature of the moss sequence, no characteristic retention motif is present, but there is an abundance of acidic residues downstream of the conserved C-terminal cysteine (not shown, terminated early in Figure 1), one feature common in luminal ER proteins. The N-terminus also has features consistent with being a signal peptide and so it remains probable that *Ceratodon* ABP1 enters the ER and is retained there. The possibility that it is secreted cannot be ruled out, however. The data for the fern *Ceratopteris* (not shown) is incomplete, being derived from an expressed sequence tag clone and neither N- or C-termini are represented. Such ER targeting has added a layer of complexity to the cell biology of ABP1 in that most of the physiological data demonstrating activity as a receptor also place the site of action ABP1 on the plasma membrane.

Despite the ER retention motif, data suggest that ABP1 does pass along the constitutive secretion pathway to the plasma membrane and cell surface (Jones and Herman, 1993; Diekmann *et al.*, 1995; Henderson *et al.*, 1997). The bulk of the protein remains in the ER and this is consistent with all the biochemical data showing characteristic ER-type glycosylation (Henderson *et al.*, 1997) and localization (Ray, 1977b; Tian *et al.*, 1995; Henderson *et al.*, 1997). That detected at the cell surface is only a small fraction of the total (Bauly *et al.*, 2000; Henderson *et al.*, 1997). Yet, a number of different approaches have suggested that ABP1, and peptides based on the C-terminus of ABP1, added to intact cells without auxin complement reconstitute auxin-like physiological responses (Barbier-Brygoo *et al.*, 1991; Thiel *et al.*, 1993; Gehring *et al.*, 1998; Leblanc *et al.*, 1999b). These experiments add to other earlier work demonstrating that maize ABP1 was able to potentiate the auxin responsiveness of tobacco protoplasts when added in the medium at low concentrations, whereas polyclonal antibodies to the maize protein were shown to impair the same auxin-dependent response (Barbier-Brygoo *et al.*, 1989; Rück *et al.*, 1993). More recently, the use of a large panel of 9 distinct monoclonal antibodies were used on the functional tobacco protoplast assay to confirm that ABP1, and not a related protein, is involved in early auxin responses at the plasma membrane (Leblanc *et al.*, 1999a). In addition, polyclonal antibodies to ABP1 have been used to show that the auxin signal for protoplast swelling is perceived by extracellular ABP1 (Steffens *et al.*, 2001).

All together, the data have shown that ABP1 is active in auxin responsiveness at the surface of the plasma membrane despite carrying a targeting motif (Figure 3). Other ER-directed proteins are also being shown to escape and have biological activities outside this compartment (Xiao *et al.*, 1999; Okazaki *et al.*, 2000); As such, ABP1 is not unusual in being both targeted to the ER and found to have activities elsewhere. The KDEL motif has been shown to be involved in the stability of ABP1 but not to be important for auxin binding and interaction with the plasma membrane (David *et al.*, 2001). Up to now, no experimental data have indicated a functional role of ABP1 inside of the ER, apart from supplying the cell surface with ABP1 (Figure 3). The possible influence of auxin on the targeting of ABP1 at the cell surface has been investigated but no significant change in ABP1 exportation has been observed consecutively to auxin treatment (Henderson *et al.*, 1997).

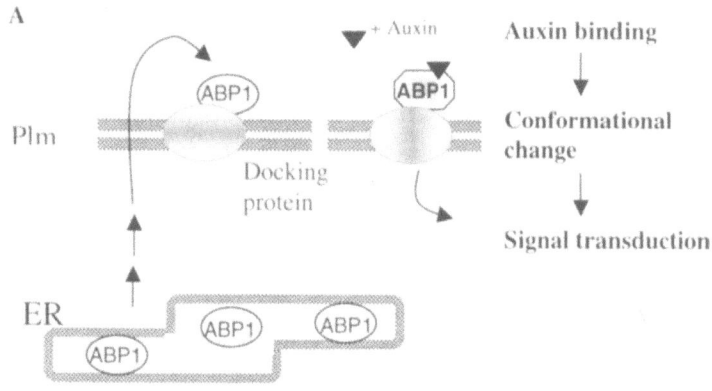

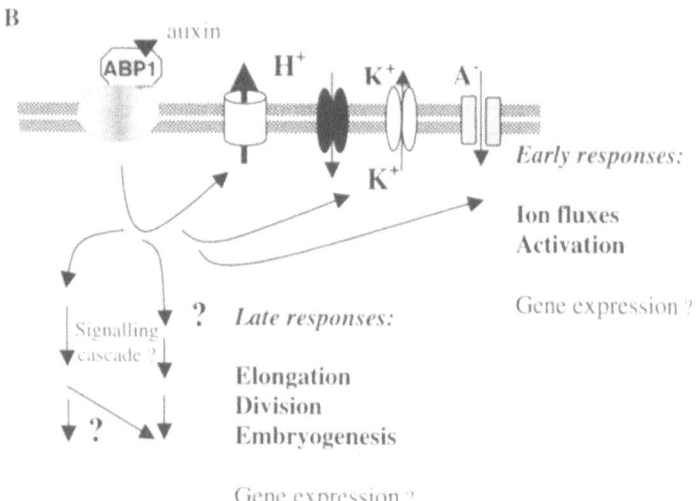

Figure 3. A. ABP1 is located both in the ER and at the outer face of the plasma membrane. Auxin binding induces a conformational change of ABP1 modifying the interaction with the plasma membrane (Plm) resulting in the activation of the signalling cascade. B. Experimental evidences support that the fraction located at the plasma membrane is involved in the control of early auxin electrophysiological responses. Correlation between such early responses and late cellular or developmental responses is not established yet. ABP1 seems to be implicated directly or indirectly in both cell elongation and cell division. No evidence has been provided so far that ABP1 mediates gene expression of early or late auxin-responsive genes.

The possibility that ABP1 has a role as chaperone when in the ER has recently been evoked (Chen *et al.*, 2001). Expression data showed that one other ER chaperone, BiP, was up-regulated in transcription by applied stresses, conditions likely to require more chaperone activity (Oliver *et al.*, 1995). ABP1 was not; indeed, heat stress down-regulated translation. However, chaperone-like interactive with specific partners cannot be ruled out. Polypeptides have been co-precipitated with ABP1, but these remain unidentified. Identification of ABP1 interacting components, at both the plasma membrane and in the ER, is one of the challenges of the years to come, but a challenge that will need to be met before we can fully understand ABP1 behaviour.

As for most phytohormones, the site of auxin perception has long been contentious (Vesper and Kuss, 1990; Claussen *et al.*, 1996). The role of ABP1 in the perception of auxin at the plasma membrane is no longer controversial. The existence of an intracellular auxin receptor remains an open question, such internal receptor could act either in concert or independently to ABP1 in response to changes in cellular auxin content. None can exclude the possibility that auxin carriers (Friml and Palme, this issue), which control auxin fluxes through plant cells and tissues, could also act as auxin receptors. Among the large number of auxin binding proteins identified in the plast 20 years, none is presently considered as a possible intracellular auxin receptor.

ABP1 and downstream auxin responses

The involvement of ABP1 in early auxin responses at the plasma membrane has already been mentioned above. These early responses consist of modifications of ion fluxes across the plasma membrane, reflecting activation or deactivation of ion channels or transporters in response to auxin (Figure 3). The electrical responses are monitored by electrophysiological measurements which reflect either the overall magnitude of ion gradients (e.g. hyperpolarization) or, more specifically, the flux of a given ion (e.g. H^+, K^+ or anions). Auxin has been shown to act via ABP1 on the activation of the proton pump ATPase (Rück et al., 1993), potassium channels (Thiel et al., 1993) and voltage-dependent anion channels (Zimmermann et al., 1994; Barbier-Brygoo et al., 1996). An enhanced auxin sensitivity of the inward and outward rectifying K^+ channels was reported in guard cells of tobacco plants overexpressing different forms of ABP1 (Bauly et al., 2000). Mesophyll protoplasts isolated from transgenic plants over-expressing the maize ABP1 were also shown to exhibit a shift of the hyperpolarization response to the lower auxin concentrations (H. Barbier-Brygoo and K. Palme, personal communication). These responses correspond to initial steps of a signalling pathway, but their correlation with other molecular or cellular auxin responses is not established yet.

Another early auxin cellular response, namely protoplast swelling, was shown to be mediated by extracellular ABP1, reinforcing the idea that the functional role of the protein in auxin responses involves the plasma membrane located protein (Steffens et al., 2001). Protoplast swelling is a rapid and specific cellular response to auxin and shares common features with auxin-mediated growth at the organ level (Steffens and Lüthen, 2000). The study of transgenic tobacco plants over-expressing Arabidopsis ABP1 under the control of an inducible promoter has provided other experimental evidence supporting ABP1-mediated, auxin-dependent cell expansion (Jones et al., 1998). This has been achieved by measuring the degree of curvature of leaf strips, or the size of mesophyll protoplasts, in samples treated or not with the inducer and auxin. No phenotype was observed at the whole-plant level when grown in the presence of the inducer (Jones et al., 1998).

Constitutive expression of antisense ABP1 in BY2 cells was shown to reduce cell growth (Chen et al., 2001). The reduction was interpreted as evidence for the role of ABP1 in both auxin-induced cell expansion and, directly or indirectly, cell proliferation. However, further experiments will be necessary to elucidate distinct roles for ABP1 in the control of cell expansion versus cell division. It is not clear whether more severe growth alteration could be observed on BY2 cells when ABP1 expression is completely abolished. Interestingly, an ABP1 knock-out mutant has been identified in Arabidopsis (Chen et al., 2001). Plants heterozygous for the mutation in the single Arabidopsis ABP1 gene do not exhibit any phenotype whereas homozygous plants stop develop and die at the early globular stage of embryogenesis. The general architecture of abp1 early embryos is severely altered, with a loss of symmetry, a fault in cell elongation and aberrant cell divisions affecting both the suspensor and the embryo. The identification of this null mutant for ABP1 demonstrates that the protein is essential during embryogenesis and, although difficult to work with, this resource could play an invaluable role in determining ABP1's place during auxin signalling during both elongation and cell division.

Challenges

Undoubtedly, ABP1 is an essential protein and is likely to be an auxin receptor. The challenges in the years to come will be to understand its mode of action and its relationship, if any, with already identified auxin signalling elements such as the transcription factors mediating the expression of early auxin-responsive genes or elements of the proteolytic pathway (Dharmasiri and Estelle, Hagen and Guilfoyle, and Liscum and Reed, this issue).

Acknowledgements

References

Barbier-Brygoo, H., Ephritikhine, G., Klambt, D., Ghislain, M. and Guern, J. 1989. Functional evidence for an auxin receptor at the plasmalemma of tobacco mesophyll protoplasts. Proc. Natl. Acad. Sci. USA 86: 891–895.
Barbier-Brygoo, H., Ephritikhine, G., Klambt, D., Maurel, C., Palme, K., Schell, J. and Guern, J. 1991. Perception of the auxin

signal at the plasma membrane of tobacco mesophyll protoplasts. Plant J. 1: 83–93.

Barbier-Brygoo, H., Zimmermann, S., Thomine, S., White, I.R., Millner, P. and Guern, J. 1996. Elementary auxin response chains at the plasma membrane involve external ABP1 and multiple electrogenic ion transport proteins. Plant Growth Regul. 18: 23–28.

Batt, S., Wilkins, M.B. and Venis, M.A. 1976. Auxin binding to corn coleoptile membranes: kinetics and specificity. Planta 130: 7–13.

Bauly, J.M., Sealy, I.M., Macdonald, H., Brearley, J., Droge, S., Hillmer, S., Robinson, D.G., Venis, M.A., Blatt, M.R., Lazarus, C.M. and Napier, R.M. 2000. Overexpression of auxin-binding protein enhances the sensitivity of guard cells to auxin. Plant Physiol. 124: 1229–1238.

Brown, J.C. and Jones, A.M. 1994. Mapping the auxin-binding site of auxin-binding protein 1. J. Biol. Chem. 269: 21136–21140.

Chen, J.G., Ullah, H., Young, J.C., Sussman, M.R. and Jones, A.M. 2001. ABP1 is required for organized cell elongation and division in *Arabidopsis* embryogenesis. Genes Dev. 15: 902–911.

Claussen, M., Lüthen, H. and Böttger, M. 1996. Inside or outside? Localization of the receptor relevant to auxin-induced growth. Physiol. Plant. 98: 861–867.

David, K.M., Carnero-Diaz, E., Leblanc, N., Monestiez, M., Grosclaude, J. and Perrot-Rechenmann, C. 2001. Conformational dynamics underlie the activity of the auxin-binding protein, Nt-abp1. J. Biol. Chem. 276: 34517–34123.

Davies, R.T., Goetz, D.H., Lasswell, J., Anderson, M.N. and Bartel, B. 1999. *IAR3* encodes an auxin conjugate hydrolase from *Arabidopsis*. Plant Cell 11: 365–376.

Delong, A., Mockaitis, K. and Christensen, S. 2001. Protein phosphorylation in the delivery of and response to auxin signals. Plant Mol Biol., 49: 285–303.

Dharmasiri, S. and Estelle, M. 2001. The role of regulated protein degradation in auxin response. Plant Mol Biol., 49: 401–408.

Dibb-Fuller, J.E and Morris, D.A. 1992. Studies on the evolution of auxin carriers and phytotropin receptors: transmembrane auxin transport in unicellular and multicellular Chlorophyta. Planta 186: 219–226.

Diekmann, W., Venis, M.A. and Robinson, D.G. 1995. Auxins induce clustering of the auxin-binding protein at the surface of maize coleoptile protoplasts. Proc. Natl. Acad. Sci. USA 92: 3425–3429.

Dunwell, J.M., Khuri, S. and Gane, P.J. 2000. Microbial relatives of the seed storage proteins of higher plants: conservation of structure and diversification of function during evolution of the cupin superfamily. Microbiol. Mol. Biol. Rev. 64: 153–179.

Edgerton, M.D., Tropsha, A. and Jones, A.M. 1994. Modelling the auxin-binding site of auxin-binding protein1 of maize. Phytochemistry 35: 1111–1123.

Farrimond, J.A., Elliot, M.C. and Clack, D.W. 1978. Charge separation as a component of the structural requirements for hormone activity. Nature 274: 401–402.

Feckler, C., Muster, M., Feser, W., Römer, A. and Palme, K. 2001. Mass spectrometric analysis reveals a cysteine bridge between residues 2 and 61 of the auxin-binding protein from *Zea mays* L. FEBS Lett., 509: 446–450.

Friml, J. and Palme, K. 2001. Polar auxin transport: old concepts and new questions. Plant Mol Biol., 49: 273–284.

Gehring, C.A., McConchie, R.M., Venis, M.A. and Parish, R.W. 1998. Auxin-binding-protein antibodies and peptides influence stomatal opening and alter cytoplasmic pH. Planta 205: 581–586.

Hagen, G. and Guilfoyle, T. 2001. Auxin-responsive gene expression: genes, promoters and regulatory factors. Plant Mol. Biol., 49: 373–385.

Henderson, J., Bauly, J.M., Ashford, D.A., Oliver, S.C., Hawes, C.R., Lazarus, C.M. and Venis, M.A. 1997. Retention of maize auxin-binding protein in the endoplasmic reticulum: quantifying escape and the role of auxin. Planta 202: 313–323.

Hertel, R., Thomson, K. and Russo, V.E.A. 1972. *In vitro* auxin binding to particulate cell fractions from corn coleoptiles. Planta 107: 325–340.

Hobbie, L. and Estelle, M. 1994. Genetic approaches to auxin action. Plant Cell Environ. 17: 525–540.

Jones, A.M. 1994. Auxin-binding proteins. Annu. Rev. Plant Physiol. Plant Mol. Biol. 45: 393–420.

Jones, A.M. and Herman, E.M. 1993. KDEL-containing auxin-binding protein is secreted to the plasma membrane and cell wall. Plant Physiol. 101: 595–606.

Jones, A.M., Im, K.H., Savka, M.A., Wu, M.J., Dewitt, N.G., Shillito, R. and Binns, A.N. 1998. Auxin-dependent cell expansion mediated by overexpressed auxin-binding protein 1. Science 282: 1114–1117.

Kaethner, T.M., 1977. Conformational change theory for auxin structure-activity relationships. Nature 267: 19–23.

Katekar, G.F. 1979. Auxins: on the nature of the receptor site and molecular requirements for auxin activity. Phytochemistry 18: 223–233.

Kim, Y.S., Kim, D.H. and Jung, J. 1998. Isolation of a novel auxin receptor from soluble fractions of rice (*Oryza sativa* L.) shoots. FEBS Lett. 438: 241–244.

Kim, Y.S., Min, J.K., Kim, D. and Jung, J. 2001. A soluble auxin-binding protein, ABP57. Purification with anti-bovine serum albumin antibody and characterization of its mechanistic role in the auxin effect on plant plasma membrane H^+-ATPase. J. Biol. Chem. 276: 10730–10736.

Laporte, K. and Rossignol, M. 1997. Auxin control of the sensitivity to auxin of the proton translocation catalyzed by the tobacco plasma membrane H^+-ATPase. Plant Growth Regul. 21: 19–25.

Lazarus, C.M. and Macdonald, H. 1996. Characterization of a strawberry gene for auxin-binding protein, and its expression in insect cells. Plant Mol. Biol. 31: 267–277.

Leblanc, N., David, K., Grosclaude, J., Pradier, J.M., Barbier-Brygoo, H., Labiau, S. and Perrot-Rechenmann, C. 1999a. A novel immunological approach establishes that the auxin-binding protein, Nt-abp1, is an element involved in auxin signaling at the plasma membrane. J. Biol. Chem. 274: 28314–28320.

Leblanc, N., Perrot-Rechenmann, C. and Barbier-Brygoo, H. 1999b. The auxin-binding protein Nt-ERabp1 alone activates an auxin-like transduction pathway. FEBS Lett. 449. 57–60.

Leyser, O. 1997. Auxins: lessons from a mutant weed. Physiol. Plant. 100: 407–414.

Liscum, E. and Reed, J.W. 2001. Genetics of Aux/IAA and ARF action in plant growth and development. Plant Mol. Biol., this issue.

Löbler, M., Klämbt, D. 1985. Auxin-binding protein from coleoptile membranes of corn (*Zea mays* L.): purification by immunological methods and characterization. J. Biol. Chem. 260: 9848–9853.

Luschnig, C. and Fink, G.R. 1999. Two pieces of the auxin puzzle. Trends Plant Sci. 4: 162–164.

Macdonald, H., Henderson, J., Napier, R.M., Venis, M.A., Hawes, C. and Lazarus, C.M. 1994. Authentic processing and targeting of active maize auxin-binding protein in the baculovirus expression system. Plant Physiol. 105: 1049–1057.

348

Massotte, D., Fleig, U. and Palme, K. 1995. Purification and characterization of an auxin-binding protein from *Arabidopsis thaliana* expressed in baculovirus-infected insect cells. Protein Expr. Purif. 6: 220–227.

Melhado, L.L., Pearce, C.J., D'Alarco M. and Leonard, N.J. 1982. Specifically deuterated and tritiated auxins. Phytochemistry 21: 2879–2885.

Napier, R.M. 2001. Models of auxin binding. J. Plant Growth Regul., 20: 244–254.

Napier, R.M. and Venis, M.A. 1995. Tansley review No 79: Auxin action and auxin-binding proteins. New Phytol. 129: 167–201.

Navé, J.F. and Benveniste, P. 1984. Inactivation by phenylglyoxal of the specific binding of 1-naphthylacetic acid with membrane-bound auxin binding sites from maize coleoptiles. Plant Physiol. 74: 1035–1040.

Ohmiya, A., Kikuchi, M., Sakai, S. and Hayashi, T. 1993. Purification and properties of an auxin-binding protein from the shoot apex of peach tree. Plant Cell Physiol. 34: 177–183.

Okazaki, Y., Ohno, H., Takase, K., Ochiai, T. and Saito, T. 2000. Cell surface expression of calnexin, a molecular chaperone in the endoplasmic reticulum. J. Biol. Chem. 275: 35751–35758.

Oliver, S.C., Venis, M.A., Freedman, R.B. and Napier, R.M. 1995. Regulation of synthesis and turnover of maize auxin-binding protein and observations on its passage to the plasma membrane: comparisons to maize immunoglobulin-binding protein cognate. Planta 197: 465–474.

Primm, T.P. and Gilbert, H.F. 2001. Hormone binding by protein disulfide isomerase, a high capacity hormone reservoir of the endoplasmic reticulum. J. Biol. Chem. 276: 281–286.

Ray, P.M. 1977a. Auxin-binding sites of maize coleoptiles are localized on membranes of the endoplasmic reticulum. Plant Physiol. 59: 594–599.

Ray, P.M. 1977b. Specificity of auxin-binding sites on maize coleoptile membranes as possible receptor sites for auxin action. Plant Physiol. 60: 585–591.

Ray, P.M., Dohrman, U. and Hertel, R. 1977. Characterization of naphthaleneacetic acid binding to receptor sites on cellular membranes of maize coleoptile tissue. Plant Physiol. 59: 357–364.

Reinard, T., Achmus, H., Walther, A., Rescher, U., Klämbt, D. and Jacobsen, H.J. 1998. Assignment of the auxin binding abilities of ABP$_{44}$ in gel. Plant Cell Physiol. 39: 874–878.

Reinard, T., Janke, V., Willard, J., Buck, F., Jacobsen, H.J. and Vockley, J. 2000. Cloning of a gene for an acyl-CoA dehydrogenase from *Pisum sativum* L. and purification and characterization of its product as an isovaleryl-CoA dehydrogenase. J. Biol. Chem. 275: 33738–33743.

Rück, A., Palme, K., Venis, M.A., Napier, R.M. and Felle, R.H. 1993. Patch-clamp analysis establishes a role for an auxin binding protein in the auxin stimulation of plasma membrane current in *Zea mays* protoplasts. Plant J. 4: 41–46.

Savitsky, P.A., Gazaryan, I.G., Tishkov, V.I., Lagrimini, L.M., Ruzgas, T. and Gorton, L. 1999. Oxidation of indole-3-acetic acid by dioxygen catalysed by plant peroxidases: specificity for the enzyme structure. Biochem. J. 340: 579–583.

Steffens, B., Feckler, C., Palme, K., Christian, M., Böttger, M. and Lüthen, H. 2001. The auxin signal for protoplast swelling is perceived by extracellular ABP1. Plant J. 27: 591–999.

Steffens, B. and Lüthen, H. 2000. New methods to analyse auxin-induced growth. II. The swelling reaction of protoplasts: a model system for the analysis of auxin signal transduction? Plant Growth Regul. 32: 115–122.

Sugaya, S., Ohmiya, A., Kikuchi, M. and Hayashi, T. 2000. Isolation and characterization of a 60 kDa 2,4-D-binding protein from the shoot apices of peach trees (*Prunus persica* L.); it is a homologue of protein disulfide isomerase. Plant Cell Physiol. 41: 503–508.

Sugaya, S. and Sakai, S. 1996. Identification of a soluble auxin-binding protein as a glutathione-dependent formaldehyde dehydrogenase. Plant Sci. 114: 1–9.

Sztein, A.E., Cohen, J.D., Slovin, J.P. and Cooke, T.J. 1995. Auxin metabolism in representative land plants. Am. J. Bot. 82: 1514–1521.

Thiel, G., Blatt, M.R., Fricker, M.D., White, I.R. and Millner, P. 1993. Modulation of K$^+$ channels in *Vicia* stomatal guard cells by peptide homologs to the auxin-binding protein C terminus. Proc. Natl. Acad. Sci. USA 90: 11493–11497.

Thimann, K.V. 1963. Plant growth substances; past present and future. Annu. Rev. Plant Physiol. 14: 1–18.

Tian, H., Klämbt, D. and Jones, A.M. 1995. Auxin-binding protein 1 does not bind auxin within the endoplasmic reticulum despite this being the predominant subcellular location for this hormone receptor. J. Biol. Chem. 270: 26962–26969.

Tomic, S., Gabdoulline, R.R., Kojic-Prodic, B. and Wade, R.C. 1998. Classification of auxin plant hormones by interaction property similarity indices. J. Comput. Aided Mol. Des. 12: 63–79.

Venis, M.A. 1977. Affinity labels for auxin binding sites in corn coleoptiles membranes. Planta 164: 145–149.

Venis, M.A. and Napier, R.M. 1995. Auxin receptors and auxin binding proteins. Crit. Rev. Plant Sci 14: 27–47.

Venis, M.A., Napier, R.M., Barbier-Brygoo, H., Maurel, C., Perrot-Rechenmann, C. and Guern, J. 1992. Antibodies to a peptide from the maize auxin-binding protein have auxin agonist activity. Proc. Natl. Acad. Sci. USA 89: 7208–7212.

Vesper, M.J. and Kuss, C.L. 1990. Physiological evidence that the primary site of auxin action in maize coleoptiles is an intracellular site. Planta 182: 486–491.

Walther, A., Rescher, U., Schiebl, C. and Klämbt, D. 1997. Antibodies against distinct ABP1 regions modify auxin binding to ABP1 and change the physiological auxin response of maize coleoptile sections. J. Plant Physiol. 150: 110–114.

Warwicker, J. 2001. Modelling of auxin-binding protein 1 suggests that its C-terminus and auxin could compete for a binding site that incorporates a metal ion and tryptophan residue 44. Planta 212: 343–347.

Woo, E.J., Bauly, J., Chen, J.G., Marshall, J., Macdonald, H., Lazarus, C., Goodenough, P., Venis, M., Napier, R. and Pickersgill, R. 2000. Crystallization and preliminary X-ray analysis of the auxin receptor ABP1. Acta Crystallog. D Biol. Crystallog. 56: 1476–1478.

Xiao, G.Q., Chung, T.F., Pyun, H.Y., Fine, R.E. and Johnson, R.J. 1999. KDEL proteins are found on the surface of NG108-15 cells. Mol. Brain Res. 72: 121–128.

Zimmermann, S., Thomine, S., Guern, J. and Barbier-Brygoo, H. 1994. An anion current at the plasma membrane of tobacco protoplasts shows ATP-dependent voltage regulation and is modulated by auxin. Plant J. 6: 707–716.

Plant Molecular Biology **49**: 349–356, 2002.
Perrot-Rechenmann and Hagen (Eds.), Auxin Molecular Biology.
© 2002 *Kluwer Academic Publishers.*

349

Channelling auxin action: modulation of ion transport by indole-3-acetic acid

Dirk Becker* and Rainer Hedrich
*Biocenter, Julius-von-Sachs-Institut for Biosciences, Department of Plant Molecular Physiology and Biophysics, Julius-von-Sachs-Platz 2, 97082 Würzburg, Germany (*author for correspondence; e-mail dbecker@botanik.uni-wuerzburg.de)*

Received 14 June 2001; accepted in revised form 9 August 2001

Key words: auxin, early auxin response gene, gravitropism, ion channels

Abstract

The growth hormone auxin is a key regulator of plant cell division and elongation. Since plants lack muscles, processes involved in growth and movements rely on turgor formation, and thus on the transport of solutes and water. Modern electrophysiological techniques and molecular genetics have shed new light on the regulation of plant ion transporters in response to auxin. Guard cells, hypocotyls and coleoptiles have advanced to major model systems in studying auxin action. This review will therefore focus on the molecular mechanism by which auxin modulates ion transport and cell expansion in these model cell types.

Introduction

A plant's life cycle from germination to reproduction is under the control of phytohormones. Among the growth hormones, the action of auxin (IAA) was first to be discovered by plant physiologist more than 100 years ago (Darwin, 1880; Sachs, 1887; Went and Thimann, 1937). Based on the introduction of modern molecular techniques in combination with plant genetics and single-cell biophysics within the past decade the auxin field has advanced enormously. These studies have led to the identification of auxin-response mutants (Firn *et al.*, 2000), the characterization of auxin-responsive genes and recently the isolation of auxin uptake and efflux carriers (Bennett *et al.*, 1996, 1998; Walker and Estelle, 1998; Gälweiler *et al.*, 1998; Palme and Gälweiler, 1999). A dominant role for auxin in the control of plant growth and morphogenesis has been elaborated and preliminary models are emerging to describe the complex molecular relation between auxin signalling and that of other hormones, such as ethylene (Burg and Burg, 1966; Pitts *et al.*, 1998; Hansen and Grossmann, 2000; Rahman *et al.*, 2001; Merritt *et al.*, 2001). Auxin-controlled processes like growth, movements or tropism in-

volve cell expansion and cell division. Although these processes significantly rely on the uptake and distribution of nutrients, only very recently insights into the molecular mechanisms of these auxin-regulated processes have been gained.

Auxin-induced K^+ channels in coleoptiles

It is well known that certain plant organs, such as the coleoptiles, hypocotyls, roots, or even guard cells, are characterized by their pronounced sensitivity to auxin. These systems have thus advanced to models for exploring the molecular events underlying auxin action. Among them, the growing coleoptile is characterized by an auxin-specific growth phenotype. The growth rates of this fast-growing organ can be accelerated by auxin application to the decapitated coleoptile tip. Since the tip represents an essential site for IAA synthesis and redistribution (cf. Ostin *et al.*, 1999; Philippar *et al.*, 1999), its removal leads to a complete loss of growth and phototropic as well as gravitropic responses (Went and Thimann, 1937).

Coleoptile growth as for plant growth in general represents a turgor-driven process based on the accumulation of osmolytes into the expanding cells. The potassium ion represents the major cation in plants (Marschner, 1996). Previous studies have linked auxin-stimulated proton efflux and potassium uptake into expanding coleoptile cells to the up-regulation of the H^+-ATPase (Hager et al., 1971; Rayle and Cleland, 1977; Senn and Goldsmith, 1988; Frias et al., 1996) and activation of K^+ uptake (Claussen et al., 1997). Using auxanometers Claussen et al. (1997) showed that in auxin-depleted coleoptile segments cell expansion could be re-initiated by addition of IAA or its physiologically active derivatives (e.g. 1-NAA). Growth, however, was abolished by the removal of potassium ions from the incubation buffer or in the presence of the K^+-channel blockers TEA^+ or Cs^+ (Hedrich and Becker, 1994; Becker et al., 1996). Re-addition of K^+ or removal of K^+-channel blockers in the presence of auxin led to a rapid enhancement of coleoptile growth rates.

Based on these observations Philippar et al. (1999) tested whether or not the activity of K^+-uptake channels in maize coleoptiles represents a rate-limiting step in auxin-dependent growth. In their studies two K^+-uptake channel genes, zmk1 and zmk2, were isolated from a coleoptile cDNA library and their expression in the growing coleoptile was documented. While zmk1 mRNA could be localized to cortex cells only, zmk2 expression was strongly enhanced in the vascular bundles (Bauer et al., 2000). Besides their differences in primary structure and localization, ZMK1 and ZMK2 exhibited different functional properties in the different coleoptile protoplast populations and after heterologous expression in Xenopus oocytes (Philippar et al., 1999; Bauer et al., 2000). ZMK2, a member of the AKT2/3 subfamily of Shaker-like plant potassium channels (see Roelfsema and Hedrich, 1999 for review), represents a weakly voltage-dependent, proton-blocked, K^+ channel, mediating potassium fluxes across the plasma membrane of vascular cells (cf. Marten et al., 1999; Lacombe et al., 2000). In contrast, ZMK1 exhibits the typical features of a voltage-dependent, proton-stimulated K^+ channel. Activated at hyperpolarizing membrane potentials and by apoplastic protons, ZMK1 mediates K^+ uptake into the expanding cortex cells. In line with the acid-growth theory, ZMK1 was thus proposed to influence turgor and cell expansion. Indeed, when probing the transcript levels of zmk1 and zmk2 upon auxin treatment of maize coleoptile segments, it was found that the zmk1 gene was up-regulated (Philippar et al., 1999). Consistent with these data, patch-clamp analy-

ses of coleoptile protoplasts revealed a time-dependent increase in the number of active K^+-uptake channels upon auxin treatment. The fact that up-regulation of zmk1 was insensitive to treatment with cycloheximide (an inhibitor of protein synthesis) indicated that all the proteins necessary to drive auxin-dependent zmk1 expression were present prior to auxin treatment. Hence zmk1 was classified as an 'early auxin-induced gene' (Abel and Theologis, 1996). As for the time course of coleoptile growth stimulation by auxin (see Figure 1), the zmk1 induction was characterized by a lag-phase of about 30 min and a peak after 60–90 min. Based on the auxin specificity and sensitivity of zmk1 towards physiologically active auxins, as well as the pH dependence of this K^+ channel, a tight coupling of auxin action and channel activation was proposed.

In response to gravistimulation, horizontally displaced coleoptiles bend upward. This process is accompanied by an increase in free IAA in the coleoptile tip and polar auxin transport towards the base of this organ (Palme and Gälweiler, 1999). In addition to basipetal auxin transport, free IAA is re-distributed from the upper half of the gravistimulated coleoptile to accumulate on the lower side (Evans, 1991; Philippar et al., 1999). Increased auxin levels initiate cell expansion resulting in differential growth in the two coleoptile halves. Within 3–4 h the coleoptile reaches a vertical position again. Consistent with a differential distribution of auxin, within 60–90 min zmk1 expression is up-regulated on the lower side of the coleoptile and down-regulated on the upper side.

Current models of the molecular events during gravitropic bending of the maize coleoptile would thus include auxin influx and efflux carriers, the H^+-ATPase, the K^+-uptake channel ZMK1 and as yet unidentified transcription factors (Figure 2). Upon gravistimulation, an increase in free IAA followed by IAA redistribution towards the lower coleoptile half would stimulate proton pumping of the H^+-ATPase by an as yet unknown mechanism. Recent studies, however, have demonstrated similar growth patterns for fusicoccin- or auxin-stimulated maize coleoptiles and thus might point to a role of 14-3-3 proteins (Sze et al., 1999; Camoni et al., 2000; Tode and Lüthen, 2001). Increased proton pumping would hyperpolarize the membrane potential (Felle et al., 1991) and thereby activate the voltage-dependent K^+-uptake channel ZMK1. The simultaneous acidification of the apoplast during H^+ extrusion will feed forward on the open probability of ZMK1 channels and thus further stimulate K^+ uptake (Hedrich and Becker, 1994; Hoth

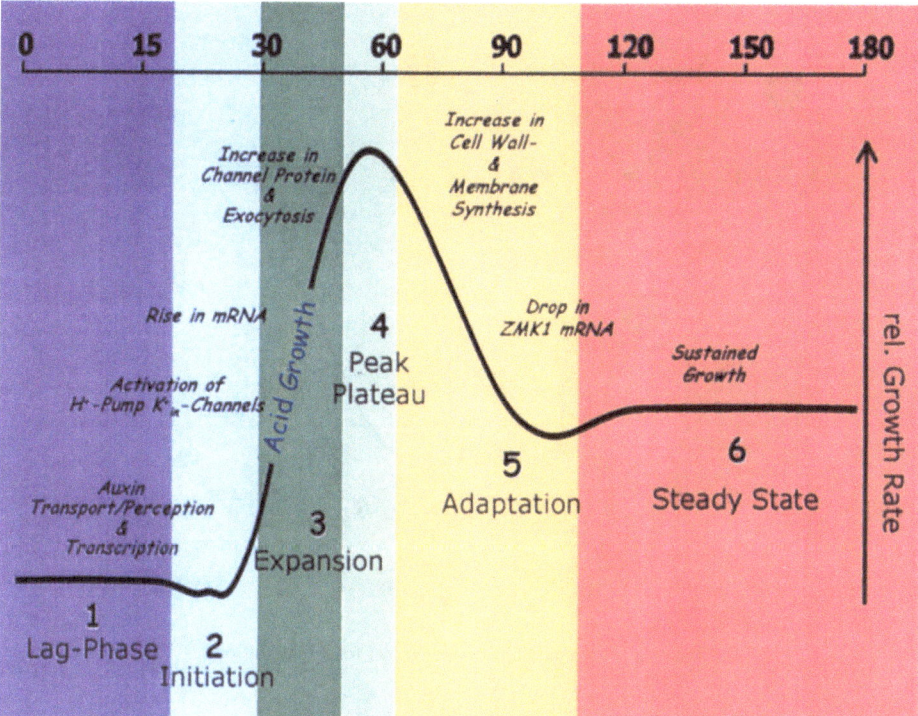

Figure 1. Time course of auxin action in coleoptiles.

et al., 1997; Hedrich *et al.*, 1998). Thus the stimulation of the H^+-ATPase and activation of ZMK1 very likely contributes to the initial growth phase of auxin-treated coleoptiles (Figures 1 and 2). In the long term, sustained growth at elevated growth rates requires incorporation of additional membrane components such as pumps, channels and carriers as well as lipids into the plasma membrane surface of the expanding cells (see also Frias *et al.*, 1996; Homann and Thiel, 1999). In this context the auxin-dependent up-regulation of *zmk1* and the coleoptile H^+-ATPase gene will maintain the density of pumps and channels.

Auxin-modulated ion transport in guard cells

Guard cells, which are sensory, turgor-driven micro-valves, control CO_2 uptake and water loss from plant surfaces. Like coleoptiles, guard cells exhibit an auxin-dependent phenotype and thus represent another model system for studying auxin action (Marten *et al.*, 1991; Hedrich and Jeromin, 1992; Blatt and Thiel, 1994; Gehring *et al.*, 1998; Bauly *et al.*, 2000). In response to exogenous auxin, the two guard cells that surround a stoma increase their volume and thereby open the stomatal pore. In contrast to the situation in

fast growing organs such as coleoptiles or hypocotyls, however, changes in guard cell volume are reversible. While signals like high CO_2 concentrations, darkness or abscisic acid initiate guard cell shrinkage and thus stomatal closure (Felle *et al.*, 2000), red and blue light (Roelfsema *et al.*, 2001), low CO_2 concentrations as well as auxin trigger stomatal opening (Hedrich *et al.*, 2001; Marton *et al.*, 1992). Stomatal opening is a turgor-driven process and relies on the accumulation of potassium salts and/or sugars (Schroeder and Hedrich, 1989; Talbott and Zeiger, 1996; Schroeder *et al.*, 2001).

The electrical response of plant plasma membranes to auxin is characterized by a transient depolarization, followed by a sustained hyperpolarization, indicating the involvement of H^+-ATPases (Felle *et al.*, 1991). The initial depolarization was thought to result from the activation of a Ca^{2+}- and nucleotide-dependent anion channel, GCAC1, in the plasma membrane of guard cells (Hedrich *et al.*, 1990). GCAC1 provides a recognition site for extracellular auxin since upon auxin application, its activation threshold is shifted towards the resting potential of the cell (Marten *et al.*, 1991; Hedrich, 1994), resulting in auxin-induced channel opening and membrane depolarization. Upon

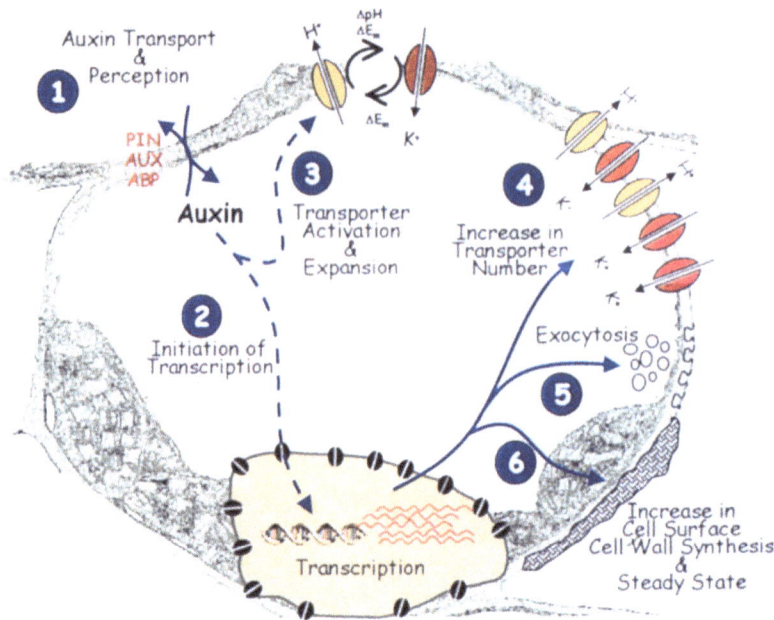

Figure 2. Auxin tunes the osmotic motor for cell expansion.

prolonged depolarization, however, GCAC1 inactivates and allows membrane repolarization. The importance of a transient membrane depolarization of the guard cell plasma membrane in response to auxin remains elusive, but GCAC1 represents the only ion channel so far, which has been shown to be modulated by auxin directly. After a lag phase of several minutes (Lohse and Hedrich, 1992), auxin treatment increases pump currents catalysed by plasma membrane H^+-ATPases which appear to be highly expressed in this sensory cell type (Lohse and Hedrich, 1992; Becker et al., 1993).

Stomatal opening is accompanied by the accumulation of potassium salts (KCl and K_2-malate) and sugar by the individual guard cells (Humble and Raschke, 1971; Lu et al., 1995; Ritte et al., 1999). This uptake of osmotically active substances is driven by a H^+ gradient and the membrane potential generated by the plasma-membrane H^+-ATPases. An auxin-induced activation of the guard cell H^+-ATPases would hyperpolarize the membrane and thereby allow K^+ uptake into the guard cells through voltage-dependent, inward-rectifying K^+ channels. In addition, the H^+ gradient would provide the driving force for the uptake of Cl^- and/or sugars via H^+-based symporters in the guard cell plasma membrane. In contrast to the activation of GCAC1, however, the stimulation of the guard cell H^+-ATPases by auxin

is characterized by a lag phase of about 20 min after stimulus onset, indicating an indirect mechanism of activation. The catalytic activity of plant H^+-ATPases has been shown to be regulated by phosphorylation/dephosphorylation mechanisms (Sze et al., 1999; Zeiger, 2000; Morsomme and Boutry, 2000). In the context of blue-light- or fusicoccin-dependent stomatal opening, the phosphorylation level of an auto-inhibitory domain in the C-terminus of this protein is critical for the activation state of the H^+-ATPase. Binding of 14-3-3 proteins to the phosphorylated C-terminal region (Kinoshita and Shimazaki, 1999; Fuglsang et al., 1999; Camoni et al., 2000; Roberts, 2000; Emi et al., 2001) leads to the displacement of the auto-inhibitory domain and thus activation of the H^+-ATPase. The fungal toxin fusicoccin, a potent stomatal opener, stabilizes this complex and thereby irreversibly activates the H^+-ATPase (Dambly and Boutry, 2000). Although auxin, like blue light, leads to an activation of the plasma membrane H^+-ATPase in guard cells (Assmann et al., 1985; Shimazaki et al., 1986; Lohse and Hedrich, 1992; Kinoshita and Shimazaki, 1999), there is no evidence so far for a common signalling pathway. Enhanced proton pumping, however, results in hyperpolarization of the guard cell plasma membrane potential and in turn activation of voltage-dependent, inward-rectifying K^+ channels (K_{in}^+) (Schroeder et al., 1984, 1987; Hedrich and

Schroeder, 1989) and, finally, potassium uptake. Furthermore, activation of the proton pump results in acidification of the apoplast. Extracellular protons have been shown to increase K_{in}^+ currents in guard cells (Blatt, 1992; Dietrich *et al.*, 1997) and thus represent modulators of auxin-mediated stomatal opening.

In 1992, the first guard cell potassium channel, KAT1, was identified in *Arabidopsis* (Anderson *et al.*, 1992; Nakamura *et al.*, 1995). Together with the potato guard cell K_{in}^+-channel KST1 (Müller-Röber *et al.*, 1995), heterologous expression studies have identified this channel type as a voltage-dependent, K^+-selective inward rectifier (Schachtman *et al.*, 1994; Müller-Röber *et al.*, 1995; Hedrich *et al.*, 1995; Hoshi, 1995). In addition to their response towards hyperpolarized voltages, these guard cell K^+-uptake channels are sensitive to extracellular protons (Hedrich and Becker, 1994; Müller-Röber *et al.*, 1995). In a feed-forward loop, acidification of the apoplast by the H^+-ATPases shifts the activation threshold of these potassium channels towards positive membrane potentials (Hoth *et al.*, 1997; Hoth and Hedrich, 1999). As a result the open probability of the K^+-uptake channels increases and potassium accumulation is further facilitated. Since heterologous expression studies with guard cell K^+ channels in *Xenopus* oocytes revealed no response to auxin, a direct regulation of these K_{in}^+ channels by auxin can be excluded. Recent studies, however, provide evidence for the expression of KAT2 α-subunits in guard cells (Pilot *et al.*, 2001). In a detailed analysis of the *Arabidopsis kat1-1* knock-out mutant, Szyroki *et al.* (2001) demonstrated the expression of at least 4 additional K^+-channel α-subunits. Therefore, it was not unexpected that stomatal opening in the *kat1-1* mutant was rescued by K^+ channels other than KAT1. Analysis of additional K^+-channel knock-out plants, together with co-expression and domain-swapping studies will now prove whether these subunits modify the response of the inward rectifier towards auxin, protons and calcium (cf. Dreyer *et al.*, 1997; Hoth *et al.*, 2001).

A model for auxin-induced growth

The response to auxin of the to model systems can be decomposed into at least six sequential phases. The first phase is considered a lag-phase because it precedes observable changes in cell volume in guard cells and coleoptiles. Although a growth response is not occurring, this phase is characterized by the percep-

tion of auxin and changes in ion transport (see Figures 1 and 2) including electrical responses at the plasma membrane. Transporters for auxin uptake and efflux have been identified at the molecular level (Bennett *et al.*, 1998; Palme and Gälweiler, 1999). Their functional characterization, however, is lacking and this represents a future challenge to explain how auxin transport is energized and how auxin gradients are established and maintained. The auxin-binding protein (ABP), the best known candidate for an auxin receptor, has been identified since nearly 25 years, but its biochemical and molecular characterization has led to a controversy about its localization and function. Antibodies directed against ABP as well as synthetic ABP peptides are able to mimic early events in auxin action such as membrane hyperpolarisation, K^+-channel activation or cell wall loosening (Barbier-Brygoo *et al.*, 1989; Thiel *et al.*, 1993; Rück *et al.*, 1993; Leblanc *et al.*, 1999). Recent studies have demonstrated that the disruption of the single-copy *abp1* gene in *Arabidopsis* correlates with reduced auxin-sensitivity (Chen *et al.*, 2001).

Auxin action is 'initiated' in phase 2 with the transcription, via auxin-activated transcription factors (cf. Ulmasov *et al.*, 1999), of very early genes of unknown function, such as the SAURs (Gee *et al.*, 1991) but also ZMK1 and H^+-ATPases. At the same time, H^+-ATPases are activated post-translationally. Enhanced proton pumping hyperpolarizes the membrane, acidifies the apoplast and thereby activates voltage- and proton-activated K_{in}^+ channels such as KAT1 and KST1 (guard cells) or ZMK1 (coleoptiles). The increase in turgor as well as cell wall loosening drives cell expansion (Nicol and Höfte, 1998; Cosgrove, 2000; see Figure 1), which is typical for phase 3, and is explained by the acid-growth theory (Rayle and Cleland, 1977).

During phase 3 the mRNA of auxin-sensitive genes rises and thus an increase in the number of channel proteins can be detected. The latter process is accompanied by enhanced exocytotic activity (Homann and Thiel, 1999; Weise *et al.*, 2000). The increase in transporter activity and density as well as the activation of processes involved in membrane and cell-wall synthesis[1] feed back on each other. This results in a peak or plateau phase (phase 4) followed by an adaptation phase (phase 5) which is characterized by a drop in, for example, *zmk1* mRNA. The final, steady-

---

[1] It is questionable whether membrane- and cell-wall synthesis processes are triggered in auxin-stimulated guard cells.

state phase (phase 6) in coleoptiles is characterized by sustained but elevated growth rates.

Ongoing auxin genomics and proteomics (see the paper by Dharmasiri and Estelle in this issue) as well as the functional analysis of the auxin transporters and/or receptors, together with the auxin-responsive genes of unknown function, will complete step-by-step our very basic model on the 'channelling' of auxin action. Likewise, related studies on plant membrane transport will provide new insight into the very early auxin-induced membrane polarization events and the tuning of the osmotic motor which drives cell expansion.

References

Abel, S. and Theologis, A. 1996. Early genes and auxin action. Plant Physiol 111: 9–17.

Anderson, J.A., Huprikar, S.S., Kochian, L.V., Lucas, W.J. and Gaber, R.F. 1992. Functional expression of a probable *Arabidopsis thaliana* potassium channel in *Saccharomyces cerevisiae*. Proc. Natl. Acad. Sci. USA 89: 3736–3740.

Assmann, S.M., Simoncini, L. and Schroeder, J.I. 1985. Blue light activates electrogenic ion pumping in guard cell protoplasts of *Vicia faba*. Nature 318: 285–287.

Barbier-Brygoo, H., Ephritikhine, G., Klämbt, D., Gishlan, M. and Guern, J. 1989. Functional evidence for an auxin receptor at the plasmalemma of tobacco protoplasts. Proc. Natl. Acad. Sci. USA 86: 891–895.

Bauer, C.S., Hoth, S., Haga, K., Philippar, K., Aoki, N. and Hedrich, R. 2000. Differential expression and regulation of K^+ channels in the maize coleoptile: molecular and biophysical analysis of cells isolated from cortex and vasculature. Plant J. 24: 139–145.

Bauly, J.M., Sealy, I.M., Macdonald, H., Brearley, J., Droge, S., Hillmer, S., Robinson, D.G., Venis, M.A., Blatt, M.R., Lazarus, C.M. and Napier, R.M. 2000. Overexpression of auxin-binding protein enhances the sensitivity of guard cells to auxin. Plant Physiol. 124: 1229–1238.

Becker, D., Zeilinger, C., Lohse, G., Depta, H. and Hedrich, R. 1993. Identification and biochemical characterization of the plasma-membrane proton ATPase in guard cells of *Vicia faba* L. Planta 190: 44–50.

Becker, D., Dreyer, I., Hoth, S., Reid, J.D., Busch, H., Lehnen, M., Palme, K. and Hedrich, R. 1996. Changes in voltage activation, Cs^+ sensitivity, and ion permeability in H5 mutants of the plant K^+ channel KAT1. Proc. Natl. Acad. Sci. USA 93: 8123–8128.

Bennett, M.J., Marchant, A., Green, H.G., May, S.T., Ward, S.P., Millner, P.A., Walker, A.R., Schulz, B. and Feldmann, K.A. 1996. *Arabidopsis* AUX1 gene: a permease-like regulator of root gravitropism. Science 273: 948–950.

Bennett, M.J., Marchant, A., May, S.T. and Swarup, R. 1998. Going the distance with auxin: unravelling the molecular basis of auxin transport. Phil. Trans. R. Soc. Lond. B Biol. Sci. 353: 1511–1515.

Blatt, M.R. 1992. K^+ channels of stomatal guard cells. Characteristics of the inward rectifier and its control by pH. J. Gen. Physiol. 99: 615–644.

Blatt, M.R. and Thiel, G. 1994. K^+ channels of stomatal guard cells: bimodal control of the K^+ inward-rectifier evoked by auxin. Plant J. 5: 55–68.

Burg, S.P. and Burg, E.A. 1966. The interaction between auxin and ethylene and its role in plant growth. Proc. Natl. Acad. Sci. USA 55: 262–269.

Camoni, L., Iori, V., Marra, M. and Aducci, P. 2000. Phosphorylation-dependent interaction between plant plasma membrane H^+-ATPase and 14-3-3 proteins. J. Biol. Chem. 275: 9919–9923.

Chen, J.G., Ullah, H., Young, J.C., Sussman, M.R. and Jones, A.M. 2001. ABP1 is required for organized cell elongation and division in *Arabidopsis* embryogenesis. Genes Dev. 15: 902–911.

Claussen, M., Lüthen, H., Blatt, M.R. and Böttger, M. 1997. Auxin-induced growth and its linkage to potassium channels. Planta 201: 227–234.

Cosgrove, D.J. 2000. Loosening of plant cell walls by expansins. Nature 407: 321–326.

Dambly, S. and Boutry, M. 2001. The two major plant plasma membrane H^+-ATPases display different regulatory properties. J. Biol. Chem. 276: 7017–7022.

Dharmasiri, S. and Estelle, M. 2002. The role of regulated protein degradation in auxin response. Plant Mol. Biol. 49: 401–408.

Darwin, C. 1880. The Power of Movement in Plants (assisted by F. Darwin). John Murray, London.

Dietrich, P., Dreyer, I., Wiesner, P. and Hedrich, R. 1997. Cation sensitivity and kinetics of guard-cell potassium channels differ among species. Planta 205: 277–287.

Dreyer, I., Antunes, S., Hoshi, T., Müller-Röber, B., Palme, K., Pongs, O., Reintanz, B. and Hedrich, R. 1997. Plant K^+ channel α-subunits assemble indiscriminately. Biophys. J. 72: 2143–2150.

Emi, T., Kinoshita, T. and Shimazaki, K. 2001. Specific binding of vf14-3-3a isoform to the plasma membrane H^+-ATPase in response to blue light and fusicoccin in guard cells of broad bean. Plant Physiol. 125: 1115–1125.

Evans, M.L. 1991. Gravitropism: Interaction of sensitivity modulation and effector redistribution. Plant Physiol. 95: 1–5.

Felle, H., Peters, W. and Palme, K. 1991. The electrical response of maize to auxins. Biochim. Biophys. Acta 1064: 199–204.

Felle, H.H., Hanstein, S., Steinmeyer, R. and Hedrich, R. 2000. Dynamics of ionic activities in the apoplast of the sub-stomatal cavity of intact *Vicia faba* leaves during stomatal closure evoked by ABA and darkness. Plant J. 24: 297–304.

Firn, R.D., Wagstaff, C. and Digby, J. 2000. The use of mutants to probe models of gravitropism. J. Exp. Bot. 51: 1323–1340.

Frias, I., Caldeira, M.T., Perez-Castineira, J.R., Navarro-Avino, J.P., Culianez-Macia, F.A., Kuppinger, O., Stransky, H., Pagès, M., Hager, A. and Serrano, R. 1996. A major isoform of the maize plasma membrane H^+-ATPase: characterization and induction by auxin in coleoptiles. Plant Cell 8: 1533–1544.

Fuglsang, A.T., Visconti, S., Drumm, K., Jahn, T., Stensballe, A., Mattei, B., Jensen, O.N., Aducci, P. and Palmgren, M.G. 1999. Binding of 14-3-3 protein to the plasma membrane H^+-ATPase AHA2 involves the three C-terminal residues Tyr(946)-Thr-Val and requires phosphorylation of Thr(947). J. Biol. Chem. 274: 36774–36780.

Gälweiler, L., Guan, C., Iler, M., Wisman, E., Mendgen, K., Yephremov, A. and Palme, K. 1998. Regulation of polar auxin transport by AtPIN1 in *Arabidopsis* vascular tissue. Science 282: 2226–2230.

Gee, M.A., Hagen, G. and Guilfoyle, T.J. 1991. Tissue-specific and organ-specific expression of soybean auxin-responsive transcripts GH3 and SAURs. Plant Cell 3: 419–430.

Gehring, C.A., McConchie, R.M., Venis, M.A. and Parish, R.W. 1998. Auxin-binding-protein antibodies and peptides influence stomatal opening and alter cytoplasmic pH. Planta 205: 581–586.

Hager, A., Menzel, H. and Krauss, A. 1971. Versuche und Hypothese zur Primärwirkung des Auxins beim Streckungswachstum. Planta 100: 47–75.

Hansen, H. and Grossmann, K. 2000. Auxin-induced ethylene triggers abscisic acid biosynthesis and growth inhibition. Plant Physiol. 124: 1437–1448.

Hedrich, R. 1994. Voltage-dependent chloride channels in plant cells: identification, characterization, and regulation of a guard cell anion channel. In: W.B. Guggino (Ed.) Chloride Channels, Academic Press, San Diego, CA, pp. 1–33.

Hedrich, R. and Becker, D. 1994. Green circuits: the potential of plant specific ion channels. Plant Mol. Biol. 26: 1637–1650.

Hedrich, R. and Jeromin, A. 1992. A new scheme of symbiosis: ligand- and voltage-gated anion channels in plants and animals. Phil. Trans. R. Soc. Lond. B Biol. Sci. 338: 31–38.

Hedrich, R. and Schroeder, J.I. 1989. The physiology of ion channels and electrogenic pumps in higher plants. Annu. Rev. Plant Physiol. Plant Mol. Biol. 40: 539–569.

Hedrich, R., Busch, H. and Raschke, K. 1990. Calcium ion and nucleotide dependent regulation of voltage dependent anion channels in the plasma membrane of guard cells. EMBO J. 9: 3889–3892.

Hedrich, R., Moran, O., Conti, F., Busch, H., Becker, D., Gambale, F., Dreyer, I., Küch, A., Neuwinger, K. and Palme, K. 1995. Inward rectifier potassium channels in plants differ from their animal counterparts in response to voltage and channel modulators. Eur. Biophys. J. 24: 107–115.

Hedrich, R., Hoth, S., Becker, D., Dreyer, I. and Dietrich, P. 1998. On the structure and function of plant K^+ channels. In: F. LoSchiavo, R.L. Last, G. Morelli and N.V. Raikhel (Eds.) Cellular Integration of Signalling Pathways in Plant Development, Springer-Verlag, Berlin/Heidelberg, pp. 35–45.

Hedrich, R., Neimanis, S., Savchenko, G., Felle, H.H., Kaiser, W.M. and Heber, U. 2001. Changes in apoplastic pH and membrane potential in leaves in relation to stomatal responses to CO_2, malate, abscisic acid or interruption of water supply. Planta 213(4): 594–601.

Homann, U. and Thiel, G. 1999. Unitary exocytotic and endocytotic events in guard-cell protoplasts during osmotically driven volume changes. FEBS Lett. 460: 495–499.

Hoshi, T. 1995. Regulation of voltage dependence of the KAT1 channel by intracellular factors. J. Gen. Physiol. 105: 309–328.

Hoth, S. and Hedrich, R. 1999. Distinct molecular bases for pH sensitivity of the guard cell K^+ channels KST1 and KAT1. J. Biol. Chem. 274: 11599–11603.

Hoth, S., Dreyer, I., Dietrich, P., Becker, D., Müller-Röber, B. and Hedrich, R. 1997. Molecular basis of plant-specific acid activation of K^+ uptake channels. Proc. Natl. Acad. Sci. USA 94: 4806–4810.

Hoth, S., Geiger, D., Becker, D. and Hedrich, R. 2001. The pore of plant K^+ channels is involved in voltage and pH sensing. Domain-swapping between different K^+ channel α-subunits. Plant Cell 13: 943–952.

Humble, G.D. and Raschke, K. 1971. Stomatal opening quantitatively related to potassium transport. Evidence from electron probe analysis. Plant Physiol. 48: 447–453.

Kinoshita, T. and Shimazaki, K. 1999. Blue light activates the plasma membrane H^+-ATPase by phosphorylation of the C-terminus in stomatal guard cells. EMBO J. 18: 5548–5558.

Lacombe, B., Pilot, G., Michard, E., Gaymard, F., Sentenac, H. and Thibaud, J.B. 2000. A shaker-like K^+ channel with weak rectification is expressed in both source and sink phloem tissues of Arabidopsis. Plant Cell 12: 837–851.

Leblanc, N., David, K., Grosclaude, J., Pradier, J.M., Barbier-Brygoo, H., Labiau, S. and Perrot-Rechenmann, C. 1999. A novel immunological approach establishes that the auxin-binding protein, Nt-abp1, is an element involved in auxin signaling at the plasma membrane. J. Biol. Chem. 274: 28314–28320.

Lohse, G. and Hedrich, R. 1992. Characterisation of the plasma-membrane H^+-ATPase from Vicia faba guard cells. Planta 188: 206–214.

Lu, P., Zhang, S.Q., Outlaw, W.H. Jr. and Riddle, K.A. 1995. Sucrose: a solute that accumulates in the guard-cell apoplast and guard- cell symplast of open stomata. FEBS Lett. 362: 180–184.

Marschner, H. 1996. Plant Nutrition of Higher Plants. Academic Press, London.

Marten, I., Lohse, G. and Hedrich, R. 1991. Plant growth hormones control voltage-dependent activity of anion channels in the plasma membrane of guard cells. Nature 353: 758–762.

Marten, I., Zeilinger, C., Redhead, C., Landry, D.W., al-Awqati, Q. and Hedrich, R. 1992. Identification and modulation of a voltage-dependent anion channel in the plasma membrane of guard cells by high-affinity ligands. Embo. J. 11(10): 3569–3575.

Marten, I., Hoth, S., Deeken, R., Ache, P., Ketchum, K.A., Hoshi, T. and Hedrich, R. 1999. AKT3, a phloem-localized K^+ channel, is blocked by protons. Proc. Natl. Acad. Sci. USA 96: 7581–7586.

Merritt, F., Kemper, A. and Tallman, G. 2001. Inhibitors of ethylene synthesis inhibit auxin-induced stomatal opening in epidermis detached from leaves of Vicia faba L. Plant Cell Physiol. 42: 223–230.

Morsomme, P. and Boutry, M. 2000. The plant plasma membrane H^+-ATPase: structure, function and regulation. Biochim. Biophys. Acta 1465: 1–16.

Müller-Röber, B., Ellenberg, J., Provart, N., Willmitzer, L., Busch, H., Becker, D., Dietrich, P., Hoth, S. and Hedrich, R. 1995. Cloning and electrophysiological analysis of KST1, an inward rectifying K^+ channel expressed in potato guard cells. EMBO J. 14: 2409–2416.

Nakamura, R.L., McKendree, W.L., Hirsch, R.E., Sedbrook, J.C., Gaber, R.F. and Sussman, M.R. 1995. Expression of an Arabidopsis potassium channel gene in guard cells. Plant Physiol. 109: 371–374.

Nicol, F. and Höfte, H. 1998. Plant cell expansion: scaling the wall. Curr. Opin. Plant Biol. 1: 12–17.

Ostin, A., Ilic, N. and Cohen, J.D. 1999. An in vitro system from maize seedlings for tryptophan-independent indole-3-acetic acid biosynthesis. Plant Physiol. 119: 173–178.

Palme, K. and Gälweiler, L. 1999. PIN-pointing the molecular basis of auxin transport. Curr. Opin. Plant Biol. 2: 375–381.

Philippar, K., Fuchs, I., Luthen, H., Hoth, S., Bauer, C.S., Haga, K., Thiel, G., Ljung, K., Sandberg, G., Bottger, M., Becker, D. and Hedrich, R. 1999. Auxin-induced K^+ channel expression represents an essential step in coleoptile growth and gravitropism. Proc. Natl. Acad. Sci. USA 96: 12186–12191.

Pilot, G., Lacombe, B., Gaymard, F., Cherel, I., Boucherez, J., Thibaud, J.B. and Sentenac, H. 2001. Guard cell inward K^+ channel activity in Arabidopsis involves expression of the twin channel subunits KAT1 and KAT2. J. Biol. Chem. 276: 3215–3221.

Pitts, R.J., Cernac, A. and Estelle, M. 1998. Auxin and ethylene promote root hair elongation in Arabidopsis. Plant J. 16: 553–560.

Rahman, A., Amakawa, T., Goto, N. and Tsurumi, S. 2001. Auxin is a positive regulator for ethylene-mediated response in the growth of Arabidopsis roots. Plant Cell Physiol. 42: 301–307.

356

Rayle, D.L. and Cleland, R. 1977. Control of plant cell enlargement by hydrogen ions. Curr. Top. Dev. Biol. 11: 187–214.

Ritte, G., Rosenfeld, J., Rohrig, K. and Raschke, K. 1999. Rates of sugar uptake by guard cell protoplasts of *Pisum sativum* L. related to the solute requirement for stomatal opening. Plant Physiol. 121: 647–656.

Roberts, M.R. 2000. Regulatory 14-3-3 protein-protein interactions in plant cells. Curr. Opin. Plant Biol. 3: 400–405.

Roelfsema, M.R.G. and Hedrich, R. 1999. Plant ion transport. In: Encyclopedia of Life Sciences, Macmillan Reference, London.

Roelfsema, M.R., Steinmeyer, R., Staal, M. and Hedrich, R. 2001. Single guard cell recordings in intact plants: light-induced hyperpolarization of the plasma membrane. Plant J. 26: 1–13.

Rück, A., Palme, K., Venis, M.A., Napier, R.M. and Felle, H.H. 1993. Patch-clamp analysis establishes a role for an auxin-binding protein in the auxin stimulation of plasma membrane current in *Zea mays* protoplasts. Plant J. 4: 41–46.

Sachs, J. 1887. Lectures in Plant Physiology. Clarendon Press, Oxford.

Schachtman, D.P., Schroeder, J.I., Lucas, W.J., Anderson, J.A. and Gaber, R.F. 1994. Expression of an inward-rectifying potassium channel by the *Arabidopsis* KAT1 cDNA. Science 258: 1654–1658.

Schroeder, J.I., Allen, G.J., Hugouvieux, V., Kwak, J.M. and Waner, D. 2001. Guard cell signal transduction. Annu. Rev. Plant Physiol. Plant Mol. Biol. 52: 627–658.

Schroeder, J.I. and Hedrich, R. 1989. Involvement of ion channels and active transport in osmoregulation and signaling of higher plant cells. Trends Biochem. Sci. 14: 187–192.

Schroeder, J.I., Hedrich, R. and Fernandez, J.M. 1984. Potassium-selective single channels in guard cell protoplasts of *Vicia faba*. Nature 312: 361–362.

Schroeder, J.I., Raschke, K. and Neher, E. 1987. Voltage dependence of K$^+$ channels in guard cell protoplasts. Proc. Natl. Acad. Sci. USA 84: 4108–4112.

Senn, A.P. and Goldsmith, M.-H.M. 1988. Regulation of electrogenic proton pumping by auxin and fusicoccin as related to the growth of *Avena* coleoptiles. Plant Physiol. 88: 131–138.

Shimazaki, K., Iino, M. and Zeiger, E. 1986. Blue light-dependent proton extrusion by guard cell protoplasts of *Vicia faba*. Nature 319: 324–326.

Sze, H., Li, X. and Palmgren, M.G. 1999. Energization of plant cell membranes by H$^+$-pumping ATPases. Regulation and biosynthesis. Plant Cell 11: 677–690.

Szyroki, A., Ivashikina, N., Dietrich, P., Roelfsema, M.R., Ache, P., Reintanz, B., Deeken, R., Godde, M., Felle, H., Steinmeyer, R., Palme, K. and Hedrich, R. 2001. KAT1 is not essential for stomatal opening. Proc. Natl. Acad. Sci. USA 98: 2917–2921.

Talbott, L.D. and Zeiger, E. 1996. Central roles for potassium and sucrose in guard-cell osmoregulation. Plant Physiol. 111: 1051–1057.

Thiel, G., Blatt, M.R., Fricker, M.D., White, I.R. and Millner, P. 1993. Modulation of K$^+$ channels in *Vicia* stomatal guard cells by peptide homologs to the auxin-binding protein C terminus. Proc. Natl. Acad. Sci. USA 90: 11493–11497.

Tode, K. and Lüthen, H. 2001. Fusicoccin- and IAA-induced elongation growth share the same pattern of K$^+$ dependence. J. Exp. Bot. 52: 251–255.

Ulmasov, T., Hagen, G. and Guilfoyle, T.J. 1999. Activation and repression of transcription by auxin-response factors. Proc. Natl. Acad. Sci. USA 96: 5844–5849.

Walker, L. and Estelle, M. 1998. Molecular mechanisms of auxin action. Curr. Opin. Plant Biol. 1: 434–439.

Weise, R., Kreft, M., Zorec, R., Homann, U. and Thiel, G. 2000. Transient and permanent fusion of vesicles in *Zea mays* coleoptile protoplasts measured in the cell-attached configuration. J. Membr. Biol. 174: 15–20.

Went, F.W. and Thimann, K.V. 1937. Phytohormones. Macmillan, New York.

Zeiger, E. 2000. Sensory transduction of blue light in guard cells. Trends Plant Sci. 5: 183–185.

Plant Molecular Biology **49**: 357–372, 2002.
Perrot-Rechenmann and Hagen (Eds.), Auxin Molecular Biology.
© 2002 *Kluwer Academic Publishers.*

357

Secondary messengers and phospholipase A₂ in auxin signal transduction

Günther F.E. Scherer
Universität Hannover, Institut für Zierpflanzenbau, Baumschule und Pflanzenzüchtung, Abt. Spezielle Ertragsphysiologie, Herrenhäuser Strasse 2, 30419 Hannover, Germany (e-mail scherer@zier.uni-hannover.de)

Received 20 April 2001; accepted in revised form 6 August 2001

Key words: auxin, fatty acid, phospholipase A₂, second messenger, signal transduction

Abstract

Despite recent progress auxin signal transduction remains largely scetchy and enigmatic. A good body of evidence supports the notion that the ABP1 could be a functional receptor or part of a receptor, respectively, but this is not generally accepted. Evidence for other functional receptors is lacking, as is any clearcut evidence for a function of G proteins. Protons may serve as second messengers in guard cells but the existing evidence for a role of calcium remains to be clearified. Phospholipases C and D seem not to have a function in auxin signal transduction whereas the indications for a role of phospholipase A₂ in auxin signal transduction accumulated recently. Mitogen-activated protein kinase (MAPK) is modulated by auxin and the protein kinase PINOID has a role in auxin transport modulation even though their functional linkage to other signalling molecules is ill-defined. It is hypothesized that signal transduction precedes activation of early genes such as IAA genes and that ubiquitination and the proteasome are a mechanism to integrate signal duration and signal strength in plants and act as major regulators of hormone sensitivity.

Abbreviations: AACOCF₃, arachidonyltrifluoromethyl carbon; ETYA, 5,8,11,14-eicosatetraynoic acid; NDGA, nordihydroguajaretic acid

Introduction: What is cellular signal transduction?

When looking into textbooks signal transduction is defined as the steps and reactions leading from a receptor to gene regulation. Usually, gene regulation by transcription factors already is described in a separate chapter. Hence, the border line between signal transduction reactions and gene regulation is the modification of transcription factors. However, many authors, especially in plant biology, include gene regulation into signalling so that the border line rather is a border zone.

The auxin receptor – or auxin receptors?

There is much literature on diverse hypothetical auxin receptors and another review in this issue on ABP1, the protein highly suspicious to have an auxin receptor function (Napier, 1995; Napier and Venis, 1995; Venis and Napier, 1995; Macdonald, 1997; Lüthen et al., 1999). It is philosophical to ask whether there could be several genetically non-homologous types of receptors for a given hormone – not a gene family as the phytochromes are. There could be, but that does not seems to be, a concept of wide distribution in nature. To evoke the complexity of auxin responses it seems to be sufficient to postulate several trans-membrane proteins as interaction partners for the ABP1. These could be more receptor-like in the classical sense, transducing the message across the membrane to a protein which carries out the next step of action (Klämbt, 1990). Different cell types can have different downstream signal transduction components to generate further complexity. The enigma is the relatively high concentration of ABP1 in the ER as compared to the very low amount in the plasma membrane. Indeed, there could start a signal transduction pathway from the ER different

358

from those at the plasma membrane even though auxin was not found to bind to ABP1 in the ER in the cell (although it certainly does so *in vitro*) (Tian *et al.*, 1995). ABP1 meets the criteria for a receptor in that modulating its amount by over-expressing it modulates hormone responses predictably (Jones *et al.*, 1998; Bauly *et al.*, 2000), knocking it out is lethal (Chen *et al.*, 2001) and antibodies against it can either inhibit or mimick responses, depending on the antibody (Barbier-Brygoo *et al.*, 1989, 1991, Venis *et al.*, 1992; Rück *et al.*, 1993; Thiel *et al.*, 1993; Leblanc *et al.*, 1999). None of the many other postulated additional receptors for auxin is more than a postulate at this stage, inspired by the fascination for auxin but not supported by such a number of facts as is the receptor function for ABP1.

What is important to recall is that auxin is outstanding in that it is polarly transported (Palme and Gälweiler, 1999) which leads to unequal concentrations of auxin in different tissues in close proximity (Jones, 1980). This in itself would be expected to lead to unequal responses to this hormone in those different tissues, even without the possibilities that different tissues could respond differentially to the same hormone concentration. Creating concentration differences by transport proteins thus will always look like influencing function even though the receptor could be a different entity. Another reason why transport proteins look like recepors is that the transport proteins certainly are hormone-binding proteins. A good example is the 'making' of the pattern of vascularization by 'draining' auxin from the surrounding tissue (Berleth *et al.*, 2000), root tip development (Sabatini *et al.*, 1999) and embryo development (Steinmann *et al.*, 1999). The next set of hormone-binding proteins are the enzymes of auxin metabolism which also change hormone concentrations by their actions. Since we cannot yet analyse all of them in enough detail, the philosophical question of whether there are one or several genetically unrelated auxin receptors will still take a while to be answered.

G proteins and auxin

The main problem in writing down opinions on auxin and G proteins is that the receptor for auxin is not unequivocally identified. Even if one accepts the concept that ABP1 has receptor function, and probably then by binding to a transmembrane protein (Klämbt, 1990), one has to realize that the only plant receptor group

where we have clear speculations about coupling to G proteins is the seven-transmembrane receptor type. All the subunits of trimeric G proteins (Weiss *et al.*, 1993; 1994; Mason *et al.*, 2000) have been found in plants. However, the known corresponding seven-transmembrane receptors have been only tentatively identified (Plackidou-Dymock *et al.*, 1998; Devoto *et al.*, 1999) and the coupling to G proteins is then – reasonably – assumed to be homologous to animal or yeast systems (Bockaert and Pin, 1999). G protein subunits might then be coupled to ion channel regulation as in animal systems (Li and Assmann, 1993; Wu and Assmann, 1994; Armstrong and Blatt, 1995). As for the other types of known plant receptors, the soluble cryptochromes and phytochromes (Quail *et al.*, 1995; Lin, 2000), leucine-rich repeat receptors and receptor kinases (Lease *et al.*, 1998; Becraft, 1998), and two-component receptors (Urao *et al.*, 2000), we do not understand the coupling to the protein carrying out the next step of signal transduction. Therefore, models for plants analogous to ras-coupled signalling (Marshall, 1996), i.e. receptor coupling to small G proteins, are attractive but pure speculation at this point (Zheng and Yang, 2000; Fu *et al.*, 2001). Thus, the weak earlier evidence of G protein involvement in auxin signal transduction (Zaina *et al.*, 1990; Scherer and André, 1993; Millner *et al.*, 1996) has to be viewed together with this lack of knowledge on the supposed or expected transmembrane receptor, the hypothetical binding partner to the extracytosolic ABP1 and to cytosolic (small or trimeric) G proteins, or yet another receptor constellation for an auxin receptor. Stronger evidence is needed to support the case for small or trimeric G proteins in auxin signalling.

Strong evidence for trimeric G proteins in plant signal transduction has been provided in several papers on mutants of Gα in rice and in *Arabidopsis* but the characterization of the phenotypes supported a function of Gα subunit in several pathways, positive in gibberellin signalling, negative in abscisic acid signalling, and positive in auxin-induced cell division (Ashikari *et al.*, 1999; Ueguchi-Tanaka *et al.*, 2000; Ullah *et al.*, 2001; Wang *et al.*, 2001). The only knockout plant for the one Gα identified in *Arabidopsis*, so far, showed decreased cell division as a major trait in its phenotype which is part of the function of auxin – but auxin is not the only player in cell division (den Boer and Murray, 2000). It remains also to be seen whether the known G protein subunits in the *Arabidopsis* genome are really the only ones and whether perhaps the underlying theme is multi-signal triggering of the cell cycle affected by auxin, gibberellin, abscisic acid, and basic signals like the nutritional status (Bögre *et al.*, 2000; Ullah *et al.*, 2001). There is evidence for a larger type of Gα protein in plants which could be a second type of α subunit and more might still be discovered (Lee and Assman, 1999; Kaydamov *et al.*, 2000). Altogether, this leaves the function of either trimeric or small G proteins in auxin signalling rather open. Undoubtedly, reverse genetics and the complete knowledge of the *Arabidopsis* genome will allow rapid progress in this area soon.

Second messengers

Ionic second messengers: pH and Ca^{2+}

Regulation of pH which may both act as a cytosolic second messenger and in the regulation of cell wall pH is described in detail by Becker and Hedrich (2002) in this issue and a number of previous reviews (Lüthen *et al.*, 1999; Roos, 2000; and references therein). What it boils down to is that either the proton pump might be down-regulated to cause cytosolic acidification (and then perhaps up-regulated after a few minutes to cause cell wall acidification) or ion channels in the plasma membrane (or even in the tonoplast) might be blocked or opened and, depending on the direction and charge of ion flow, cytosolic acidification would be the consequence, according to the strong ion difference theory (Stewart *et al.*, 1983). In the case of low auxin concentrations rapid cytosolic acidification was found, whereas alkalinization occurred in the case of very high auxin or of physiological abscisic acid concentrations (Blatt and Thiel, 1994). Rapid stimulation of potassium influx (<5 min) was also observed (Thiel *et al.*, 1993; Blatt and Thiel, 1994; Grabov and Blatt, 1997; Thiel and Weise, 1999) but is not considered as a second messenger response and is also a necessity for long-term cell expansion (Claussen *et al.*, 1997; Philippar *et al.*, 1999). A change of the cytosolic pH clearly might activate enzymes inducing the next step(s) (Tena and Renaudin, 1998). From the signal transduction point of view this means that either ion channels or the H^{+}-ATPase may be receptors for auxin (or for the ABP1) or that some signal transduction step(s) from a receptor to an ion channel or the H^{+}-ATPase have to occur. Direct evidence for either one, ion channel or H^{+}-ATPase, as auxin or ABP1 receptors is not available but the time lag between auxin application and the cytosolic pH reaction is 1–2 min (Thiel *et al.*, 1993; Rück *et al.*, 1993; Blatt

and Thiel, 1994; Shishova and Lindberg, 1997; S. Gilroy, personal communication) which certainly excludes gene regulation as a cause. In conclusion, at least one branch of auxin signal transduction leads directly to activation changes of proteins (ion channels or the proton pump or both) and does not need or use gene regulation. The resulting change in the cytosolic pH may be part of the mechanism of auxin signalling.

Although up-regulation of the biosynthesis of some isoforms of the H^{+}-ATPase was described (Hager *et al.*, 1991; Frias *et al.*, 1996) and up-regulation of at least one potassium channel is necessary for sustained growth and the graviresponse (Philippar *et al.*, 1999) this is related to extracytosolic acidification and potassium uptake. The slower up-regulation of genes for H^{+}-ATPase and channels is not a contradiction to an initially more rapid regulation of ion gradients without gene regulation. The regulation of cell wall pH is not further discussed here (see Lüthen *et al.*, 1999).

Whether or not cytosolic calcium participates in auxin signal transduction is still an enigma. There are a surprisingly low number of reports regarding the importance of the issue, and the methods used in those positive reports where direct measurements were tried are not up to the highest possible technical standard which is available now. The early report by Felle (1988) showed slow oscillations of cytosolic calcium and pH on a time scale of 10–20 min which may reflect rather a cellular system property and does not fit to the time scale measured for pH changes found by other authors mentioned above. The work by Parish's group was done with the non-ratiometric dye fluo3 (Gehring *et al.*, 1990) and was at best partially supported by others working with the ratiometric indo1 and fura 2 (Ayling *et al.*, 1994). Non-ratiometric dyes are hard to calibrate and deliver only qualitative results since precise concentration determinations are not possible, besides the loading problems and bleaching problems associated with them (S. Gilroy, personal communication). More recent experiments with indo1, a ratiometric dye which was injected into *Arabidopsis* root cells, showed no calcium response to auxin but did show a calcium response to touch as a positive control. Also in *Arabidopsis* transformed by a chameleon Ca^{2+}-sensing GFP no calcium response to auxin could be detected but an acidification response could be found in plants transformed by a pH-indicator type of GFP f (S. Gilroy, personal communication).

Several papers provided indirect evidence that calcium could be involved in auxin signalling. Induc-

360

tion of a calmodulin-related gene by auxin may be support for this idea but gene regulation is a downstream consequence of signal transduction so that any calmodulin-dependent function late in auxin action would explain its induction by auxin (Yang and Poovaiah, 2000). The dependence of rapid intracellular acidification on extracellular calcium (Shishova and Lindberg, 1997) may be only indirectly related to auxin since it was measured in protoplasts and protoplasting elicits strongly so that acidification might have been a consequence of elicitation further influenced by auxin (Roos, 2000). Other indirect clues are rapid inhibitory effects of auxin on plasma streaming (Sweeney and Thimann, 1942; Sweeney, 1944; Ayling et al., 1994). The rationale for the plasma streaming effects is that elevated calcium concentrations inhibit plasma streaming (Shimmen and Yokota, 1994; Yokota et al., 1999). However, the most recent report drew inconclusive conclusions on calcium and auxin (Ayling et al., 1994) and elevated cytosolic calcium would be difficult to reconcile with a stimulation of the proton pump and of potassium influx (Kinoshita et al., 1995; Grabov and Blatt, 1997; de Nisi et al., 1999; Thiel and Weise, 1999). Even though I am inclined to find the evidence for an involvement of cytosolic calcium inconclusive, one has to keep in mind that rapid calcium changes in small compartments (e.g. between plasma membrane and cortical ER) still could go undetected by current methods.

Phospholipase C and phospholipase D

The reactions catalysed by phospholipases are shown in Figure 1. There are only few reports on auxin activation of phosphatidylinositol-splitting phospholipase C (PLC) *in vitro* and *in vivo* (Ettlinger and Lehle, 1988; Zbell and Walter-Back, 1989). Knowing now the primary structures of plant PLCs one can conclude on that basis that they probably are not activated by G proteins because they do not possess the corresponding domain structure (Munnik et al., 1998). But, they could be activated by calcium since they possess EF hands even though other activation mechanisms, such as phosphorylation, cannot be excluded yet. For the report on *in vitro* activation of PLC by auxin in isolated vesicles an activation by modulation of calcium seems very unlikely as an explanation and, furthermore, the identification of the products of the reaction was insufficient and the effects were small (Zbell and Walter-Back, 1989). For instance, lysolipids were not excluded as the quantified reaction products. Another

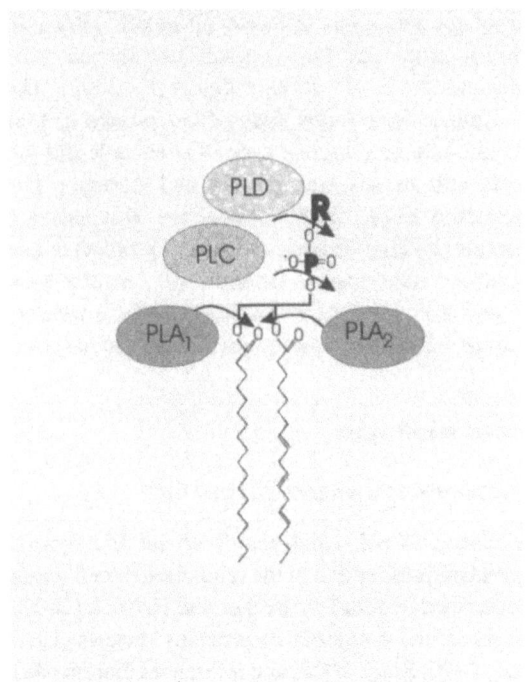

Figure 1. Scheme of hydrolysis of phospholipids by phospholipase A_1 (PLA$_1$), phospholipase A_2 (PLA$_2$), phospholipase C (PLC), and phospholipase D (PLD).

report described the activation of PLC by auxin *in vivo* (Ettlinger and Lehle, 1988) but this work was discontinued.

Even though the existing reports on PLC activation by auxin must be considered as insufficient, future work could still support this concept. Better methods, like the ones used to show activation of PLC by nod factor and elicitors, are needed to reinvestigate this topic for clarity (den Hartog et al., 2000; van der Luit et al., 2000). In recent publications a diacylglycerol-activated protein kinase was convincingly described (Subramanian et al., 1997; Chandok and Sopory, 1999) so that the classical pathway, receptor-(G protein?)-PLC-diacylglycerol + inositol trisphosphate, can occur in plants although, in the auxin field, it still awaits convincing experimental support.

There is only one report on auxin and phospholipase D: auxin had no effect on it (Scherer and André, 1993). This also awaits confirmation.

Phospolipase A$_2$

Presence and functions of phospolipase A$_2$ in animals, plants, fungi and bacteria

Since PLA$_2$ (phospolipase A$_2$) is probably the least recognized enzyme in plant signal transduction it seems appropriate here to explain the 'taxonomy' of this group of enzymes. In a recent series of reviews the diverse structural groups and functions of phospholipases A$_2$ have been described (Dennis, 1994; Balsinde and Dennis, 1997; Dessen, 2000; Six and Dennis, 2000; Winsteadt et al., 2000). The major groups are the small secreted PLA$_2$s (sPLA$_2$s) in animals and plants, the calcium-independent PLA$_2$s (iPLA$_2$s) in protists, plants and animals, the calcium-dependent (or cytosolic) PLA$_2$s (cPLA$_2$s) in animals and lower fungi (yeasts), and the low-density lipoprotein and platelet-activating-factor-hydrolysing PLA$_2$s in animals (Stafforini et al., 1997). This last group seems to have undergone a change in function into an epoxide hydrolase in plants (Bennedetti et al., 1998). Finally, bacteria possess a PLA$_2$ which is an intrinsic membrane protein having no sequence homology to any of the above isoforms (Snijder et al., 1999). Finer distinctions into more subgroups have been made but are not pertinent here (Six and Dennis, 2000).

A number of functions are known for higher-animal PLA$_2$s which include all four eukaryotic groups. A function in signal transduction is carried out mostly by the cPLA$_2$ which is activated by a number of hormones and releases arachidonic acid, the precursor to biologically active oxylipins in animals, leucotrienes and prostaglandins. Arachidonic acid also activates certain isoforms of protein kinase C as a lipidic second messenger (Oishi et al., 1990; Khan et al., 1993). Activation of cPLA$_2$ is achieved by elevated cytosolic calcium binding the enzyme to the membrane and by phosphorylation of cPLA$_2$ by mitogen-activated protein kinase (MAPK) and by protein kinase C (Nemenoff et al., 1993). In yeasts the enzyme homologous to cPLA$_2$ is often called phospholipase B because it hydrolyses both fatty acids efficiently. The yeast enzymes have switched their compartment and are exoenzymes found in the extracytosolic space. We could not find cPLA$_2$ sequences in plants and none in *Caenorhabditis* by BLAST search even though BLAST searches using cPLA$_2$ as a template never yielded any iPLA$_2$ sequences.

The second group of cytosolic PLA$_2$s are the iPLA$_2$s which are assumed in animals to mainly play a role in reshuffling of fatty acids (Winstead et al.,

2000) although they may also be activated by signals (Satoshi et al., 1999). They are related to plant patatins (vacuolar storage proteins and iPLA$_2$s) by sequence homology but also to cytosolic plant iPLA$_2$s (see below). They are found in *Caenorhabditis* and in protists but not in yeasts.

The sPLA$_2$ in animals has a function in digestion as an enzyme secreted from the pancreas but certain isoforms may also liberate arachidonic acid from the outer leaflet of the plasma membrane (Six and Dennis, 2000). Many sPLA$_2$s serve as haemolytic toxins in bee, scorpion and snake venoms. The sPLA$_2$ has been found in plants (Ståhl et al., 1998) and, as judged by sequence analysis, should be a secreted enzyme.

PLA$_2$ in auxin signal transduction: work with isolated vesicles

Rapid activation by auxin of PLA$_2$ was first shown by experiments with auxin-treated isolated microsomes and cell cultures (Scherer and André, 1989, 1993; Scherer et al., 1990; André and Scherer, 1991) and was continued later with microsomes isolated from hypocotyl segments (Scherer, 1995). Pre-labelling the cells or tissue by radioactive choline or ethanolamine provided that only one substrate phospholipid was labelled and accumulation of the respective lysophospholipid in response to auxin could be demonstrated, despite a background of PLA$_2$ activity which did not respond to hormone. The membrane vesicle system showed several of the properties to be expected from an auxin-responsive system: (1) activation was rapid within 1 min; (2) low concentrations (down to 0.02 μM 1-NAA) activated PLA$_2$; (3) inactive auxins did not stimulate PLA$_2$, while active auxins including IAA did; (4) antibodies inhibitory in the electrophysiological response to auxin, made against the ABP1 (Barbier-Brygoo et al., 1989, 1991), inhibited the *in vitro* PLA$_2$ response, suggesting that ABP1 could be the receptor for this response; (5) GDP and ADP, when added prior to isolation, i.e. inside the vesicles, inhibited the PLA$_2$ response but not when given to the outside of the vesicles. There was no effect of applying UDP and UTP (inside) or ATPγS and GTPγS (neither inside nor outside) (Scherer and André, 1993). This might be explained as an effect of G protein necessary inside the vesicles (see previous section) but other, less obvious explanations might also explain the findings. The accessibility of the PLA$_2$ activation to an antibody from outside and to GDP from inside suggest that outside-out plasma membrane or inside-out ER vesicles might have been the right compartment, but this was not investigated by cell fractionation. More important than these details is the consideration that auxin activation of PLA_2 in vesicles did not require energy-rich co-factors and was observed within 1 min in the presence of 20 mM EGTA. This indicates that only binding reactions caused the observed effects and neither phosphorylation nor calcium release played a role in this membrane vesicle system although phosphorylation prior to isolation of one or several components might have been necessary. Though not conclusive beyond this vague formulation, it means that interactions of only very few components seem possible in such a system, i.e. the PLA_2 was close to the receptor and perhaps bound directly to a transmembrane protein receptor which could be, in turn, binding the ABP1.

Taken together, work on PLA_2 activation by auxin in vesicles led to the breakthrough concept of a PLA_2 being active in plant signal transduction and acting upstream of gene regulation. Much remains to be done to verify this working hypothesis.

Activation of PLA_2 in vivo: pointing out the role of fatty acids and role of inhibitors to identify the relevant enzyme group

Measuring the activation of PLA_2 by auxins *in vivo* was achieved by two methods, selective labelling of a single lipid substrate, phosphatidylcholine or phosphatidylethanolamine, in the living cell and quantification of the labelled lysophospholipids as the breakdown products (Scherer and André, 1989; Scherer *et al.*, 1990; Scherer, 1990, 1992, 1995; Paul *et al.*, 1998) or by feeding the cell cultures phosphatidylcholine labelled by two fluorescent fatty acids as an artificial indicator substrate (Paul *et al.*, 1998; Paul, 1999; Scherer *et al.*, 2000). The second method revealed that only fatty acids and not lysophospholipids accumulated in living cells after treatment by physiological auxin concentrations, the threshold level being 1 μM auxin. Activation was observed as soon as 1 min after auxin application (Scherer, 1990; Paul *et al.*, 1998) or 5 min (Paul *et al.*, 1998) depending on the labelling method and was specific for active auxins.

Support for the concept of PLA_2 activation by auxin came from several pieces of work showing that PLA_2 inhibitors blocked downstream auxin responses. Yi *et al.* (1996) showed that fatty acids and lysophosphatidylcholine stimulated auxin-induced acidification and growth in maize coleoptiles and that these responses were inhibited by PLA_2 inhibitors. Similarly, it was shown that PLA_2 inhibitors inhibited activation

of PLA_2 by auxin (Paul *et al.*, 1998; Paul, 1999) and auxin-induced growth, but not growth induced by fusicoccin, gibberellin or cytokinin (Scherer and Arnold, 1997). The lipid platelet-activating factor, which is similar to lysophosphatidylcholine, stimulated proton extrusion in cultured cells as did fusicoccin (Scherer and Nickel, 1988; Nickel *et al.*, 1991). Polyunsaturated fatty acids modulated proton extrusion in guard cells (Lee *et al.*, 1994) and PLA_2 inhibitors inhibited auxin-stimulated stomatal opening but not blue light-stimulated stomatal opening (H. Macdonald, personal communication). Studies with purified bean $iPLA_2$ (Jung and Kim, 2000) or $iPLA_2$ expressed and purified from isolated *Arabidopsis* genes in *Escherichia coli* (Rietz, Holk and Scherer, unpublished) showed that the same inhibitors that inhibited purified enzymes also inhibited the activation of PLA_2 by auxin or the other described downstream responses.

The disadvantage of work with vesicles was that only a single reaction could be observed and subsequent metabolic reactions linked to this reaction in living cells remained unaccounted for. Lysophospholipids were found to accumulate in isolated vesicles but not in living cells, unless unphysiologically high auxin concentrations were used. This disproved the first interpretation of lysophospholipids being second messengers in auxin action and showed that lysophospholipids were down-regulated in cells, most likely by re-acylation of lysophospholipids, unless overridden by very strong and toxic signal concentrations (Paul *et al.*, 1998). This first and probably erroneous interpretation was influenced by earlier work with vesicles, which shared a stimulation of ATPase activity (Scherer, 1981; Cho and Hong, 1995) or proton pumping activity in plasma membrane vesicles by auxin as a common theme (Santoni *et al.*, 1991; François *et al.*, 1992; Bellamine *et al.*, 1993; Peltier and Rossignal, 1996). It seems likely that in all those experiments accumulation of lysophosphatidylcholine was the cause of ATPase activity increases, since this has been shown independently by addition of lysophosphatidylcholine to plasma membrane vesicles (Scherer and Stoffel, 1987; Scherer *et al.*, 1988; Palmgren and Sommarin, 1989). However, since lysophospholipids did not accumulate after application of physiological auxin concentrations in the living cells, different models are necessary, emphasizing the role of fatty acids.

The most likely next step(s) seems to be carried out by a lipid-activated protein kinase (Scherer, 1996 and references therein) but whether fatty acids themselves or rather some of their many possible metabolites generated from them (Farmer, 1992; Farmer *et al.*, 1998; Blée, 1998; Doss *et al.*, 2000) are second messengers remains to be seen.

PLA$_2$ and plant defence
Because jasmonate is synthesized from linolenic acid as a precursor, and because it is a wound and defence hormone, one function for a PLA$_2$ might be to release this fatty acid from phospholipids. However, since the jasmonate biosynthetic enzymes are localized to the chloroplasts (Creelman and Mullet, 1997) other acyl hydrolases or lipases could also fulfil this function, possibly hydrolysing galactolipids, the most abundant lipids in chloroplasts, as a source for linolenic acid. Despite this, a function for PLA$_2$ in defence signalling is indicated by a rapid activation of phosphatidylcholine hydrolysis by elicitors or wounding observed by a number of groups (Lee *et al.*, 1992, 1997; Roy *et al.*, 1995; Chandra *et al.*, 1996; Roos *et al.*, 1999; Paul, 1999; Narvaez-Vasquez *et al.*, 2000; Scherer *et al.*, 2000). Activation of PLA$_2$ was as rapid as 2 min (Chandra *et al.*, 1996; Lee *et al.*, 1997; Roos *et al.*, 1999) so that a signal function seems most likely as an explanation. Other reports of rapid accumulation of fatty acid after elicitor challenge or wounding may either relate to signal transduction by PLA$_2$ or to linolenic acid release for jasmonate biosynthesis (Mueller *et al.*, 1993; Conconi *et al.*, 1996; Ryan, 2000). To unravel this complexity of functions of plant iPLA$_2$s in auxin and pathogen signalling it will be necessary to know the compartments of both the fatty acid-releasing enzymes and their substrates and the compartments of the those enzymes which perform the subsequent conversions of the fatty acids to oxylipins.

Identification of iPLA$_2$ genes in plants
For a better understanding of the functions of iPLA$_2$ in plants a molecular approach is needed. Therefore, several laboratories, including my laboratory, are engaged in characterizing iPLA$_2$ genes and proteins (Senda *et al.*, 1996; Dhondt *et al.*, 2000; Jung and Kim, 2000; Huang *et al.*, 2001). Ten iPLA$_2$ genes are annotated in the *Arabidopsis* genome which can be grouped into three subgroups by sequence comparisons (Holk, Rietz, Zahn, Quader and Scherer, unpublished). One unique iPLA$_2$ has two additional domains, one containing a leucine-rich repeat and another domain with no homology to other proteins. The next subgroup comprises five genes having six introns and the third subgroup has only one intron each. The three sub-

groups can also be defined on the basis of amino acid sequence homology. We have expressed several of the respective proteins in *E. coli* and verified that they have indeed PLA$_2$ activity towards the artificial fluorescent phosphatidylcholine that we used in the work with cell cultures (Paul *et al.*, 1998; Scherer *et al.*, 2000). When the same inhibitors (NDGA, ETYA, AACOCF$_3$) were used that inhibited the auxin stimulation of PLA$_2$ or auxin-induced elongation growth (Scherer and Arnold, 1997; Paul *et al.*, 1998) they also inhibited the purified enzyme expressed in *E. coli* (Rietz, Holk, Oppermann, Scherer, unpublished). One of these inhibitors, ETYA, also inhibited a soluble enzyme PLA$_2$ purified from bean (Jung and Kim, 2000) and other PLA$_2$ inhibitors inhibited proton extrusion and elongation growth in coleoptiles (Yi *et al.*, 1996). The finding that HELSS, an inhibitor thought to be diagnostic of the iPLA$_2$ type of enzyme (Street *et al.*, 1993), was a strong inhibitor of PLA$_2$ activation by auxin in cell cultures (Paul, 1999), had prompted us to search for iPLA$_2$s in *Arabidopsis* and to abandon the idea that a cPLA$_2$ was the enzyme relevant for plant signal transduction, contrary to the animal signal transduction paradigm. Using the BLAST program we have not been able to find any plant cPLA$_2$ up to now. Moreover, an iPLA$_2$ from *Hevea brasiliensis* was shown to be a cytosolic enzyme (Sowka *et al.*, 1998) and all iPLA$_2$-GFP hybrid proteins expressed so far in tobacco leaf cells proved to be cytosolic (Holk, Zahn, Rietz, Quader, Scherer, unpublished). It appears that secreted iPLA$_2$s found as patatins in vacuoles of *Solanum*, *Nicotiana* and *Cucumis* may be rather the exception in a small group of related plant families, because all *Arabidopsis* sequences and other plant sequences we checked (*Sorghum*, *Hevea*) did not contain secretory signal peptides. The enzymes purified from tobacco and bean were soluble but their compartment is unclear (Senda *et al.*, 1996; Jung and Kim, 2000; Dhondt *et al.*, 2000). In *Arabidopsis* the mutation STURDY was caused by activation tagging of an iPLA$_2$ gene and showed slightly enhanced apical dominance, thicker stems and increased vascularization consistent with an auxin-related function (Huang *et al.*, 2001).

Molecular biology of iPLA$_2$ in plants is at an early stage and the availability of the molecular tools provided by the *Arabidopsis* genome project should help to unravel the complexity to be expected from a ten-member gene family and to assign functions to the individual genes. Functional redundancy is to be expected as in other gene families. This functional redundancy likely is a reason why mutations in an $iPLA_2$ gene were never found in an auxin signal transduction mutant.

What could be the next step downstream? Lipid-activated protein kinase and other protein kinases in auxin signal transduction

Since the review on phospholipid signalling and on lipid-activated protein kinase in plants five years ago (Scherer, 1996), very little happened in the field of lipid-activated protein kinase. The evidence for the long-sought-after diacylglcerol-activated protein kinase C-like kinase in plants hardened (Subramanian *et al.*, 1997; Chandok and Sopory, 1998). In addition, phosphatidylserine-activated (Szczegielniak *et al.*, 2000) and phosphatidic acid-activated protein kinases were described (Farmer and Choi, 1999) both of which seem not to be pertinent for auxin because of a lack of clear evidence for a phospholipase C in auxin action (see above).

Protein kinases activated by fatty acids and lysophospholipids *in vitro* have been described, most likely members of the CDPK family (Klucis and Polya, 1987; Lucantoni and Polya, 1987; Scherer *et al.*, 1988; Martiny-Baron and Scherer, 1989; Martiny-Baron *et al.*, 1992; Schaller *et al.*, 1992; Harper *et al.*, 1993), but their substrates remained largely unknown and, therefore, their clear functions. Actually, it is also not exactly clear which fatty acids or which possible metabolites of fatty acids could initiate the next hypothetical step(s) of auxin signal transduction *in vivo*. Schweizer *et al.* (1996) showed that certain hydroxylated fatty acids regulate protein phosphorylation and could relate this to an elicitor effect of these derivatized fatty acids in plant cells. Although a 20-fold activation of kinase activity by arachidonic acid in the absence of Ca^{2+} was shown by Polya's group, arachidonic acid is not present in higher plants. There seems to be no immediate molecular or mutant approach to the field of lipid-activated protein kinases in plants.

Another hint that protein kinase could be involved in auxin action was that staurosporine repressed proton extrusion in coleoptiles (Yi *et al.*, 1996) but this inhibitor does not identify the type of protein kinase. Both phosphorylation and dephoshorylation can activate the proton pump (Desbrooes *et al.*, 1998; Kinoshita and Shimazaki, 1999) in addition to similar actions of fusicoccin, where the exact C-terminal phosphorylated amino acid has been identified (Fuglsang *et al.*, 1999). Conceivably, the plasma membrane H^+-ATPase could be a target of a lipid-activated protein kinase but the gap between hypothesis and facts has to be closed and, as explained above, channel regulation could lead to enhanced proton extrusion as well.

Two other protein kinases are known to play a role in auxin signalling, MAPK (Mizoguchi *et al.*, 1994; Suzuki and Shinshi, 1995; Kovtun *et al.*, 1998, 2000; Mockaitis and Howell, 2000) and the kinase PINOID (Christensen *et al.*, 2000). A review in this issue will give the reader more details about them (Delong *et al.*, 2002). The auxin-activated MAPK showed a rapid activation peaking at about 5–10 min which is slower than PLA_2 activation. This can be taken as an indication that it may act downstream (or independently) of PLA_2 which would be contrary to the animal $cPLA_2$ which is phosphorylated by MAPK (Nemenoff *et al.*, 1993). MAPK would be expected to act upstream of gene regulation, by analogy to other MAPKs.

PINOID was defined as a mutant morphological phenotype. Mutants exhibit no time course so that PINOID could act anywhere upstream or downstream of other known signal elements in auxin signalling. If it affects auxin transport, as the phenotype suggests, one would suppose it to act downstream. This would be an exciting possibility, a link between auxin action and auxin transport. As a conclusion, one would like to know a lot more about lipid-activated protein kinases, MAPK and PINOID and probably some more as yet unknown kinases and phosphatases in auxin signal transduction.

How to integrate auxin signal transduction with auxin physiology

Of all the postulated secondary messengers or mediators in auxin signal transduction described by various groups (G proteins, cytosolic pH, cytosolic calcium, breakdown products of lipid hydrolysis by phospholipases A_2, C or D), only cytosolic pH and activation of PLA_2 seem to be supported by several groups independently. As explained under 'Ionic second messengers', regulation of cytosolic pH requires signal transduction reactions and an involvement of PLA_2 in this has not (yet?) been shown. This and other unknown pieces of the auxin puzzle could mean that a PLA_2-dependent and a PLA_2-independent one could, or rather should, exist which could lead to downstream reactions (Figure 2). This is also assumed as the basis for the scheme on the transition of signal transduction to gene regulation (Figure 3). The action of additional

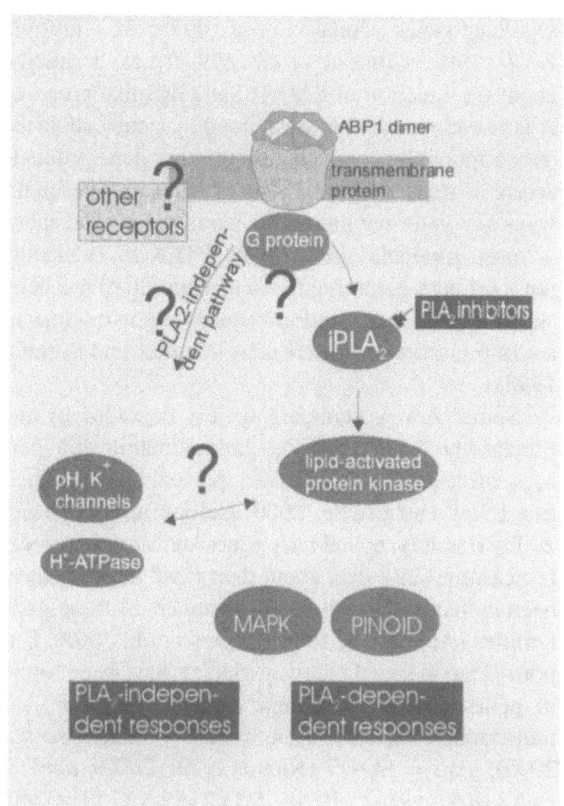

Figure 2. Signal transduction components and supposed interactions of cellular auxin signal transduction. For explanations, see text (ABP1, auxin-binding protein 1; iPLA₂, calcium-independent phospholipase A₂; MAPK, mitogen-activated protein kinase).

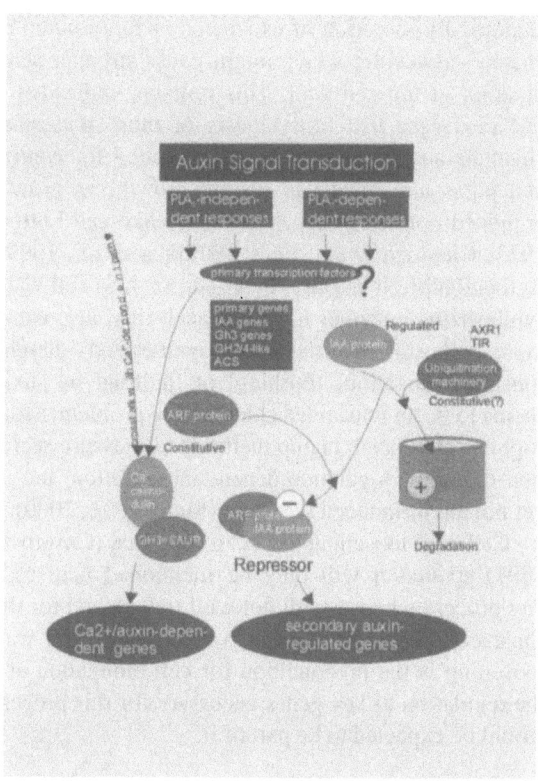

Figure 3. Transition of cellular signal transduction to gene regulation. One (or more) PLA₂-independent and a PLA₂-dependent signal transduction pathway are assumed which feed into the IAA gene-dependent pathway. Other auxin-regulated transcription factors probably exist (Hardte and Berleth, 1998) but are not included into the scheme.

auxin receptors besides the ABP1 cannot and should not be excluded so that both schemes are definitely simplifications. How actually other responses like redox balance, secretion, cell wall loosening and gene regulation are linked to upstream signal transduction reactions remains largely unknown (see below).

A recent review tried to tie up all different ends in auxin signal transduction and classic auxin physiology in an excellent manner so that this short section highlights only a few topics (Lüthen *et al.*, 1999). Redox-balance regulation is an event in auxin physiology which may be the explanation for the stimulation of an NADH oxidase by 2,4-D (Morré, 1994). This NADH oxidase is thought to regulate exchange of electrons from reduced SH groups to oxidized S-S bridges, perhaps in cell wall proteins (Morré, 1995; Morré *et al.*, 1995). It was not shown that this enzyme has any relationship to an auxin-binding protein, nor that it might be an auxin-binding protein itself (Sugaya *et al.*, 2000). Only 2,3-D and 2,4-D were tested so that auxin specificity remained unclear. This makes

it difficult to connect this work to any other known work in auxin physiology like proton extrusion, cell wall loosening, secretion, or gene activation.

Stimulation of secretion by auxin was implied by a necessity for continued cell wall biosynthesis during elongation (Ray, 1987; Phillips *et al.*, 1988; Bret-Harte *et al.*, 1991). Lipid biosynthesis as part of secretion was shown to be activated by auxin within 10 min (Hager *et al.*, 1989, 1991) and increased secretion of the H⁺-ATPase can also be viewed as secretion activation (Hager *et al.*, 1991; Frias *et al.*, 1996) as can be the up-regulation of gene activity for an inward-rectifiying potassium channel (Philippar *et al.*, 1999). Even though secretion is a rapid event and Brefeldin A, an inhibitor of secretion, inhibited auxin-induced elongation (Schindler *et al.*, 1994), it is not entirely clear whether secretion is being 'pulled' by events of cell wall loosening and cell expansion, requiring the biosynthesis of more secreted cell wall material, or whether secretion is 'pushing' cell wall loosening, for example by secretion of expansins or more pump or channel molecules which might cause stronger acidification of the cell wall. This remains undecided if one envisages that biosynthesis of more potassium channels and pump molecules is needed for continued potassium uptake to sustain growth, as growth is indeed potassium-dependent (Haschke and Lüttge, 1973; Claussen et al., 1997; Philippar et al., 1999). Activation of other genes by auxin, such as cell wall-synthesizing enzymes might, conceivably, also cause more cell wall material to be synthesized, thereby 'pushing' secretion. 'Pushing' or 'pulling' by auxin seems to be an undecided chicken-egg problem. More sophisticated patch clamp methods to measure secretion did show a calcium-dependent secretion, but as yet not auxin-induced secretion (Sutter et al., 2000).

Cell wall loosening has been reviewed (Cosgrove, 1997) so that it will only be mentioned here as a key process which is still not well understood for the shear complexity of the cell wall. Clearly, cell wall loosening is the precondition for cell elongation and the regulation of key genes necessary for this process would be expected to be part of it.

Gene regulation by auxin

This leaves gene regulation, the key word of our times. Part of it is realizing that RNA and DNA methods lead to quicker and clearer results and finding the relevant genes activated or down-regulated by auxin could give the new clues needed to integrate such knowledge into classic physiology (Sitbon and Perrot-Rechenmann, 1997). It seems that gene regulation by auxin may be organized in two waves or steps, since some early auxin-regulated genes are transcription factors and their rapid degradation is part of auxin action (Abel et al., 1994; Figure 3).

Primary genes, by definition, are regulated rapidly by auxin and comprise the Aux/IAA genes, the SAURs, GH3-like genes, aminocyclopropane-1-carboxylic acid synthase (ACS) and glutathione-S-transferase (GH2/4-like genes) (Abel and Theologis, 1996). A list of primary and secondary (slowly regulated) auxin-regulated genes can be found in Sitbon and Perrot-Rechenmann (1997). Whereas GH2/4-like genes and ACS are clearly enzymes not involved in gene regulation, the IAA proteins bind to constitutively expressed ARF proteins to form complexes, probably on the DNA, which are thought to control negatively the transcription of secondary auxin-

regulated genes (Ulmasov et al., 1997a, b; Guilfoyle et al., 1998; Ulmasov et al., 1999). Less is known about the function of SAURs and GH3-like proteins. It is not clear whether all secondary genes could be transcriptionally controlled by primary gene products acting as transcription factors, or whether other pathways for gene regulation by auxin exist. The latter is more probable since MONOPTEROS (which is not a primary auxin-regulated gene product) has been shown to be a transcription factor acting in or close to auxin-transport-competent cells (Hardtke and Berleth, 1998).

Some IAA proteins are rapidly degraded by the proteasome which requires ubiquitination and perhaps protein phosphorylation, preceding ubiquitination (Gray and Estelle, 2000; Colon-Carmona et al., 2000). Possibly, not all IAA genes are degraded by the proteasome since data about their rapid turnover have been collected for only a few members of these gene families (Abel et al., 1994; Worley et al., 2000). Important auxin signal pathway mutants have been found in genes coding for components of the ARF/IAA transcription factors NPH4 = ARF7 (Harper et al., 2000), axr2 = IAA17 (Nagpal et al., 2000), axr3 = IAA7 (Rouse et al., 1998), SHY2 = IAA3 (Tian and Reed, 1999) and in the ubiquitination machinery, axr1 (Leyser et al., 1993), TIRs and SAR (Gray and Estelle, 2000). Since these mutations alter many different auxin responses this, in summary, suggests that the combined action of ARF/IAA transcription factors and ubiquitination controls many auxin responses, probably more than are currently identified. Thus, it seems that most auxin responses that require gene activation are channelled through this machinery (Leyser, 1998). Only this can explain that this machinery is a process which can determine auxin sensitivity as the axr mutants demonstrate.

Hormone sensitivity: determined by receptor and/or transcriptional repressors as 'midway' checkpoint

In theoretical articles on hormone action and sensitivity, Trewavas developed the concept that hormone sensitivity is not determined by receptor occupancy alone but is the result of the interactions of several components (Trewavas, 1981a, b). New features of plant signal transduction in several pathways (phytochrome, cytokinin, auxin, gibberellin, pathogens) point out a two-step signal transduction machinery:

initial signal transduction and induction of early genes may use protease-sensitive proteins so that only an extended or strong signal will lead to a developmental response. This means the initial response must be integrated over a certain amount of time to become apparent. Hormone/signal sensitivity may be determined more by the amount of protease-sensitive components, or other post-translational modifications of critical proteins, than by receptor occupancy and integration over time may be more decisive for hormone action than receptor occupancy alone. The protease-sensitive step in the case of auxin is carried out with some (all?) IAA proteins, as described above. Phytochrome signal transduction involves protein phosphorylation (Yeh and Lagarias, 1998; Fankhauser *et al.*, 1999), which is probably very quick, and proteolytic degradation of phytochrome A (Casal *et al.*, 1997) and migration into the nucleus where transcriptional regulation occurs (Gil *et al.*, 2000; Hisada et al. 2000; Kim *et al.*, 2000). Proteolysis of phytochrome A is a critical step to enable the high-irradiance responses, i.e. to extend the sensitivity range greatly for far-red light and thus determine sensitivity. Although we know little about cytokinin signal transduction, the rapid up-regulation and rapid turnover of ARR genes and products is reminiscent of the general functional setup of the IAA/ARF genes and auxin (D'Agostino and Kieber, 1999; D'Agostino *et al.*, 2000). In the case of gibberellins the SPINDLY and GAI proteins control hormone sensitivity both of which are likely downstream elements in signalling (Peng *et al.*, 1997, 1999; Robertson *et al.*, 1998; Silverstone *et al.*, 1998; Thornton *et al.*, 1999) so that the final height of plants is determined by the action and amounts of downstream elements. Elicitors can also rapidly up-regulate early genes which turned out to be transcription factors that are turned over rapidly, again reminiscent of auxin and the IAA/ARF genes (Becker *et al.*, 2000; Eulgem *et al.*, 2000). Thus, the recurring theme is a downstream step which is metabolically sensitive and, thus, controls stimulus output, probably as a means of stimulus-time integration. It will be interesting to see how general such a concept for plant signal transduction might become.

Acknowledgements

This work was supported by grants from the Deutsche Forschungsgemeinschaft, the European Union (ER-BIC 15 CT98 0118), the Bundesland Niedersachsen ('Agrarbiotechnologie') and the DLR (project 50WB0010).

References

Abel, S. and Theologis, A. 1996. Early genes and auxin action. Plant Physiol. 111: 9–17.

Abel, S., Oeller, P.W. and Theologis, A. 1994. Early auxin-induced genes encode short-lived nuclear proteins. Biochemistry 33: 326–330.

André, B. and Scherer, G.F.E. 1991. Stimulation by auxin of phospholipase A in membrane vesicles from an auxin-sensitive tissue is mediated by an auxin receptor. Planta 185: 209–214.

Armstrong, F. and Blatt, M.R. 1995. Evidence for K^+ channel control in *Vicia* guard cells coupled by G-proteins to a 7TMS receptor. Plant J. 8: 187–198.

Ashikari, M., Wu, J., Yano, M., Sasaki, T. and Yoshimura, A. 1999. Rice gibberellin-insensitive dwarf mutant gene Dwarf 1 encodes the alpha-subunit of GTP-binding protein. Proc. Natl. Acad. Sci. USA 96: 10284–10289.

AylIng, S.M., Brownlee, C. and Clarkson, D.T. 1994. The cytoplasmic streaming response of tomato root hairs to auxin; observations of cytosolic calcium levels. J. Plant Physiol. 143: 184–188.

Balsinde, J. and Dennis, E.A. 1997. Function and inhibition of intracellular calcium-independent phospholipase A_2. J. Biol. Chem. 272: 16069–16072.

Barbier-Brygoo, H., Ephritikhine, G., Klämbt, D., Ghislain, M. and Guern, J. 1989. Functional evidence for an auxin receptor at the plasmalemma of tobacco mesophyll protoplasts. Proc. Natl. Acad. Sci. USA 86: 891–895.

Barbier-Brygoo, H., Ephritikhine, G., Klämbt, D., Maurel, C., Palme, K., Schell, J. and Guern, J. 1991. Perception of the auxin signal at the plasma membrane of tobacco mesophyll protoplasts. Plant J. 1: 83–93.

Bauly, J.M., Sealy, I.M., Macdonald, H., Brearley, J., Droge, S., Hillmer, S., Robinson, D.G., Venis, M.A., Blatt, M.R., Lazarus, C.M. and Napier, R.M. 2000. Overexpression of auxin-binding protein enhances the sensitivity of guard cells to auxin. Plant Physiol. 124: 1229–1238.

Becker, D. and Hedrich, R. 2002. Channeling auxin action: modulation of ion transport by indole-3-acetic acid. Plant Mol. Biol. 49: 349–356.

Becker, J., Kempf, R., Jeblick, W. and Kauss, H. 2000. Induction of competence for elicitation of defense response in cucumber hypocotyls requires proteasome activity. Plant J. 21: 311–316.

Becraft, P.W. 1998 Receptor kinases in plant development. Trends Plant Sci. 3: 384–388.

Bellamine, J., Penel, C. and Greppin, H. 1993. Proton pump and IAA sensitivity changes in spinach leaves during the flowering induction. Plant Physiol. Biochem. 31: 197–203.

Benedetti, C.E., Costa, C.L., Turcinelli, C.R. and Arruda, P. 1998. Differential expression of a novel gene in response to coronatine, methyl jasmonate, and wounding in the *coi1* mutant of *Arabidopsis*. Plant Physiol. 116: 1037–1042.

Berleth, T., Mattson, J. and Hardtke, C.S. 2000. Vascular continuity and auxin signals. Trends Plant Sci. 5: 387–394.

Blatt, M.R. and Thiel, G. 1994. K^+ channels of stomatal guard cells: bimodal control of the K^+ inward-rectifier evoked by auxin. Plant J. 5: 55–68.

Blee, E. 1998. Phytooxylipins and plant defense reactions. Prog. Lipid Res. 37: 33–72.

Bockaert, J. and Pin. J.P. 1999. Molecular tinkering of G protein-coupled receptors: an evolutionary success. EMBO J. 18: 1723–1729.

Bögre, L., Meskiene, I., Heberle-Bors, E. and Hirt, H. 2000. Stressing the role of MAP kinases in mitogenic stimulation. Plant Mol. Biol. 43: 705–718.

Bret-Harte, M.S., Baskin, T.I., and Green, P.B. 1991. Auxin-stimulated deposition and breakdown in material in the pea outer epidermal cell wall, as measured interferometrically. Planta 185: 462–470.

Casal, J.J., Sánchez, R.A., and Yanowsky, R.A. 1997. The function of phytochrome A. Plant Cell Environ. 20: 813–819.

Chandok M.R. and Sopory, S.K. 1998. ZmcPKC70, a protein kinase C-type enzyme from maize. Biochemical characterization, regulation by phorbol 12-myristate 13-acetate and its possible involvement in nitrate reductase gene expression. J. Biol. Chem. 273: 19235–19242.

Chandra, S., Heinstein, P.F. and Low, P.S. 1996. Activation of phospholipase A by plant defense elicitors. Plant Physiol. 110: 979–986.

Chen, J.-G., Ullah, H., Young, Y.C., Sussman, M.R. and Jones, A.M. 2001. ABP1 is required for organized cell elongation and division in *Arabidopsis* embryogenesis. Genes Dev. 15: 902–911.

Cho, H.-T. and Hong, Y.-N. 1995. Effect of IAA on synthesis and activity of the plasma membrane H^+-ATPase of sunflower hypocotyls, in relation to IAA-induced cell elongation and H^+ excretion. J. Plant Physiol. 145: 717–725.

Christensen, S.K., Dagenais, N., Chory, J. and Weigel, D. 2000. Regulation of auxin response by the protein kinase PINOID. Cell 100: 469–478.

Claussen, M., Lüthen, H., Blatt, M. and Böttger, M. 1997. Auxin-induced growth and its linkage to potassium channels. Planta 201: 227–234.

Colon-Carmona, A., Chen, D.L., Yeh, K.C. and Abel, S. 2000. Aux/IAA proteins are phosphorylated by phytochrome *in vitro*. Plant Physiol. 124: 1728–1738.

Conconi, A., Miquel, M., Browse, J.A. and Ryan, C.A. 1996. Intracellular levels of free linolenic acid and linoleic acids increase in tomato leaves in response to wounding. Plant Physiol. 111: 797–803.

Cosgrove, D.J. 1997. Assembly and enlargement of the primary cell wall in plants. Annu. Rev. Cell. Dev. Biol. 13: 171–201.

Creelman, R.A. and Mullet, J.E. 1997. Biosynthesis and action of jasmonates in plants. Annu. Rev. Plant Physiol. Plant Mol. Biol. 48: 355–381.

D'Agostino, I.B. and Kieber, J.J. 1999. Molecular mechanisms of cytokinin action. Curr. Opin. Plant Biol. 2: 359–364.

D'Agostino, I.B., Deruere, J. and Kieber, J.J. 2000. Characterization of the response of the *Arabidopsis* response regulator gene family to cytokinin. Plant Physiol. 124: 1706–1717.

den Boer, B.G. and Murray, J.A. 2000. Triggering the cell cycle in plants. Trends Cell Biol. 10: 245–250.

den Hartog, M., Musgrave, A. and Munnik, T. 2000. Nod factor-induced phosphatidic acid and diacylglycerol pyrophosphate formation: a role for phospholipase C and D in root hair deformation. Plant J. 25: 55–65.

DeLong, A., Mockaitis, K. and Christensen, S. 2001. Protein phosphorylation in the delivery of and response to auxin signals. Plant Mol. Biol. 49: 285–303.

De Nisi, P., Dell'Orto, M., Pirovano, L., and Zocchi, G. 1999. Calcium-dependent phosphorylation regulates the plasma-membrane H^+-ATPase activity of maize (*Zea mays* L.) roots. Planta 209: 187–194.

Dennis, E. 1994. Diversity of group types, regulation, and function of phospholipase A_2. J. Biol. Chem. 269: 13057–13060.

Dessen, A. 2000. Structure and mechanism of human cytosolic phospholipase A_2. Biochim. Biophys. Acta 1488: 40–47.

Desbrooses, G., Stelling, J. and Renaudin, J.P. 1998. Dephosphorylation activates the purified plasma membrane H^+-ATPase: possible function of phosphothreonine residues in a mechanism not involving the regulatory C-terminal domain of the enzyme. Eur. J. Biochem. 251: 496–503.

Devoto, A., Piffanelli, P., Nilsson, I., Wallin, E., Panstruga, R., von Heijne, G. and Schulze-Lefert, P. 1999. Topology, subcellular localization, and sequence diversity of the Mlo family in plants. J. Biol. Chem. 274: 34993–35004.

Dhondt, S., Geoffroy, P., Stelmach, B.A., Legrand, M. and Heitz, T. 2000. Soluble phospholipase A_2 activity is induced before oxylipin accumulation in tobacco mosaic virus-infected tobacco leaves and is contributed by patatin-like enzymes. Plant J. 23: 431–440.

Doss, R.P., Oliver, J.E., Proebsting, W.M., Potter, S.W., Kuy, S., Clement, S.L., Williamson, R.T., Carney, J.R. and DeVilbiss, E.D. 2000. Bruchins: insect-derived plant regulators that stimulate neoplasm formation. Proc. Natl. Acad. Sci. USA 97: 6128–6223.

Ettlinger, C. and Lehle, L. 1988. Auxin induces rapid changes in phosphatidylinositol metabolites. Nature 331: 176–178.

Eulgem, T., Rushton, P.J., Robatzek, S. and Somssich, I. 2000. The WRKY superfamily of plant transcription factors. Trends Plant Sci. 5: 199–206.

Fankhauser, C., Yeh, K.C., Lagarias, J.C., Zhang, H., Elich, T.D. and Chory, J. 1999. PKS1, a substrate phosphorylated by phytochrome that modulates light signaling in *Arabidopsis*. Science 284: 1539–1541.

Farmer, E.F. 1992. Fatty acid signalling in plants and their associated microorganisms. Plant Mol. Biol. 26: 1423–1437.

Farmer, P.K. and Choi, J.H. 1999. Calcium and phospholipid activation of a recombinant calcium-dependent protein kinase (DcCPK1) from carrot (*Daucus carota* L.). Biochim. Biophys. Acta 1434: 6–17.

Farmer, E.E., Weber, H. and Vollenweider, S. 1998. Fatty acid signaling in *Arabidopsis*. Planta 206: 167–174.

Felle, H. 1988. Auxin causes oscillations of cytosolic free calcium and pH in *Zea mays* coleoptiles. Planta 174: 495–499.

François, J.M., Bervilé, A. and Rossignol, M. 1992. Development and line variations of *Petunia* plasma membrane H^+-ATPase sensitivity to auxin. Plant Sci. 87: 19–27.

Frias, I., Caldeira, M.T., Perez-Castineira, J.R., Navarro-Avino, J.P., Culianez-Mazia, F.A., Kuppinger, O., Stransky, H., Montserrat, P., Hager, A. and Serrano, R. 1996. A major isoform of the maize plasma membrane ATPase: characterization and induction by auxin in coleoptiles. Plant Cell 8: 1533–1544.

Fu, Y., Wu, G. and Yang, Z. 2001. Rop gtpase-dependent dynamics of tip-localized f-actin controls tip growth in pollen tubes. J. Cell Biol. 152: 1019–1032.

Fuglsang, A.T., Visconti, S., Drumm, K., Jahn, T., Stensballe, A., Mattei, B., Jensen, O.N., Aducci, P. and Palmgren M.G. 1999. Binding of 14-3-3 protein to the plasma membrane H^+-ATPase AHA2 involves the threeC-terminal residues Tyr^{946}-Thr-Val and requires phosphorylation of Thr^{947}. J. Biol. Chem. 274: 36774–36780.

Gehring, C.A., Irving, H.R. and Parish, R.W. 1990. Effects of auxin and abscisic acid on cytosolic calcium and pH in plant cells. Proc. Natl. Acad. Sci. USA 87: 9645–9649.

Gil, P., Kircher, S., Adam, E., Bury, E., Kozma-Bognar, L., Schäfer, E. and Nagy, F. 2000. Photocontrol of subcellular partitioning of phytochrome-B:GFP fusion protein in tobacco seedlings. Plant J. 22: 135–145.

Grabov, A. and Blatt, M.R. 1997. Parallel control of the inward-rectifier K$^+$ channel by cytosolic-free Ca^{2+} and pH in *Vicia* guard cells. Planta 201: 84–95.

Gray, W.M. and Estelle, I. 2000. Function of the ubiquitin-proteasome pathway in auxin response. Trends Biochem. Sci. 25: 133–138.

Guilfoyle, T., Hagen, G., Ulmasov, T. and Murfett, J. 1998. How does auxin turn on genes? Plant Physiol. 118: 341–347.

Hager, A., Debus, G., Edel, H.-G., Stransky, H. and Serrano, R. 1991. Auxin induces exocytosis and the rapid synthesis of a high-turnover pool of plasma-membrane H$^+$-ATPase. Planta 185: 527–537.

Hager, A., Brich, M., Debus, G., Edel, H.G. and Priester, G. 1989. Membrane metabolism and growth. Phospholipases, protein kinases and exocytotic processes in coleoptiles in *Zea mays*. In: Plant Water Relations and Growth under Stress, Yamada Science Foundation, Osaka, Tokyo, pp. 275–282.

Haschke, H.P. and Lüttge, U. 1973. β–Indolylessigsäure (IES)-abhängiger K$^+$-H$^+$-Austauschmechanismus und Streckungswachstum bei *Avena*-Koleoptilen. Z. Naturforsch. 28: 555–558.

Hardtke, C.S. and Berleth, T. 1998. The *Arabidopsis* gene MONOPTEROS encodes a transcription factor mediating embryo axis formation and vascular development. EMBO J. 17: 1405–1411.

Harper, J.F., Binder, B.M. and Sussman, M.R. 1993. Calcium and lipid regulation of an *Arabidopsis* protein kinase expressed in *Escherichia coli*. Biochemistry 32: 3282–3290.

Harper, R.M., Stowe-Evans, E.L., Luesse, D.R., Muto, H., Tatematsu, K., Watahiki, M.K., Yamamoto, K. and Liscum, E. 2000. The NPH4 locus encodes the auxin response factor ARF7, a conditional regulator of differential growth in aerial *Arabidopsis* tissue. Plant Cell 12: 757–770.

Hisada, A., Hanzawa, H., Weller, J.L., Nagatani, A., Reid, J.B. and Furuya, M. 2000. Light-induced nuclear translocation of endogenous pea phytochrome A visualized by immunocytochemical procedures. Plant Cell 12: 1063–1078.

Huang, S., Cerny, R.E., Bhat, D.S. and Brown, S.M. 2001. Cloning of an *Arabidopsis* patatin-like gene, STURDY, by activation tagging. Plant Physiol. 125: 573–584.

Jones, A.M. 1980. Location of transported auxin in etiolated maize shoots using 5-azidoindol-3-acetic acid. Plant Physiol. 93: 1154–1161.

Jones, A.M., Im, K.H., Savka, M.A., Wu, M.J., DeWitt, N.G., Shillito, R. and Binns, A.N. 1998. Auxin-dependent cell expansion mediated by overexpressed auxin-binding protein 1. Science 282: 1114–1117.

Jung, K.M. and Kim, D.K. 2000. Purification and characterization of a membrane-associated 48-kilodalton phospholipase A$_2$ in leaves of broad bean. Plant Physiol. 123: 1057–1067.

Kaydamov, C., Tewes, A,, Adler, K. and Manteuffel, R. 2000. Molecular characterization of cDNAs encoding G protein alpha and beta subunits and study of their temporal and spatial expression patterns in *Nicotiana plumbaginifolia* Viv. Biochim. Biophys. Acta 1491: 143–160.

Khan, W.A., Blobe, C., Halpern, A., Taylor, W., Wetsel, W.C., Burns, D., Loomis, C. and Hannun, Y.A. 1993. Selective regulation by protein kinase C isoenzymes by oleic acid in human platelets. J. Biol. Chem. 268: 5063–5068.

Kim, L., Kircher, S., Toth, R., Adam, E., Schäfer, E. and Nagy, F. 2000. Light-induced nuclear import of phytochrome-A:GFP fusion proteins is differentially regulated in transgenic tobacco and *Arabidopsis*. Plant J. 22: 125–123.

Kinoshita, T., Nishimura, M. and Shimazaki, K.I. 1995. Cytosolic concentration of Ca^{2+} regulates the plasma membrane H$^+$-ATPase in guard cells of fava bean. Plant Cell 7: 1333–1342.

Kinoshita, T. and Shimazaki, K. 1999. Blue light activates plasma membrane H$^+$-ATPase by phosphorylation of the C-terminus in stomatal guard cells. EMBO J. 18: 5548–5558.

Klämbt, D. 1990. A view about the function of auxin-binding proteins at the plasma membranes. Plant Mol. Biol. 14: 1045–1050.

Klucis, E. and Polya, G.M. 1987. Calcium-independent activation of two plant leaf calcium-regulated protein kinases by fatty acids. Biochem. Biophys. Res. Commun. 147: 1041–1047.

Kovtun, Y., Chiu, W.L., Zeng, W. and Sheen, J. 1998. Suppression of auxin signal transduction by a MAPK cascade in higher plants. Nature 395: 716–720.

Kovtun, Y., Chiu, W.L., Tena, G. and Sheen, J. 2000. Functional analysis of oxidative stress-activated mitogen-activated protein kinase cascade in plants. Proc. Natl. Acad. Sci. USA 97: 2940–2945.

Lease, K., Ingham, E. and Walker, J.C. 1998. Challenges in understanding RLK function. Curr. Opin. Plant Biol. 5: 388–392.

Leblanc, N., David, K., Grosclaude, J., Pradier, J.M., Barbier-Brygoo, H., Labiau, S. and Perrot-Rechenmann, C. 1999. A novel immunological approach establishes that the auxin-binding protein, Nt-abp1, is an element involved in auxin signaling at the plasma membrane. J. Biol. Chem. 274: 28314–28320.

Lee, Y.R. and Assmann, S.M. 1999. *Arabidopsis thaliana* 'extra-large GTP-binding protein' (AtXLG1): a new class of G-protein. Plant Mol. Biol. 40: 55–64.

Lee, S.-S., Kawakita, K., Tsuge, T. and Doke, N. 1992. Stimulation of phospholipase A$_2$ in strawberry cells treated with AF-toxin 1 produced by *Alternaria alternata* strawberry phenotype. Physiol. Mol. Plant Path. 41: 283–294.

Lee, Y., Lee, H.J., Crain, R.C., Lee, A. and Korn, S.J. 1994. Polyunsaturated fatty acids modulate stomatal aperture and two distinct K$^+$ channel currents in guard cells. Cell. Sign. 6: 181–186.

Lee, S., Suh, S., Kim, S., Crain, R.C., Kwak, J.M., Nam, H.-G. and Lee, Y. 1997. Systemic elevation of phosphatidic acid and lysophospholipid levels in wounded plants. Plant J. 12: 547–556.

Leyser, H.M., Lincoln, C.A., Timpte, C., Lammer, D., Turner, J. and Estelle, M. 1993. *Arabidopsis* auxin-resistance gene AXR1 encodes a protein related to ubiquitin-activating enzyme E1. Nature 364: 161–164.

Leyser, O. 1998. Auxin signalling: protein stability as a versatile control target. Curr. Biol. 8: 305–307.

Li, W. and Assmann, S.M. 1993. Characterization of a G-protein-regulated outward K$^+$ current in mesophyll cells of *Vicia faba* L. Proc. Natl. Acad. Sci. USA 90: 262–266.

Lin, C. 2000. Plant blue light receptors. Trends Plant Sci. 5: 337–342.

Lucantoni, A. and Polya, G.M. 1987. Activation of wheat embryo Ca^{2+}-regulated protein kinase by unsaturated fatty acids in the presence and absence of calcium. FEBS Lett. 221: 33–36.

Lüthen, H., Claussen, M. and Böttger, M. 1999. Growth: progress in auxin research. Prog. Bot. 60: 315–340.

Marshall, C.J. 1996. Ras effectors. Curr. Opin. Cell Biol. 8: 197–204.

Martiny-Baron, G., Hecker, D., Manolson, M.F., Poole, R.J. and Scherer, G.F.E. 1992. Proton transport and phosphorylation of tonoplast polypeptides from zucchini are stimulated by the phospholipid platelet-activating factor. Plant Physiol. 99: 1635–1641.

370

Martiny-Baron, G. and Scherer, G.F.E. 1989. Phospholipid-stimulated protein kinase in plants. J. Biol. Chem. 264: 18052–18059.

Mason, M.G. and Botellan, J.R. 2000. Completing the heterotrimer: isolation and characterization of an *Arabidopsis thaliana* G protein γ-subunit cDNA Proc. Natl. Acad. Sci. USA 97: 14784–14788.

McDonald, H. 1997. Auxin perception and signal transduction. Physiol. Plant. 100: 423–430.

Millner, P.A., Groarke, D.A. and White, I.R. 1996. Synthetic peptides as probes of plant cell signalling: G-proteins and the auxin signalling pathway. Plant Growth Regul. 18: 143–147.

Mizoguchi, T., Gotoh, Y., Nishida, E., Yamaguchi-Shinozaki, K., Hayashida, N., Iwasaki, T., Kamada, H., and Shinozaki, K. 1994. Characterization of two cDNAs that encode MAP kinase homologues in *Arabidopsis thaliana* and analysis of the possible role of auxin in activating such kinase activities in cultured cells. Plant J. 5: 111–122.

Mockaitis, K. and Howell, S.H. 2000. Auxin induces mitogenic activated protein kinase (MAPK) activation in roots of *Arabidopsis* seedlings. Plant J. 24: 785–796.

Morré, D.J. 1994. Hormone- and growth factor-stimulated NADH oxidase. J. Bioenerg. Biomembr. 26: 421–433.

Morré, D.J. 1995. The role of NADH oxidase in growth and physical membrane displacement. Protoplasma 184: 14–21.

Morré, D.J., de Cabo, R., Jacobs, E. and Morré, D.M. 1995. Auxin-modulated protein disulfide-thiol-interchange activity from soybean plasma membranes. Plant Physiol. 109: 573–578.

Mueller, M.J., Brodschelm, W., Spannagl, E.and Zenk, M.H. 1993. Signalling in the elicitation process is mediated through the octadecenoid pathway leading to jasmonic acid. Proc. Natl. Acad. Sci. USA 90: 7490–7494.

Munnik, T., Irvine, R.F. and Musgrave, A. 1998. Phospholipid signalling in plants. Biochim. Biophys. Acta 1389: 222–272.

Nagpal, P., Walker, L.M., Young, J.C., Sonawala, A., Timpte, C., Estelle, M. and Reed, J.W. 2000. AXR2 encodes a member of the Aux/IAA protein family. Plant Physiol. 123: 563–574.

Napier, R.M. 1995. Towards an understanding of ABP1. J. Exp. Bot. 46: 1787–1795.

Napier, R.M. and Venis, M.A. 1995. Auxin action and auxin-binding proteins. New Phytol. 129: 167–201.

Narvaez-Vasquez, J., Florin-Christensen, J. and Ryan, C.A. 1999. Positional specificity of a phospholipase A activity induced by wounding, systemin, and oligosaccharide elicitors in tomato leaves. Plant Cell 11: 2249–2260.

Nemenoff, R.A., Winitz, S., Quian, N.-X., van Putten, V., Johnson, G.L. and Heasley, L.E. 1993. Phosphorylation and activation of a high molecular weight form of phospholipase A₂ by p42 microtubule-associated protein 2 kinase and protein kinase C. J. Biol. Chem. 268: 1960–1964.

Nickel, R., Schütte, M., Hecker, D. and Scherer, G.F.E. 1991. The phospholipid platelet-activating factor stimulates proton extrusion in cultured soybean cells and protein phosphorylation and ATPase activity in plasma membranes. J. Plant Physiol. 139: 205–211.

Oishi, K., Zheng, B. and Kuo, J.F. 1990. Inhibition of Na,K-ATPase and sodium pump by protein kinase C regulators sphingosine, lysophosphatidylcholine, and oleic acid. J. Biol. Chem. 265: 70–75.

Palme, K. and Gälweiler, L. 1999. PIN-pointing the molecular basis of auxin transport. Curr. Opin. Plant Biol. 5: 375–381.

Palmgren, M.G. and Sommarin, M. 1989. Lysophosphatidylcholine stimulates ATP-dependent proton accumulation in isolated oat root plasma membrane vesicles. Plant Physiol. 90: 1009–1014.

Paul, R. 1999. Untersuchungen zur Funktion von Phospholipase A₂ und Phospholipase C im Signaltransduktionsweg von Auxin und Pilzelicitor in Petersiliezellkulturen. Doctoral thesis, University of Hannover, Germany.

Paul, R., Holk, A. and Scherer, G.F.E. 1998. Fatty acids and lysophospholipids as potential second messengers in auxin action. Rapid activation of phospholipase A₂ activity by auxin in suspension-cultured parsley and soybean cells. Plant J. 16: 601–611.

Peltier, J.B. and Rossignol, M. 1996. Auxin-induced differential sensitivity of the H⁺-ATPase in plasma membrane subfractions from tobacco cells. Biochem. Biophys. Res. Commun. 219: 492–496.

Peng, J. Carol, P., Richards, D.E., King, K.E., Cowling, R.J., Murphy, G.P. and Harberd, N.P. 1997. The *Arabidopsis GAI* gene defines a signalling pathway that negatively regulates gibberellin responses. Genes Dev. 11: 3194–3205.

Peng, J., Richards, D.E., Hartley, N.M., Murphy, G.P., Devos, K.M., Flintham, J.E., Beales, J., Fish, L.J., Worland, A.J., Pelica, F., Sudhakar, D., Christou, P., Snape, J.W., Gale, M.D., and Harberd, N.P. 1999. 'Green Revolution' genes encode mutant gibberellin response modulators. Nature 400: 265–261.

Philippar, K., Fuchs, I., Lüthen, H., Hoth, S., Bauer, C.S., Haga, K., Thiel, G., Ljung, K., Sandberg, G., Böttger, M., Becker, D. and Hedrich, R. 1999. Auxin-induced K⁺ channel expression represents an essential step in coleoptile growth and gravitropism. Proc. Natl. Acad. Sci. USA 96: 12186–12191.

Phillips, G.D., Preshaw, C. and Steer, M.W. 1988. Dictyosome vesicle production and plasma membrane turnover in auxin-stimulated outer epidermal cells of coleoptile segments from *Avena sativa* (L.). Protoplasma 145: 59–65.

Plakidou-Dymock, S., Dymock, D. and Hooley, R. 1998. A higher plant seven-transmembrane receptor that influences sensitivity to cytokinins. Curr. Biol. 12: 315–324.

Quail, P.H., Boylan, M.T., Parks, B.M., Short, T.W., Xu, Y. and Wagner, D. 1995. Phytochromes: photosensory perception and signal transduction. Science 268: 675–680.

Ray, P.M. 1987. Involvement of macromolecule biosynthesis in auxin and fusicoccin enhancement of β-glucan synthase activity in pea. Plant Physiol. 85: 523–528.

Robertson, M., Swain, S.M., Chandler, P.M., Olszewski, N.E. 1998. Identification of a negative regulator of gibberellin action, *HvSPY*, in barley. Plant Cell 10: 995–1007.

Roos, W. 2000. Ion mapping in plant cells–methods and applications in signal transduction research. Planta 210: 347–370.

Roos, W., Dordschbal, B., Steighardt, J., Hieke, M., Weiss, D. and Saalbach, G. 1999. A redox-dependent, G-protein-coupled phospholipase A of the plasma membrane is involved in the elicitation of alkaloid biosynthesis in *Eschscholtzia californica*. Biochim. Biophys. Acta 1448: 390–402.

Rouse, D., Mackay, P., Stirnberg, P., Estelle, M. and Leyser, O. 1998. Changes in auxin response from mutations in an AUX/IAA gene. Science 279: 1371–1373.

Roy, S., Pouénat, M.-L., Caumont, C., Cariven, C., Prévost, M.-C. and Esquerré-Tugayé, M.-T. 1995. Phospholipase activity and phospholipid patterns in tobacco cells treated with fungal elicitor. Plant Sci. 107: 17–25.

Rück, A., Palme, K., Venis, M.A., Napier, R.A. and Felle, H.H. 1993. Patch-clamp analysis establishes a role for an auxin-binding protein in the auxin stimulation of plasma membrane currents in *Zea mays* protoplasts. Plant J. 4: 41–46.

Ryan, C.A. 2000. The systemin signaling pathway: differential activation of plant defensive genes. Biochim. Biophys. Acta 1477: 112–121.

Sabatini, S., Beis, D., Wolkenfelt, H., Murfett, J., Guilfoyle, T., Malamy, J., Benfey, P., Leyser, O., Bechtold, N., Weisbeek, P. and Scheres, B. 1999. An auxin-dependent distal organizer of pattern and polarity in the *Arabidopsis* root. Cell 99: 463–472.

Santoni, V., Vansuyt, G. and Rossignol. M. 1991. Indoleacetic acid pretreatment of tobacco plants *in vivo* increases the *in vitro* sensitivity to auxin of the plasma membrane H^+-ATPase from leaves and modifies the polypeptide composition of the membrane. FEBS Lett. 326: 17–20.

Satoshi, A., Shingo, M., Keisuke, K., Misako, H. and Takashi, S. 1999. Involvement of group VI Ca^{2+}-independent phospholipase A_2 in protein kinase C-dependent arachidonic acid liberation in zymosan-stimulated macrophage-like P388D1 cells. J. Biol. Chem. 274: 19906–19912.

Schaller, G.E., Harmon, A. and Sussman, M.A. 1992. Characterization of a calcium- and lipid-dependent protein kinase associated with the plasma membrane of oat. Biochemistry 31: 1721–1727.

Scherer, G.F.E. 1981. Auxin-stimulated ATPase in membrane fractions from pumpkin hypocotyls (*Cucurbita maxima* L.). Planta 151: 434–438.

Scherer, G.F.E. 1990. Phospholipid-activated protein kinase in plants: coupled to phospholipase A_2? In: R. Ranjeva and A.M. Boudet (Eds.) Signal Perception and Transduction in Higher Plants, NATO ASI series H vol. 47, Springer-Verlag, Berlin/Heidelberg/New York, pp. 69–82.

Scherer, G.F.E. 1992. Stimulation of growth and phospholipase A_2 by the peptides mastoparan and melittin and by the auxin 2,4-dichlorophenoxyacetic acid. Plant Growth Regul. 11: 153–157.

Scherer, G.F.E. 1995. Activation of phospholipase A by auxin and mastoparan in hypocotyl segments from zucchini and sunflower. J. Plant Physiol. 145: 483–490.

Scherer, G.F.E. 1996. Phospholipid signalling and lipid-derived second messengers in plants. Plant Growth Regul. 18: 125–133.

Scherer, G.F.E. and André, B. 1989. A rapid response to a plant hormone: auxin stimulates phospholipase A_2 *in vivo* and *in vitro*. Biochem. Biophys. Res. Commun. 163: 111–117.

Scherer, G.F.E. and André, B. 1993. Stimulation of phospholipase A_2 by auxin in microsomes from suspension-cultured soybean cells is receptor-mediated and influenced by nucleotides. Planta 191: 515–523.

Scherer, G.F.E. and Arnold, B. 1997. Auxin-induced growth is inhibited by phospholipase A_2 inhibitors. Implications for auxin-induced signal transduction. Planta 202: 462–469.

Scherer, G.F.E and Nickel, R. 1988. The animal ether phospholipid platelet-activating factor stimulates acidification of the incubation medium by cultured soybean cells. Plant Cell Rep. 7: 575–578.

Scherer, G.F.E. and Stoffel, B. 1987. A plant lipid and the platelet-activating factor stimulate ATP-dependent H^+ transport in isolated plant membrane vesicles. Planta 172: 127–130.

Scherer, G.F.E., Martiny-Baron, G. and Stoffel, B. 1988. A new set of regulatory molecules in plants: a plant phospholipid similar to platelet-activating factor stimulates protein kinase and proton-translocating ATPase in membrane vesicles. Planta 175: 241-253

Scherer, G.F.E., André, B. and Martiny-Baron, G. 1990. Hormone-activated phospholipase A_2 and lysophospholipid-activated protein kinase: a new signal transduction chain and a new second messenger system in plants? Curr. Top. Plant Biochem. Physiol. 9: 190–218.

Scherer, G.F.E., Paul, R.U. and Holk, A. 2000. Phospholipase A_2 in auxin and elicitor signal transduction in cultured parsley cells (*Petrosilenium crispum* L.). Plant Growth Regul. 32: 123–128.

Schindler, T., Bergfeld, R., Hohl, M. and Schopfer, P. 1994. Inhibition of Golgi-apparatus function by brefeldin A in maize coleoptiles and its consequences on auxin-mediated growth, cell-wall extensibility and secretion of cell-wall proteins. Planta 192: 404–413.

Schweizer, P., Felix, G., Buchala, A., Müller, C. and Métraux, J.-P. 1996. Perception of free cutin monomers by plant cells. Plant J. 10: 331–341.

Senda, K., Yoshioka, H., Doke, N. and Kawakita, K. 1996. A cytosolic phospholipase A_2 from potato tissues appears to be patatin. Plant Cell Physiol. 37: 347–353.

Shimmen, T. and Yokota, E. 1994. Physiological and biochemical aspects of cytoplasmic streaming. Int. Rev. Cytol. 155: 97–139.

Shishova, M. and Lindberg, S. 1997. Auxin-induced cytosol acidification in wheat leaf protoplasts depends on expternal concentration of Ca^{2+}. J. Plant Physiol. 155: 190–196.

Silverstone, A.L., Ciampaglio, C.N. and Sun, T. 1998. The new RGA locus encodes a negative regulator of gibberellin response in *Arabidopsis thaliana*. Genetics 146: 1087–1099.

Sitbon, F. and Perrot-Rechenmann, C. 1997. Expression of auxin-regulated genes. Plant Physiol. 100: 443–455.

Six, D.A. and Dennis, E.A. 2000. The expanding superfamily of phospholipase A_2 enzymes: classification and characterization. Biochim. Biophys. Acta 1488: 1–19.

Snijder, H.J., Ubarretxena-Belandia, I., Blaauw, M., Kalk, K.H., Verheij, H.M., Egmond, M.R., Dekker, N. and Dijkstra, B.W. 1999. Structural evidence for dimerization-regulated activation of an integral membrane phospholipase. Nature 401: 717–721.

Sowka, S., Wagner, S., Krebitz, M., Arija-Mad-Arif, S., Yusof, F., Kinaciyan, T., Brehler, R., Scheiner, O. and Breiteneder, H. 1998. cDNA cloning of the 43-kDa latex allergen Hev b 7 with sequence similarity to patatins and its expression in the yeast *Pichia pastoris*. Eur. J. Biochem. 255: 213–219.

Stafforini, D.M., McIntyre, T.M., Zimmermann, G.A. and Prescott, S.M. 1997. Platelet-activating factor acetylhydrolases. J. Biol. Chem. 272: 17895–17898.

Ståhl, U., Ek, B. and Stymme, S. 1998. Purification and characterization of low-molecular-weight phospholipase A_2 from developing seeds of elm. Plant Physiol. 117: 197–205.

Steinmann, T., Geldner, N., Grebe, M., Mangold, S., Jackson, C.L., Paris, S., Gälweiler, L., Palme, K. and Jürgens, G. 1999. Coordinated polar localization of auxin efflux carrier PIN1 by GNOM ARF GEF. Science 286: 316–318.

Stewart, P.A. 1983. Modern quantitative acid-base chemistry. Can. J. Physiol. Pharmacol. 61: 1444–1461.

Street, I.P., Lin, H.K., Lalibert, F., Ghomashchi, F., Wang, Z.Y., Perrier, H., Tremblay, N.M., Huang, Z., Weech, P.K. and Gelb, M.H. 1993. Slow-binding and tight-binding inhibitors of the 85-kDa human phospholipase A_2. Biochemistry 32: 5935–5940.

Subramanian, R., Després, C. and Brisson, N. 1997. A functional homolog of mammalian protein kinase C participates in the elicitor-induced defense response in potato. Plant Cell 9: 653–664.

Sugaya, S., Ohmiya, A., Kikuchi, M. and Hayashi, T. 2000. Isolation and characterization of a 60 kDa 2,4-D-binding protein from the shoot apices of peach trees (*Prunus persica* L.); it is a homologue of protein disulfide isomerase. Plant Cell Physiol. 41: 503–508.

Sutter, J.-U., Homann, U. and Thiel, G. 2000. Ca^{2+}-stimulated exocytosis in maize coleoptiles. Plant Cell 12: 1127–1136.

Suzuki, K. and Shinshi, H. 1995. Transient activation and tyrosine phosphorylation of a protein kinase in tobacco cells treated with a fungal elicitor. Plant Cell 7: 639–647.

Sweeney, B.M. and Thimann, K.V. 1942. The effects of auxin on protoplasmic streaming. III. J. Gen. Physiol. 25: 841–851.

Sweeney, B.M. 1944. The effect of auxin on protoplasmic streaming in root hairs of *Avena*. Am. J. Bot. 31: 78–80.

Szczegielniak, J., Liwosz, A., Jurkowski, I., Loog, M., Dobrowolska, G., Ek, P., Harmon, A.C. and Muszynska, G. 2000. Calcium-dependent protein kinase from maize seedlings activated by phospholipids. Eur. J. Biochem. 267: 3818–3827.

Tena, G. and Renaudin, J.P. 1998. Cytosolic acidification but not auxin at physiological concentration is an activator of MAP kinases in tobacco cells. Plant J. 16: 173–182.

Thiel, G. and Weise, R. 1999. Auxin augments conductance of K^+ inward rectifier in maize coleoptile protoplasts. Planta 208: 38–45.

Thiel, G., Blatt, M.R., Fricker, M.D., White, I.R. and Millner, P. 1993. Modulation of K^+ channels in *Vicia* stomatal guard cells by peptide homologs to the auxin-binding protein C terminus. Proc. Natl. Acad. Sci. USA 90: 11493–11497.

Thornton, T.M., Swain, S.M. and Olszewski, N.E. 1999. Gibberellin signal transduction presents ... the SPY who O-GlcNAc'd me. Trends Plant Sci. 4: 424–428.

Tian, Q. and Reed, J.W. 1999. Control of auxin-regulated root development by the *Arabidopsis thaliana* SHY2/IAA3 gene. Development 126: 711–721.

Tian, H., Klämbt, D. and Jones, A.M. 1995. Auxin binding protein 1 does not bind auxin within the endoplasmic reticulum despite its being the predominant subcellular location of this hormone receptor. J. Biol. Chem. 270: 26962–26969.

Trewavas, A.J. 1981a. Growth substance sensitivity: the limiting factor in plant development. Physiol. Plant. 55: 60–70.

Trewavas, A.J. 1981b. How do plant growth substances work? Plant Cell Envir. 4: 203–228.

Ueguchi-Tanaka, M., Fujisawa, Y., Kobayashi, M., Ashikari, M., Iwasaki, Y., Kitano, H. and Matsuoka, M. 2000. Rice dwarf mutant d1, which is defective in the alpha subunit of the heterotrimeric G protein, affects gibberellin signal transduction. Proc. Natl. Acad. Sci. USA 97: 11638–11643.

Ullah, H., Chen, J.-G., Young, J.C., Im, K.-H., Sussman, M.R., Jones, A.M. 2001. Modulation of cell proliferation by G-protein alpha subunits in *Arabidopsis*. Science 292: 2066–2069.

Ulmasov, T., Murfett, J., Hagen, G. and Guilfoyle, T.J. 1997a. Aux/IAA proteins repress expression of reporter genes containing natural and highly active synthetic auxin response elements. Plant Cell 9: 1963–1971.

Ulmasov, T., Hagen, G., Guilfoyle, T.J. 1997b. ARF1, a transcription factor that binds auxin response elements. Science 276: 1865–1866.

Ulmasov, T., Hagen, G. and Guilfoyle, T.J. 1999. Activation and repression of transcription by auxin-response factors. Proc. Natl. Acad. Sci. USA 96: 5844–5849.

Urao, T., Yamaguchi-Shinozaki, K. and Shinozaki, K. 2000. Two-component systems in plant signal transduction. Trends Plant Sci. 5: 67–74.

van der Luit, A.H., Piatti, T., van Doorn, A., Musgrave, A., Felix, G., Boller, T. and Munnik, T. 2000. Elicitation of suspension-cultured tomato cells triggers the formation of phosphatidic acid and diacylglycerol pyrophosphate. Plant Physiol. 123: 1507–1524.

Venis, M.A. and Napier, R.M. 1995. Auxin receptors and auxin binding proteins. Crit. Rev. Plant Sci. 14: 27–47.

Venis, M.A., Napier, R.M., Barbier-Brygoo, H., Maurel, C., Perrot-Rechenmann, C. and Guern, J. 1992. Antibodies to a peptide from the maize auxin-binding protein have auxin agonist activity. Proc. Natl. Acad. Sci. USA 89: 7208–7212.

Wang, X.-Q., Ullah, H., Jones, A.M. and Assmann, S.M. 2001. G-protein regulation of ion channels and abscisic acid signaling in *Arabidopsis* guard cells. Science 292: 2070–2072.

Weiss, C.A., Garnaat, C.W., Mukai, K., Hu, Y. and Ma, H. 1994. Isolation of cDNAs encoding guanine nucleotide-binding protein β-subunit homologues from maize (ZGB1) and *Arabidopsis* (AGB1). Proc. Natl. Acad. Sci. USA 91: 9554–9558.

Weiss, C.A., Huang, H. and Ma, H. 1993. Immunolocalization of the G protein alpha subunit encoded by the GPA1 gene in *Arabidopsis*. Plant Cell 11: 1513–1528.

Winstead, M.V., Balsinde, J. and Dennis, E.A. 2000. Calcium-independent phospholipase A_2: structure and function. Biochim. Biophys. Acta 1488: 28–39.

Worley, C.K., Zenser, N., Ramos, J., Rouse, D., Leyser, O., Theologis, A. and Callis, J. 2000. Degradation of Aux/IAA proteins is essential for normal auxin signalling. Plant J. 21: 553–562.

Wu, W.H. and Assmann, S.M. 1994. A membrane-delimited pathway of G-protein regulation of the guard-cell inward K^+ channel. Proc. Natl. Acad. Sci. USA. 91: 6310–6314.

Yang, T. and Poovaiah, B.W. 2000. Molecular and biochemical evidence for the involvement of calcium/calmodulin in auxin action. J. Biol. Chem. 275: 3137–3143.

Yeh, K.C. and Lagarias, J.C. 1998. Eukaryotic phytochromes: light-regulated serine/threonine protein kinases with histidine kinase ancestry. Proc. Natl. Acad. Sci. USA 95: 13976–13981.

Yi, H., Park, D. and Lee, Y. 1996. *In vivo* evidence for the involvement of phospholipase A and protein kinase in the signal transduction pathway for auxin-induced corn coleoptile elongation. Physiol. Plant. 96: 359–368.

Yokota, E., Muto, S. and Shimmen, T. 1999. Inhibitory regulation of higher-plant myosin by Ca^{2+} ions. Plant Physiol. 119: 231–239.

Zaina, S., Reggiani, R. and Bertani, A. 1990. Preliminary evidence for involvement of GTP-binding protein(s) in auxin signal transduction in rice (*Oryza sativa* L.) coleoptile. J. Plant Physiol. 136: 653–658.

Zbell, B. and Walter-Back, C. 1989. Signal transduction of auxin on isolated plant cell membranes: indications for a rapid polyphosphoinositide response stimulated by indoleacetic acid. J. Plant Physiol. 133: 353–360.

Zheng, Z.L. and Yang, Z. 2000. The Rop GTPase switch turns on polar growth in pollen. Trends Plant Sci. 5: 298–303.

Plant Molecular Biology **49**: 373–385, 2002.
Perrot-Rechenmann and Hagen (Eds.), Auxin Molecular Biology.
© 2002 *Kluwer Academic Publishers.*

Auxin-responsive gene expression: genes, promoters and regulatory factors

Gretchen Hagen* and Tom Guilfoyle
*Department of Biochemistry, University of Missouri, 117 Schweitzer Hall, Columbia, MO 65211, USA (*author
for correspondence; e-mail HagenG@missouri.edu)*

Received 3 May 2001; accepted in revised form 6 July 2001

Key words: auxin response elements (AuxREs), auxin response factors (ARFs), auxin response genes

Abstract

A molecular approach to investigate auxin signaling in plants has led to the identification of several classes of early/primary auxin response genes. Within the promoters of these genes, *cis* elements that confer auxin responsiveness (referred to as auxin-response elements or AuxREs) have been defined, and a family of *trans*-acting transcription factors (auxin-response factors or ARFs) that bind with specificity to AuxREs has been characterized. A family of auxin regulated proteins referred to as Aux/IAA proteins also play a key role in regulating these auxin-response genes. Auxin may regulate transcription on early response genes by influencing the types of interactions between ARFs and Aux/IAAs.

Abbreviations: ARF, auxin-response factor; AuxRE, auxin-response element; CHX, cycloheximide; DBD, DNA-binding domain; DR, direct repeat; ER, everted repeat; EST, expressed sequence tag; ORF, open reading frame

Introduction

Auxins play a critical role in most major growth responses throughout the development of a plant. Auxins are thought to regulate or influence diverse responses on a whole-plant level, such as tropisms, apical dominance and root initiation, and responses on a cellular level, such as cell extension, division and differentiation. Over the past 20 years, it has been clearly demonstrated that auxins can also exert rapid and specific effects on genes at the molecular level. Numerous sequences that are up-regulated or down-regulated by auxin have been described (for reviews, see Abel and Theologis, 1996; Sitbon and Perrot-Rechenmann, 1997; Guilfoyle, 1999). Research efforts in a number of labs are currently focused on characterizing the mechanisms involved in the regulation of genes by auxin. The genes that have been most extensively studied are those that are specifically induced by active auxins within minutes of exposure to the hormone, and are induced by auxin in the absence of protein synthesis. These genes are referred to as early, or pri-

mary auxin response genes, and fall into three major classes (*Aux/IAAs*, *SAURs* and *GH3s*). In this review, we will briefly discuss the early/primary auxin response gene families, the TGTCTC-containing auxin response promoter elements and the auxin response factor (ARF) family of transcription factors. A number of reviews that cover these areas in more detail have been published (Guilfoyle, 1999; Guilfoyle *et al.*, 1998a, b). This review will expand on these areas and focus on information that has recently emerged (for example, information derived from the publication of the genome sequence of *Arabidopsis*) or has recently been published. We also present a working model for the regulation of auxin response genes, based on the current available information.

Auxin-responsive genes

Aux/IAA genes

Aux/IAA genes have been identified in several laboratories as rapidly induced auxin response genes, using

374

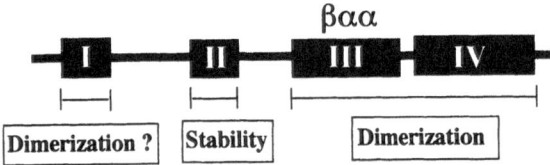

Figure 1. Diagram of an Aux/IAA protein with conserved domains I, II, III and IV. Domain I may play a role in homodimerization of Aux/IAA proteins (Ouellet *et al.*, 2001). Domain II plays a role in destabilizing Aux/IAA proteins (Colon-Carmona *et al.*, 2000; Ouellet *et al.*, 2001; Worley *et al.*, 2000). Domains III and IV function as dimerization domains, facilitating interactions among Aux/IAA and ARF proteins (Kim *et al.*, 1997; Ulmasov *et al.*, 1997b; Morgan *et al.*, 1999). Domain III contains a predicted $\beta\alpha\alpha$ fold, which has been characterized in DNA-binding domains of Arc and MetJ repressor proteins (Abel *et al.*, 1994).

differential hybridization with probes from untreated and auxin-treated hypocotyls or epicotyls (Walker and Key, 1982; Hagen *et al.*, 1984; Theologis *et al.*, 1985). The original *Aux/IAA* genes to be described (soybean *GmAux22, GmAux28, GH1* and pea *PS-IAA4/5* and *PS-IAA6*) were expressed to moderate levels in elongating regions of etiolated hypocotyls or epicotyls. When these elongating regions are excised and incubated in auxin-free medium, the Aux/IAA mRNAs are rapidly depleted, and can be rapidly induced by addition of auxin to the medium. Aux/IAA mRNAs are specifically induced by active auxins; protein synthesis inhibitors, such as cycloheximide, also induce the accumulation of Aux/IAA transcripts (for review, see Guilfoyle, 1999).

Aux/IAA genes are present as multigene families in soybean (Ainley *et al.*, 1988), pea (Oeller *et al.*, 1993), mung bean (Yamamoto *et al.*, 1992), tobacco (Dargeviciute *et al.*, 1998) and tomato (Nebenfuhr *et al.*, 2000). *Arabidopsis* contains 29 *Aux/IAA* genes (see article by Liscum and Reed, 2002). Most of the *Arabidopsis* genes are induced by auxin and show a range of induction kinetics (Abel *et al.*, 1995); however, *IAA 28* is not responsive to exogenous auxin (Rogg *et al.*, 2001). *Aux/IAA* genes are also found in monocots and gymnosperms (GenBank EST database), but are not found in organisms other than plants.

Aux/IAA proteins generally range in size from 20 to 35 kDa. They are short-lived and localize to the nucleus (Abel *et al.*, 1994; Abel and Theologis, 1995). Four conserved motifs are found in most Aux/IAA proteins, and these are referred to as domains I, II, III and IV (Figure 1; Ainley *et al.*, 1988; Abel *et al.*, 1995). Domain II plays a role in destabilizing Aux/IAA proteins, and may be a target for

ubiquitination (Worley *et al.*, 2000; Colon-Carmona *et al.*, 2000; Ouellet *et al.*, 2001). Domain III is part of a motif that is predicted to resemble the amphipathic $\beta\alpha\alpha$-fold found in the β-ribbon multimerization and DNA-binding domains of Arc and MetJ repressor proteins (Abel *et al.*, 1994). The predicted $\beta\alpha\alpha$ motif has been shown to play a role in dimerization/multimerization of Aux/IAA proteins and in heterodimerization between Aux/IAA and ARF proteins (Kim *et al.*, 1997; Ulmasov *et al.*, 1997b; Morgan *et al.*, 1999; Ouellet *et al.*, 2001); however, a role for this motif in DNA binding has not been demonstrated. The function of domains I and IV in Aux/IAA proteins is not clear, but recent experiments suggest that domain I may play a role in homodimerization of Aux/IAA proteins (Ouellet *et al.*, 2001).

A number of mutations in *Aux/IAA* genes have been identified that provide insight into the role played by these proteins in auxin responses. Some of these mutants display light-grown phenotypes when grown in the dark, suggesting that they bypass a requirement for phytochrome in selected aspects of photomorphogenesis. In this regard, recent studies have shown that phytochrome A interacts with and phosphorylates the amino-terminal half (encompassing domains I and II) of selected Aux/IAA proteins *in vitro* (Colon-Carmona *et al.*, 2000). The Aux/IAA mutants are discussed in more detail in the article by Liscum and Reed in this issue.

SAUR genes

A group of small, auxin-induced RNAs, referred to as SAURs, was identified in a differential hybridization screen of clones from auxin-treated soybean elongating hypocotyl sections (McClure and Guilfoyle, 1987). These RNAs are induced within 2–5 min of exposure to exogenous auxin. SAURs are moderately abundant in the zone of cell elongation in soybean hypocotyls, and most strongly expressed in epidermal and cortical cells; induction by auxin results in an elevation of transcripts within the same cell types (Gee *et al.*, 1991). Auxin induction of soybean SAURs is transcriptionally regulated (McClure *et al.*, 1989) and specific for active auxins (McClure and Guilfoyle, 1987). Treatment with the protein synthesis inhibitor cycloheximide (CHX) does not inhibit or enhance auxin-induced transcriptional activation of soybean SAURs, but does result in an increase in the abundance of SAUR transcripts (McClure and Guilfoyle, 1987). This induction by CHX was shown, however, not to be at the level of transcription, and must result from the stabilization of SAUR transcripts (Franco *et al.*, 1990).

Sequence analysis of three soybean SAUR cDNAs and genes revealed that the genes contain no introns (McClure *et al.*, 1989). The deduced open reading frames (ORF) encode proteins of 9–10 kDa. Five soybean *SAUR* genes are clustered in a single 7 kb locus of the nuclear genome, and each gene is oriented in an opposing orientation. The 5 ORFs show a high degree of homology, particularly in the C-terminal portion of the protein. Database searches indicate that the predicted structures of SAUR proteins are not highly homologous to any other published amino acid sequences.

Auxin-inducible SAURs have also been described from mung bean (Yamamoto *et al.*, 1992), pea (Guilfoyle *et al.*, 1993), *Arabidopsis* (Gil *et al.*, 1994), radish (Anai *et al.*, 1998) and *Zea mays* (Yang and Poovaiah, 2000). In addition to auxin, some of these SAUR mRNAs are induced by CHX (Gil *et al.*, 1994). In contrast to the soybean SAURs, the *Arabidopsis* SAUR-AC1 appears to be transcriptionally induced by CHX (Gil *et al.*, 1994); SAUR-AC1 is also induced by the plant hormone cytokinin (Timpte *et al.*, 1995). There are over 70 *SAUR* genes in *Arabidopsis* (Table 1). With one exception (*AtSAUR11*), all genes appear to lack introns (GenBank annotations for *AtSAUR26*, *-33*, *-39* and *− − 52* contain introns; however, we suspect that these annotations are incorrect and that these genes consist of single exons, because the annotated 3' exons are unrelated to conserved *SAUR* sequences). Many of the *SAUR* genes in *Arabidopsis* are found in clusters, like those originally identified in soybean. Clusters of eight, five, six and seven, and five *SAUR* genes are found on chromosomes 1, 3, 4 and 5, respectively (Figure 2). It is not known how many genes in this large gene family are expressed and are auxin-inducible.

As mentioned above, at least some *SAUR* genes are transcriptionally regulated by active auxins. There is evidence, however, that *SAURs* are also regulated post-transcriptionally. *SAURs* encode unstable mRNAs (McClure and Guilfoyle, 1989; Franco *et al.*, 1990), and their high turnover rate may be due, in part, to a conserved element (DST) in the 3'-untranslated region of the mRNA (McClure *et al.*, 1989; Newman *et al.*, 1993) and/or elements within the ORF (Li *et al.*, 1994). SAUR proteins may also be regulated posttranslationally. Based on studies using anti-SAUR antibodies, there is evidence that SAUR protein abun-

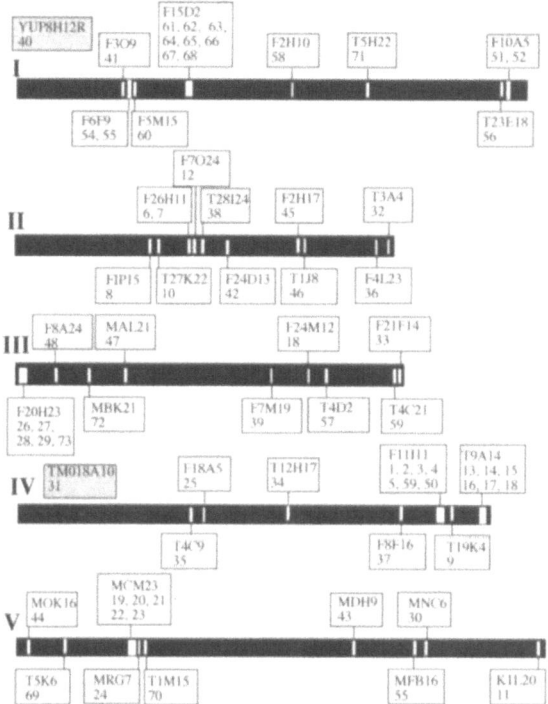

Figure 2. Chromosome positions of *SAUR* genes in *Arabidopsis*. *SAUR* genes are indicated in boxes along with the BAC clone on which they are found. Gray boxes above chromosomes 1 and 4 indicate that the chromosome position has not been determined. See Table 1 for *SAUR* gene nomenclature and GenBank Gen Info Identifier Number (Gene ID).

dance is low (Guilfoyle, 1999), suggesting that SAUR protein half-life may be very short.

The function of SAUR proteins is still unknown; however, they may play some role in an auxin signal transduction pathway that involves calcium and calmodulin. This possible role is suggested from recent experiments that demonstrate *in vitro* binding of calmodulin to an amino terminal domain in several SAUR proteins (i.e., maize ZmSAUR1, soybean SAUR 10A5 and *Arabidopsis* SAUR-AC1; Yang and Poovaiah, 2000). While the amino terminus is not highly conserved in amino acid sequence among the SAUR proteins, a putative basic α-amphipathic helix domain found in the amino terminus may provide a calmodulin-binding site in these proteins.

GH3 genes

The GH3 mRNA is one of several sequences that was recovered in a differential hybridization screen of auxin-induced cDNA sequences derived from auxin-treated, etiolated soybean seedlings (Hagen *et al.*,

376

Table 1. Arabidopsis SAUR genes.

AtSAUR			AtSAUR		
gene	chromosome	gene ID	Gene	chromosome	gene ID
1	4	5123694	37	4	2827527
2	4	5123695	38	2	4337198
3	4	5123696	39	3	7529253
4	4	5123697	40	1	3152585
5	4	5123698	41	1	4966371
6	2	4803922	42	2	AC005851
7	2	4803923	43	5	9759477
8	2	4589972	44	5	7378611; 9757782
9	4	3036815	45	2	4510355
10	2	4406817	46	2	4883619
11	5	10177420	47	3	9293978
12	2	4582443	48	3	6681332
13	4	AL035656	49	4	3096944
14	4	4490336	50	4	3096945
15/AC1	4	4490337	51	1	9369378
16	4	4490338	52	1	9369367
17	4	AC006567	53	1	10086488
18	3	6562272	54	1	10086504
19	5	9757893	55	5	9777399
20	5	9757894	56	1	6573709
21	5	9757895	57	3	6630745
22	5	9757897	58	1	12321539
23	5	9757898	59	3	7329679
24	5	9758890	60	1	8778613
25	4	4455308	61	1	12323520
26	3	6006857	62	1	12323532
27	3	6006858	63	1	12323534
28	3	6006859	64	1	12323536
29	3	6006860	65	1	12323538
30	5	9759198	66	1	12323511
31	4	2252854	67	1	12323513
32	2	3831443	68	1	12323545
33	3	6899888	69	1	8979733
34	4	2827539	70	5	AF296832
35	4	5281052	71	1	6056370
36	2	2583132	72	3	11994425

AtSAUR15 is *SAUR-AC1. AtSAUR26*, 33, 39, 52, and 68 are likely to be incorrectly annotated in GenBank. *AtSAUR13*, 17, and 70 are not annorated in GenBank. Gene ID is the
GenBank Gen Info Identifier number. The chromosome number is indicated.

1984). In the absence of exogenous auxin treatment, the steady-state mRNA level of soybean GH3 is low and largely associated with the vascular system; exogenous auxin application induces GH3 expression in every major organ and tissue type throughout the soybean plant (reviewed by Guilfoyle, 1999). GH3 mRNA is transcriptionally induced within 5 minutes of auxin treatment (Hagen and Guilfoyle, 1985), and is specifically induced only by active auxins (Hagen *et al.*, 1984). In contrast to other auxin-regulated sequences (SAURs, Aux/IAAs), soybean GH3 mRNA levels are unaffected by treatment with protein synthesis inhibitors (Hagen and Guilfoyle, 1985; Franco *et al.*, 1990). The *GH3* gene encodes a 70 kDa protein of unknown function, and is part of a small, multigene family in soybean (Hagen *et al.*, 1991; unpublished

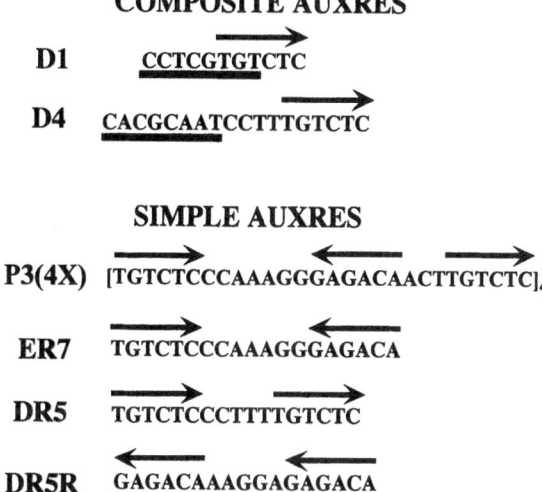

COMPOSITE AUXRES

D1   CCTCGTGTCTC

D4   CACGCAATCCTTTGTCTC

SIMPLE AUXRES

P3(4X)   [TGTCTCCCAAAGGGAGACAACTTGTCTC]4

ER7   TGTCTCCCAAAGGGAGACA

DR5   TGTCTCCCTTTTGTCTC

DR5R   GAGACAAAGGAGAGACA

Figure 3. Composite and simple AuxREs. The TGTCTC elements or inverted GAGACA elements are indicated by arrows. The coupling or constitutive elements in the D1 and D4 promoter fragments of the *GH3* gene are indicated by the underlines. P3 (4X) contains an everted repeat separated by 7 bp, which is identical to ER7, and an inverted repeat separated by 3 bp. ARF1 protein was shown to bind to the ER in a P3 (4X) probe (Ulmasov *et al.*, 1997). DR5 is a direct repeat, and DR5R is the inverse of the DR5 repeat; both are active AuxREs, indicating that the orientation of the DR has no effect on activity.

Table 2. Arabidopsis GH3 genes.

AtGH3 gene	Chromosome	Gene ID
1	2	3650037
2	4	4468805
3	2	2642446
4	1	8778768
5	4	3269287
6/DFL1	5	8885594
7	1	2829896
8	5	9758189
9	2	3738288
10	4	4262168
11/FIN219	2	4559380
12	5	7529287
13	5	7529290
14	5	7543903
15	5	7543904
16	5	7543905
17	1	9795624
18	1	12597809
19	1	12597811
20/truncated	1	12597807

data). The protein is localized to the cytoplasm and appears to be stable (Wright *et al.*, 1987).

An auxin-responsive *GH3* gene has been identified in tobacco using differential display (Roux and Perrot-Rechenmann, 1997). In contrast to soybean GH3 mRNA, the tobacco transcript is induced by the protein synthesis inhibitor CHX. A family of *GH3* genes exists in *Arabidopsis*, consisting of 19 members; an additional partial gene contains only the amino-terminal third of the protein. We refer to the *Arabidopsis* full-length genes as *AtGH3-1* to *AtGH3-19* (Table 2). The *GH3* gene family in *Arabidopsis* is composed of three subfamilies, based on sequence similarities and splicing patterns. Five of the genes are found in a cluster on chromosome 5 (*AtGH3-12* to *AtGH3-16*), and two others (*AtGH3-18* and *AtGH3-19*) are clustered on chromosome 1, along with the truncated *AtGH3-20* gene. Database searches have identified *GH3*-related genes in other dicots and monocots, the blue green algae, *Synechocystis* sp. (GenBank D90900), mouse (AC074918/AF316996), man (AF316997) and other vertebrates (GenBank EST database). *GH3*-like genes are not found in the sequenced genomes of yeast, *Caenorhabditis elegans* or *Drosophila*. All of the *GH3* genes encode proteins

with predicted molecular masses of 65–70 kDa. The GH3 proteins have no particular striking features, with the exception that putative coiled-coil domains can be found in the amino-and carboxyl-terminal regions of AtGH3-11/FIN219 (Hsieh *et al.*, 2000) and some other AtGH3 proteins (T. Guilfoyle, unpublished).

Open reading frames from several *Arabidopsis GH3* genes (*GH3-1, -2, -3, - 5, -6, -14, -17*) have been cloned and used to investigate the auxin-inducibility of this gene family. All but *GH3-17* show some level of auxin induction after a 1.5 h treatment with 1-naphthaleneacetic acid (NAA); *GH3-17* is expressed in the absence of auxin treatment, and the levels are not further elevated by auxin treatment (G. Hagen, unpublished). An additional family member (*GH3-11/FIN219*) is induced by auxin (Hsieh *et al.*, 2000). It has not been determined if the remaining *GH3* genes are expressed in *Arabidopsis*.

The recent isolation of two mutants, each involving members of the *Arabidopsis GH3* gene family, suggests a role for at least some GH3 proteins in photomorphogenesis, and strengthens the potential link between phytochrome signaling and auxin responses (Tian and Reed, 1999). The *fin219* (far-red-insensitive 219) mutant was isolated as a suppressor of a mutation in a key regulatory component of photomorphogenesis, COP1 (Hsieh *et al.*, 2000). COP1 is a repressor of photomorphogenic development, and its activity is negatively regulated by light (Osterlund *et al.*, 1999). The *fin219* mutant plants display a long hypocotyl phenotype when grown under continuous far-red light. The *FIN219* gene encodes a cytoplasmically-localized protein with homology to soybean GH3 (designated *AtGH3-11*; Table 2). FIN219 mRNA is rapidly induced by auxin and, in contrast to soybean GH3 mRNA, is also induced by cycloheximide. Hsieh *et al.* (2000) suggest that FIN219 defines a new component of the phytochrome A (phyA) pathway, and may play a specific role in the phyA-mediated far-red inactivation of COP1. The dominant *dfl1-D* (dwarf in light 1) mutant was selected in a screen for hypocotyl length mutants from *Arabidopsis* activation-tagged lines (Nakazawa *et al.*, 2001). In the light, both hypocotyl cell elongation and lateral root formation are inhibited in the *dfl1-D* mutant. The *DFL1-D* gene encodes another *GH3* family member (*GH3-6*), and, in the mutant, the transcript is highly expressed. As noted above, this gene is also auxin-inducible. We have found that *Arabidopsis* plants that overexpress a soybean *GH3* gene also have an exaggerated dwarf phenotype and show a deetiolated phenotype when germinated and grown in the dark (G. Hagen, unpublished results). Over-expression of the *FIN219* gene was reported to result in a short hypocotyl phenotype in plants grown under continuous far-red light, but dwarfing of adult plants was not reported (Hsieh *et al.*, 2000). Therefore, ectopic expression of different *GH3* genes results in a variety of phenotypes, suggesting that they may have partially redundant functions, but also possibly non-redundant roles in auxin and/or light signaling. This is similar to observations made with *Aux/IAA* mutant plants, which have some related, yet distinct phenotypes (see article by Liscum and Reed in this issue).

Auxin-responsive promoters, promoter elements and interacting factors

The promoters of several auxin-responsive genes (soybean *GH3*, soybean *SAUR15A* and pea *PS-IAA4/5*) have been analyzed in some detail, using a variety of methods (e.g. deletion analysis, linker-scanning, site directed mutagenesis, gain of function analysis; reviewed by Guilfoyle, 1999). The smallest element to be identified as auxin-responsive is a six-base pair

sequence, TGTCTC (Ulmasov *et al.*, 1997a, b). This element has been shown to function in both composite and simple auxin-response elements (AuxREs; Figure 3). In composite AuxREs, such as those found in the GH3 promoter fragments D1 and D4, the TGTCTC element is only functional if combined with a coupling or constitutive element (Figure 3; reviewed by Guilfoyle *et al.*, 1998a; Guilfoyle, 1999). Simple AuxREs, derived from the alteration of naturally occurring AuxREs, may function in the absence of a coupling element if the TGTCTC elements occur as direct or palindromic repeats that are appropriately spaced (Figure 3; P3 (4X)-palindromic repeats spaced by 3 bp; ER7-everted repeats spaced by 7 bp; DR5-direct repeats spaced by 5 bp; DR5R-direct repeats in the inverse orientation; reviewed by Guilfoyle *et al.*, 1998a; Guilfoyle, 1999). These simple, synthetic AuxREs are 5–10 times more auxin-responsive than natural AuxREs (Guilfoyle, 1999).

Natural AuxRE promoter-reporter constructs have been used to study organ and tissue expression patterns of auxin-responsive genes (Guilfoyle, 1999). These constructs have been valuable tools to follow gene expression events during growth responses associated with changes in auxin gradients or sensitivities, such as gravitropism and phototropism (Li *et al.*, 1999), and in studies of signal transduction pathways in plants (Kovtun *et al.*, 1998; Kovtun *et al.*, 2000). Synthetic AuxRE-reporter genes have been shown to respond to auxin in a wide variety of organs, tissues and cell types (Ulmasov *et al.*, 1997b; Oono *et al.*, 1998). These synthetic AuxREs, when fused to minimal promoter-reporter genes, have been used to monitor cell and/or tissue responses to endogenous auxin in wild type and mutant plants carrying the reporter gene (Sabatini *et al.*, 1999; Mockaitis and Howell, 2000; Zhao *et al.*, 2001). In addition, these constructs have provided the basis to develop genetic screens for auxin response mutants (Oono *et al.*, 1998; Murfett *et al.*, 2001).

To identify proteins that bind to the TGTCTC element, Ulmasov *et al.* (1997) used the synthetic AuxRE P3 (4X) (see Figure 3) as a bait in a yeast one-hybrid screen of an *Arabidopsis* cDNA expression library. A novel transcription factor, referred to as auxin-response actor 1 or ARF1, was identified and shown to bind with specificity to TGTCTC AuxREs. *Arabidopsis* has 23 *ARF*-related genes (Table 3). One of these genes (*ARF23*) is, however, likely to be a pseudogene, because it lacks any sequence carboxy-terminal to the DNA-binding domain (DBD; see below) and contains a stop codon within the DBD (based on our anno-

Table 3. Arabidopsis ARF genes.

ARF	Chromosome	Gene ID
1	1	5080809
2	5	10176918
3	2	3805770
4	5	12744969
5	1	10086486
6	1	12322119
7	5	8071649
8	5	9758525
9	4	4972102
10	2	4432846
11	2	4415934
12	1	5091627
13	1	10086460
14	1	8778363
15	1	8778352
16	4	4938484
17	1	16573757
18	3	6850874
19	1	8954059
20	1	12322942
21	1	12323856
22	1	12323853
23/truncated	1	8778678*

Gene ID is the GenBank Gen Info Identifier number. ARF23 is likely to be incorrectly annotated in GenBank.

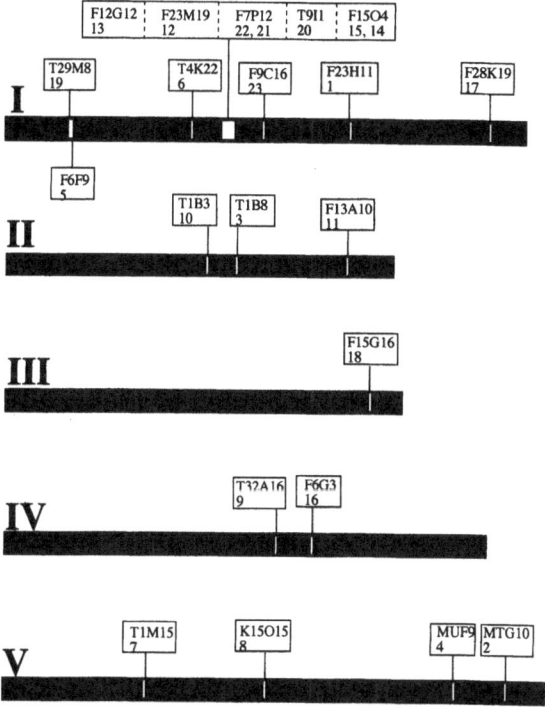

Figure 4. Chromosome positions of *ARF* genes in *Arabidopsis*. *ARF* genes are indicated in boxes along with the BAC clone on which they are found. See Table 3 for *ARF* gene nomenclature and GenBank Gen Info Identifier Number (Gene ID).

tations; the GenBank annotation of this ARF lacks portions of the conserved DBD sequence found in all other ARFs and is likely to be incorrect). In addition, there is no evidence that *ARF23* is expressed. Seven ARF genes are clustered on chromosome I (Figure 4). All of the genes in this cluster, along with *ARF23*, are more closely related to one another than to other ARFs (see the phylogenetic tree for ARFs presented in the article by Liscum and Reed, this issue). It has not been shown that any of the ARFs in this cluster are expressed (i.e., no expressed sequence tags (ESTs) or mRNAs have been reported). In this regard, we have been unsuccessful in our attempts to clone a cDNA for *ARF12* (another gene in this cluster). Interestingly, database searches have uncovered a segment of DNA in the mitochondrial genome of *Arabidopsis* that has some homology to *ARF17*. This DNA is also found associated with a large fragment of mitochondrial DNA located on *Arabidopsis* chromosome II. ARFs are found in other dicots, monocots, gymnosperms and ferns (GenBank EST database), but are not found outside of the plant kingdom.

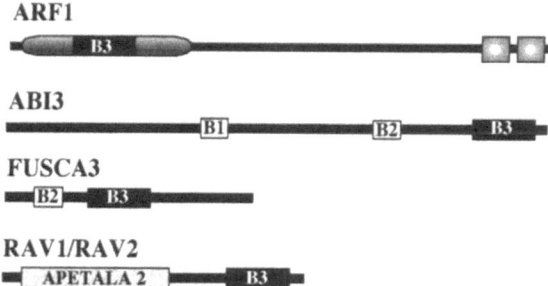

Figure 5. B3 DBD proteins. Diagrams of the B3 domains found in ARFs and other transcription factors identified in *Arabidopsis*. The B3 domain is indicated by the black boxes. The minimal DBD in ARF1 is indicated by the oval in the amino-terminal portion of the protein. The two boxes in the carboxy-terminal portion of ARF1 represent domains related to domains III and IV in Aux/IAA proteins. The transcription factors ABI3 and FUSCA3 contain two related domains, B2 and B3 (Giraudat *et al.*, 1992; Luerssen *et al.*, 1998). The B1 domain in ABI3 is also found in maize VP1. RAV1 and RAV2 transcription factors contain an APETALA2 (AP2) DBD and a B3 DBD (Kagaya *et al.*, 1999).

Analysis of *ARF1–ARF10* has shown that these ARF mRNAs are ubiquitously expressed in most major organs of *Arabidopsis* plants, as well as in *Arabidopsis* suspension culture cells (Ulmasov *et al.*, 1999b). Presently, it is not known if ARF mRNAs are expressed in a tissue-specific manner within major organs. Of the ARFs that have been tested, exogenous auxin treatment (i.e. 50 μM NAA, 5 h) did not alter ARF mRNA levels (G. Hagen, unpublished results; Ulmasov *et al.*, 1999b).

ARF proteins range in size from 70 to 130 kDa and are defined by their N-terminal DBD (Ulmasov *et al.*, 1997a,b; Guilfoyle *et al.*, 1998a, b), which is unique to plants. The central region of about 100 amino acids in ARF DBDs is related to the B3 DBD domain found in transcription factors like maize VP1 and *Arabidopsis* ABI3, FUSCA3, RAV1 and RAV2 (Figure 5; McCarty *et al.*, 1991; Giraudat *et al.*, 1992; Luerssen *et al.*, 1998; Kagaya *et al.*, 1999). The ARF DBD binds with specificity to TGTCTC AuxREs (Ulmasov *et al.*, 1997a). Gel mobility shift assays have shown that the first 4 nucleotides (TGTC or positions +1 to +4) of the TGTCTC element are required for binding of ARFs; substitutions at +5 are tolerated, and the importance of position +6 to binding shows some variation (Ulmasov *et al.*, 1997a, 1999b). ARFs bind as dimers on palindromic AuxREs that are spaced by 7 to 9 residues (Ulmasov *et al.*, 1999b). Putative nuclear localization signals (NLS) can be found within the DBD of most ARFs (Ulmasov *et al.*, 1999b). Hardtke and Berleth (1998) showed that the N-terminal (residues 4–289) portion of the MONOPTEROUS protein (MP), which is encoded by the *ARF5* gene, could confer nuclear targeting to a GUS reporter gene. Putative NLS sequences downstream of residue 289 were also identified in MP/ARF5; however, a GUS reporter construct containing residues 297–901 was unable to specifically target the reporter to the nucleus.

All ARFs but ARF3 and ARF17 contain a carboxy-terminal domain related to domains III and IV found in Aux/IAA proteins (Ulmasov *et al.*, 1997a, 1999b). The carboxy-terminal domain in both ARF and Aux/IAA proteins is a protein-protein interaction domain that allows the homo- and heterodimerization of ARFs and the heterodimerization of ARF and Aux/IAA proteins (Ulmasov *et al.*, 1997a, b, 1999b; Kim *et al.*, 1997; Ouellet *et al.*, 2001). Domains III and IV in some ARF proteins increase the *in vitro* binding (i.e., gel mobility shift assays) of these ARFs to TGTCTC AuxREs by facilitating the formation of dimers that occupy the two half-sites in palindromic AuxREs (Ulmasov *et al.*, 1999b). The middle regions of ARFs, which lie between the DBD and domains III and IV, function as activation or repression domains. Transfection assays with protoplasts indicate that ARF1 and 2, which contain middle regions rich in proline and serine, are repressors and ARF5, -6, -7, and -8, which contain middle regions rich in glutamine, are activators (Ulmasov *et al.*, 1999a; our unpublished results). ARF19 also contains a glutamine-rich middle region and is likely to be a transcriptional activator.

Arabidopsis mutants containing lesions in three different ARFs have recently been described. Mutations in the *MONOPTEROUS (MP)/ARF5* gene disrupt embryo axis formation and vascular development (Hardtke and Berleth, 1998). Lesions in the *ETTIN (ETT)/ARF3* gene lead to defects in flower development (Sessions *et al.*, 1997). The *NPH4* gene encodes the ARF7 protein, and mutations in this gene result in plants with abnormal phenotypes associated with differential growth (Harper *et al.*, 2000). These mutants provide valuable tools to study the role of ARF proteins in auxin-regulated growth responses. Additionally, they should help to elucidate the importance of the different ARF domains to ARF protein function. For a detailed discussion of these mutants, see the article by Liscum and Reed (this issue).

Mechanisms involved in auxin-responsive gene expression

Genetic and molecular/biochemical experimental approaches have provided a large body of preliminary evidence that both ARFs and Aux/IAA proteins function as transcription factors on primary auxin-response genes. The data indicate that there may be specific interactions among ARFs, and between ARFs and Aux/IAAs (Ulmasov *et al.*, 1997a; our unpublished data). As mentioned above, there is evidence that some ARFs are activators and other ARFs are repressors of auxin-response genes (Ulmasov *et al.*, 1999a); additional data (i.e. over-expression of Aux/IAA proteins in transient assay systems) implicate Aux/IAA proteins as repressors of auxin-response genes (Ulmasov *et al.*, 1997b). The precise details/mechanisms for the transcriptional regulation of these genes by 22 ARFs and 29 Aux/IAAs in *Arabidopsis* remain to be determined. Regulation of gene expression by auxin will, undoubtedly, be complex. Our understanding is lim-

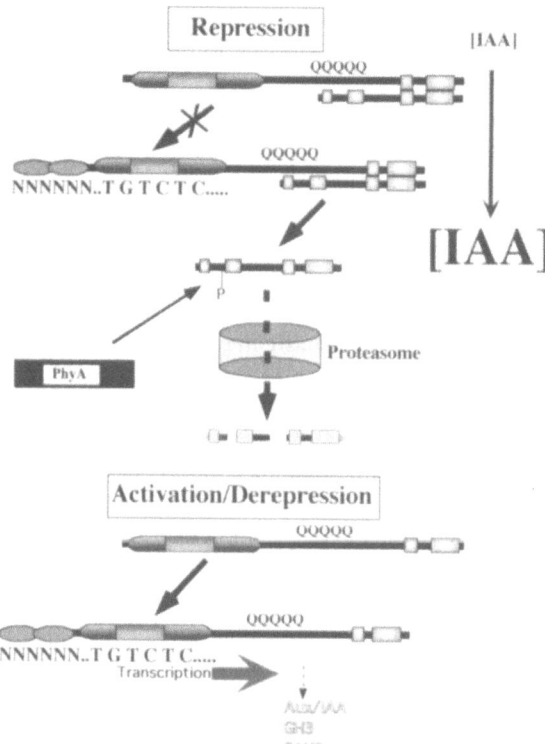

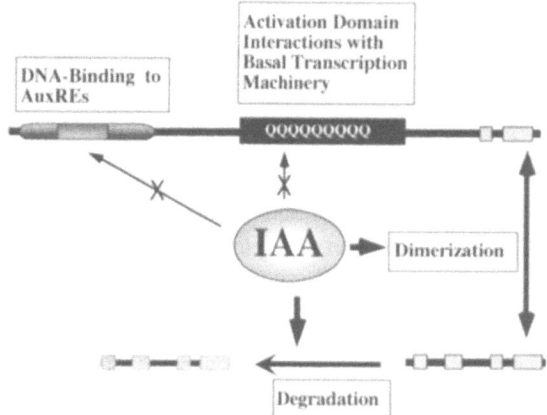

Figure 7. Possible target sites for auxin action on ARF and Aux/IAA proteins. Arrows indicate possible targets for IAA signaling on the DBD or activation domain of an ARF transcriptional activator. Xs through arrows indicated that these are unlikely target sites (see text). Other arrows indicate possible target sites that involve the dimerization of ARF and Aux/IAA proteins or the enhanced degradation of Aux/IAA proteins (which in turn would result in their failure to dimerize with ARFs).

Figure 6. Model for the roles of ARF and Aux/IAA proteins in regulating early auxin-response genes. When auxin concentrations are low, ARF activators are associated with Aux/IAA repressors through their conserved carboxy-terminal dimerization domains, and early auxin response genes are repressed. ARF and Aux/IAA dimers might form on or off AuxRE target sites. When auxin concentrations are high, Aux/IAA repressors dissociate from ARF activators, allowing the ARFs to reach their AuxRE targets or become active on their DNA targets, and early auxin response genes (e.g., *Aux/IAA*, *GH3*, and *SAUR* genes) are derepressed/activated. Early response genes are continuously transcribed as long as auxin concentrations remain high. Concomitant with or subsequent to their dissociation from ARFs, Aux/IAA proteins are possibly phosphorylated by phytochrome(s) and marked for degradation by the ubiquitin pathway (proteasome). This would prevent Aux/IAA proteins from reassociating with ARF activators as long as auxin concentrations remain elevated. When auxin concentrations become low, Aux/IAA proteins are degraded less rapidly and increase in abundance, eventually reaching levels sufficient to complex with ARF activators and repress transcription.

ited, thus far, by the knowledge of only two of the *trans*-acting factors involved.

Based on current information, we favor the following model for the regulation of primary auxin response genes that contain TGTCTC AuxREs (Figure 6). When auxin concentrations are low, auxin-response genes are repressed. This repression could result from the dimerization via domains III and IV of Aux/IAA repressors with ARF transcriptional activa-

tors. These heterodimers might form whether or not an ARF transcription factor is bound to its TGTCTC target site. Dimerization off of the DNA might prevent an ARF transactivator from binding to its target site, while dimerization to a DNA-bound ARF might result from active repression by the Aux/IAA protein. We favor the latter scenario because results of Ulmasov *et al.* (1999a) suggest that ARFs are bound to AuxREs regardless of the auxin status in cells and whether the auxin response genes are repressed or activated. The targeting of selected ARFs to an AuxRE could result from the particular coupling factor that binds to the composite AuxRE and acts to recruit selected ARFs to the TGTCTC element. Tissue-specific expression patterns of both ARF and Aux/IAA proteins (if they exist) could also determine which ARF can target an AuxRE and which Aux/IAA protein can interact with an ARF to regulate transcription.

When auxin concentrations are elevated, transcription is rapidly derepressed/activated (i.e. as early as 2 to 5 min after exogenous auxin application; McClure *et al.*, 1989; Hagen and Guilfoyle, 1985). We propose that an early event in derepression might be the dissociation of Aux/IAA repressors from their ARF counterparts and subsequent degradation of the Aux/IAA proteins by the ubiquitin/proteasome pathway (Gray and Estelle, 2000; Worley *et al.*, 2000). We favor an initial derepression effect by elevated auxin levels resulting in the dissociation of Aux/IAA

382

proteins from ARFs bound to AuxREs, because of the rapidity with which derepression/activation takes place. Phytochrome(s) might come into play by phosphorylating Aux/IAA proteins (Colon-Carmona et al., 2000) and marking them for degradation via the proteasome pathway. The phosphorylation of Aux/IAA proteins by the kinase activity associated with phytochromes might be one explanation for the apparent cross-talk between light signaling and auxin signaling pathways (Reed et al., 1998; Tian and Reed, 1999; Colon-Carmona et al., 2000). Once the Aux/IAA proteins dissociate from the ARF transcription factor and are decreased in concentration by degradation, derepression/activation of the auxin response gene can occur. Activation might be further potentiated by the binding of additional ARF transcriptional activators (i.e., by dimerization through domains III and IV) to the DNA-bound ARF (Ulmasov et al., 1999a). We expect that mutations in domain II of Aux/IAA proteins not only increase the stability of these proteins (Ouellet et al., 2001), but also increase their repressive nature because the normal regulation via degradation is altered.

In our proposed model, Aux/IAA genes that contain TGTCTC AuxREs would be activated in response to auxin, resulting in an increase in mRNA and protein. As proposed repressors of AuxRE function, Aux/IAA proteins might, at some point, feed back on their own genes to down-regulate expression. This feedback loop of repression would, however, most likely occur only when auxin levels are low, based on the observation that in the continuous presence of high auxin levels, early auxin response genes remain activated for at least several hours (Walker and Key, 1982; Theologis et al., 1985; Abel et al., 1995).

This model is clearly oversimplified because it does not take into account all the experimental details that have been published. For example, it does not explain what role ARF transcriptional repressors, like ARF1 and ARF2, might play in regulating auxin response genes. We speculate that ARF repressors could repress transcription in a similar fashion to the Aux/IAA proteins (i.e., by binding to ARF transcriptional activators and repressing their activator function). As another example, dimers are invoked in our model, but it is possible that higher order multimers of ARF and Aux/IAA proteins might be involved with the regulation of auxin response genes. That higher order multimers of Aux/IAA proteins are capable of forming is suggested from experiments of Morgan et al. (1999) and Ouellet et al. (2001).

To test the validity of this model, a number of questions need to be addressed. Do Aux/IAA proteins possess intrinsic repressor activity? Is there selectivity in the types of Aux/IAA and ARF homo- and heterodimers (or multimers) that can form? What role, if any, do Aux/IAA homodimers or Aux/IAA heterodimers play in auxin-responsive transcription? How different or similar are the temporal and tissue-specific expression patterns of the many ARF and Aux/IAA proteins?

Perhaps the most intriguing questions are where does auxin come into play and what is the signal-transduction machinery involved in auxin's activation of early response genes? Figure 7 highlights some of the possible sites of auxin action in regulating genes that bind ARFs. One possible site of auxin action is in regulating the binding of ARFs to their DNA target sites. This may be unlikely, however, based on results of Ulmasov et al. (1999a; discussed above). A second possibility is that auxin exerts an effect on the activation domain of ARF transcriptional activators by altering its ability to interact with co-activators and the general transcription machinery. This again seems unlikely, because when ARF activation domains are fused to a heterologous DNA-binding domain (e.g., yeast GAL4 DBD) and tested with reporter genes that contain GAL4 binding sites, there is no auxin response (S. Tiwari, G.Hagen and T. Guilfoyle, unpublished). A third possibility is that auxin has an effect on the ARF and Aux/IAA dimers that form, favoring the dissociation of Aux/IAAs from ARF transcriptional activators under high auxin concentrations. We speculate that this is a likely target for auxin action. A fourth possibility is that auxin regulates the degradation of Aux/IAA proteins, resulting in rapid degradation under high auxin concentrations. While there is some evidence suggesting that auxin does not alter the life times of Aux/IAA proteins (Abel et al., 1994), we, nevertheless, feel that this is still a likely possibility because of the connection between auxin action and the ubiquitin pathway (Gray and Estelle, 2000). Obviously, the third and fourth possibilities are related to one another, in that depletion of Aux/IAA proteins upon auxin treatment would likely upset the balance of interactions that take place among Aux/IAA and ARF proteins.

Conclusions and perspectives

While our understanding of the signal transduction pathway from auxin perception to gene expression is still rudimentary, progress is being made and interesting challenges remain. The models (Figures 6 and 7) we have presented for the regulation of primary auxin response genes are derived from the characterization of promoters containing TGTCTC or related elements that function as AuxREs, and a transcription factor (ARF) that binds with specificity to these AuxREs. Naturally occurring AuxREs that contain TGTCTC elements have been shown, however, to be composite AuxREs, and a future challenge will be to identify additional transcription factors (e.g., coupling factors) involved with the targeting of ARFs to different types of composite AuxREs. These factors might be recovered in yeast one-hybrid screens (Ulmasov *et al.*, 1997), using a composite element as bait, for example, or from genetic screens for auxin signal transduction mutants (Oono *et al.*, 1998; Murfett *et al.*, 2001). Yeast two-hybrid screens might identify other proteins that specifically interact with ARF and Aux/IAA protein domains. These proteins will be crucial to the characterization of higher order transcriptional complexes that regulate auxin response genes.

Another goal of the studies of auxin response genes will be to identify additional genes, and to characterize their promoter elements. It is reasonable to assume that some auxin response genes may contain auxin-responsive promoter elements other than TGTCTC. An AuxRE that has received considerable attention is the ocs/as-1 element found in a variety of genes, including plant glutathione *S*-transferases (for a recent review see Guilfoyle, 1999). This AuxRE is a 20 bp DNA sequence with the consensus sequence of TGACG(T/C)AAG(C/G)(G/A)(C/A)T(G/T)ACG(T/C)(A/C)(A/C) (Ellis *et al.*, 1993). Specific basic region-leucine zipper proteins (bZIP transcription factors) that bind to TGACG(T/C) motifs within ocs/as-1 elements have been cloned (Katagiri *et al.*, 1989; Singh *et al.*, 1990)). In some cases, this element is induced within minutes upon exogenous application of active auxins; however, at least in some cases, this element is also induced by inactive auxin analogues such as 2,3-dichlorophenoxyacetic acid, other hormones and stress-inducing compounds (Ulmasov *et al.*, 1994). Thus, this AuxRE may function in signal transduction pathways involved with stress responses. Potential targets for further analysis of functional AuxREs include the maize potassium channel *ZMK1* (Philippar *et al.*, 1999), the recently described auxin-inducible *NAC1* gene from *Arabidopsis* (Xie *et al.*, 2000) and, perhaps, a mitogen-activated protein kinase (*MAPK*) gene (Mockaitis and Howell, 2000). In the future, alternative experimental strategies involving techniques such as differential display or microarrays might uncover new primary auxin response genes.

There is now strong evidence that auxin-regulated gene expression is also influenced by other signaling pathways. This evidence has come predominantly from the characterization of auxin response mutants, and indicates that auxin signaling is impacted by protein degradation via the ubiquitin/proteasome pathway, protein phosphorylation, light and other hormones. A discussion of these interacting pathways and their connection with auxin signaling and growth responses can be found in the articles by Dharmasiri and Estelle, Delong *et al.*, Liscom and Reed, and Swarup *et al.* in this issue.

Acknowledgements

Research from our laboratory and unpublished results cited in this article have been supported by National Science Foundation Grant MCB-0080096 and University of Missouri Food for the 21st Century Program.

References

Abel, S, Oeller, P.W. and Theologis, A. 1994. Early auxin-induced genes encode short-lived nuclear proteins. Proc. Natl. Acad. Sci. USA 91: 326–330

Abel, S., Nguyen, M.D. and Theologis, A. 1995. The PS-IAA4/5-like family of early auxin-inducible mRNAs in *Arabidopsis thaliana*. J. Mol. Biol. 251: 533–549.

Abel, S. and Theologis, A. 1995. A polymorphic bipartite motif signals nuclear targeting of early auxin-inducible proteins related to PS-IAA4 from pea (*Pisum sativum*). Plant J. 8: 87–96.

Abel, S. and Theologis, A. 1996. Early genes and auxin action. Plant Physiol. 111: 9–17.

Ainley, W.M., Walker, J.C., Nagao, R.T. and Key, J.L. 1988. Se quence and characterization of two auxin-regulated genes from soybean. J. Biol. Chem. 263: 10658–10666.

Anai, T., Kono, N., Kosemura, S., Yamamura, S. and Hasegawa, K. 1998. Isolation and characterization of an auxin-inducible SAUR gene from radish. DNA Seq. 9: 329–333.

Casimiro, I., Marchant, A., Bhalerao, R.P, Beeckman, T., Dhooge, S., Swarup, R., Graham, N., Inzé, D., Sandberg, G., Casero, P.J. and Bennett, M. 2001. Auxin transport promotes *Arabidopsis* lateral root initiation. Plant Cell 13: 843–852.

Colon-Carmona, A., Chen, D.L., Yeh, K.C. and Abel, S. 2000. Aux/IAA proteins are phosphorylated by phytochrome *in vitro*. Plant Physiol. 124: 1728–1738.

Conner, T.W., Goekjian, V.L., LaFayette, P.R. and Key, J.L. 1990. Structure and expression of two auxin-inducible genes from *Arabidopsis*. Plant Mol. Biol. 15: 623–632.

Dargeviciute, A., Roux, C., Decreux, A., Sitbon, F. and Perrot-Rechenmann, C. 1998. Molecular cloning and expression of the early auxin-responive Aux/IAA gene family in *Nicotiana tabacum*. Plant Cell Physiol. 39: 993–1002.

Dharmasiri, S. and Estelle, M. 2002. The role of regulated protein degradation in auxin response. Plant Mol. Biol. 49: 401–408.

Delong, A., Mockaitis, K. and Christensen, S. 2002. Protein phosphorylation in the delivery of and response to auxin signals - Plant Mol. Biol. 49: 285–303.

Ellis, J.G., Tokuhisa, J.G., Llewellyn, D.J., Bouchez, D., Singh, K., Dennis, E.S. and Peacock, W.J. 1993. Does the ocs-element occur as a functional component of the promoters of plants? Plant J. 4: 433–443.

Franco, A., Gee, M.A. and Guilfoyle, T.J. 1990. Induction and superinduction of auxin-responsive genes with auxin and protein synthesis inhibitors. J. Biol. Chem. 265: 15845–15849.

Gee, M.A., Hagen, G. and Guilfoyle, T.J. 1991. Tissue-specific and organ-specific expression of the auxin-responsive transcripts, SAURs and GH3, in soybean. Plant Cell 3: 419–430.

Gil, P., Liu, Y., Orbovic, V., Verkamp, E., Poff, K.L. and Green, P. 1994. Characterization of the auxin-inducible SAUR-AC1 gene for use as a genetic tool in *Arabidopsis*. Plant Physiol. 104: 777–784.

Giraudat, J., Hauge, B.M., Valon, C., Smalle, J., Parcy, F. and Goodman, H.M. 1992. Isolation of the *Arabidopsis ABI3* gene by positional cloning. Plant Cell 4: 1251–1261.

Gray, W.M. and Estelle, M. 2000. Function of the ubiquitin-proteasome pathway in auxin response. Trends Biochem. Sci. 25: 133–138.

Guilfoyle, T.J. 1999. Auxin-regulated genes and promoters. In: P.J.J. Hooykaas, M. Hall and K.L. Libbenga (Eds.) Biochemistry and Molecular Biology of Plant Hormones, Elsevier, Leiden, Netherlands, pp. 423–459.

Guilfoyle, T.J., Hagen, G., Li, Y., Ulmasov, T., Liu, Z., Strabala, T. and Gee, M.A. 1993. Auxin-regulated transcription. Aust. J. Plant Physiol. 20: 489–502.

Guilfoyle, T.J., Ulmasov, T. and Hagen, G. 1998a. The ARF family of transcription factors and their role in plant hormone responsive transcription. Cell. Mol. Life Sci. 54: 619–627.

Guilfoyle, T.J., Hagen, G., Ulmasov, T. and Murfett, J. 1998b. How does auxin turn on genes? Plant Physiol. 118: 341–347.

Hagen, G. and Guilfoyle, T.J. 1985. Rapid induction of selective transcription by auxin. Mol. Cell. Biol. 5: 1197–1203.

Hagen, G., Kleinschmidt, A.J. and Guilfoyle, T.J. 1984. Auxin-regulated gene expression in intact soybean hypocotyl and excised hypocotyl sections. Planta 16: 147–153.

Hagen, G., Martin, G., Li, Y. and Guilfoyle, T.J. 1991. Auxin-induced expression of the soybean *GH3* promoter in transgenic tobacco plants. Plant Mol. Biol. 17: 567–579.

Hardtke, C.S. and Berleth, T. 1998. The *Arabidopsis* gene MONOPTEROUS encodes a transcription factor mediating embryo axis formation and vascular development. EMBO J. 17: 1405–1411.

Harper, R.M., Stowe-Evans, E.L., Luesse, D.R., Muto, H., Tatematsu, K., Watahiki, M.K., Yamamoto, K. and Liscum, E. 2000. The NPH4 locus encodes the auxin response factor ARF7, a conditional regulator of differential growth in aerial *Arabidopsis* tissue. Plant Cell 12: 757–770.

Hsieh, H.L., Okamoto, H., Wang, M., Ang, L.H., Matsui, M., Goodman, H. and Deng, X.W. 2000. *FIN219*, an auxin-regulated gene, defines a link between phytochrome A and the downstream regulator COP1 in light control of *Arabidopsis* development. Genes Devel. 14: 1958–1970.

Kagaya, Y., Ohmiya, K. and Hattori T. 1999. RAV1, a novel DNA-binding protein, binds to bipartite recognition sequence through two distinct DNA-binding domains uniquely found in plants. Nucl. Acids Res. 27: 470–478.

Katagiri, F., Lam, E. and Chua, N.-H. 1989. Two tobacco DNA-binding proteins with homology to the nuclear factor CREB. Nature 340: 727–730.

Kim, J., Harter, K. and Theologis, A. 1997. Protein-protein interactions among the Aux/IAA proteins. Proc. Natl. Acad. Sci. USA 94: 11786–11791.

Kovtun, Y., Chiu, W.-L., Zeng, W. and Sheen, J. 1998. Suppression of auxin signal transduction by a MAPK cascade in higher plants. Nature 395: 716–720.

Kovtun, Y., Chiu, W.-L., Tena, G. and Sheen J. 2000. Functional analysis of oxidative stress-activated mitogen-activated protein kinase cascade in plants. Proc. Natl. Acad. Sci. USA 97: 2940–2945.

Li, Y., Strabala, T.J., Hagen, G. and Guilfoyle, T.J. 1994. The SAUR open reading frame contains a *cis* element responsible for cycloheximide-mediated mRNA accumulation. Plant Mol. Biol. 24: 715–723.

Li, Y., Wu, Y.H., Hagen, G. and Guilfoyle, T.J. 1999. Expression of the auxin-inducible GH3 promoter GUS fusion gene as a useful molecular marker for auxin physiology. Plant Cell Physiol. 40: 675–682.

Liscum, E. and Reed, J.W. 2002. Genetics of Aux/IAA and ARF action in plant growth and development. Plant Mol. Biol. 49: 387–400.

Liu, Z.-B., Ulmasov, T., Shi, X., Hagen, G. and Guilfoyle, T.J. 1994. The soybean GH3 promoter contains multiple auxin-inducible elements. Plant Cell 6: 645–657.

Luerssen, H., Kirik, V., Herrmann, P. and Misera, S. 1998. FUSCA3 encodes a protein with a conserved VP1/ABI3-like B3 domain which is of functional importance for the regulation of seed maturation in *Arabidopsis thaliana*. Plant J. 15: 755–764.

McCarty, D.R., Hattori, T., Carson, C.B., Vasil, V., Lazar, M. and Vasil, I.K. 1991. The *viviparous-1* developmental gene of maize encodes a novel transcriptional activator. Cell 66: 895–905.

McClure, B.A. and Guilfoyle, T. 1987. Characterization of a class of small auxin-inducible soybean polyadenylated RNAs. Plant Mol. Biol. 9: 611–623.

McClure, B.A. and Guilfoyle, T.J. 1989. Rapid redistribution of auxin-regulated RNAs during gravitropism. Science 243: 91–93.

McClure, B.A., Hagen, G., Brown, C.S., Gee, M.A. and Guilfoyle, T.J. 1989. Transcription, organization, and sequence of an auxin-regulated gene cluster in soybean. Plant Cell 1: 229–239.

Mockaitis, K. and Howell, S.H. 2000. Auxin induces mitogenic activated protein kinase (MAPK) activation in roots of *Arabidopsis* seedlings. Plant J. 24: 785–794.

Morgan, K.E., Zarembinski, T.I., Theologis, A. and Abel, S. 1999. Biochemical characterization of recombinant polypeptides corresponding to the predicted $\beta\alpha\alpha$ fold in Aux IAA proteins. FEBS Lett. 454: 283–287.

Murfett, J., Wang, X-J., Hagen, G. and Guilfoyle, T.J. 2001. Identification of *Arabidopsis* histone deacetylase HDA6 mutants that affect transgene expression. Plant Cell 13: 1047–1061.

Nakazawa, M., Yabe, N., Ichikawa, T., Yamamoto, Y.Y., Yoshizumi, T. Hasunuma, K. and Matsui, M. 2001. *DFL1*, an auxin-responsive *GH3* gene homologue, negatively regulates shoot cell elongation and lateral root formation, and positively regulates the light response of hypocotyl length. Plant J. 25: 213–221.

Nebenfuhr, A., White, T.J. and Lomax, T.L. 2000. The *diageotropica* mutation alters auxin induction of a subset of the Aux/IAA gene family in tomato. Plant Mol. Biol. 44: 73–84.

Newman, T.C., Ohme-Takagi, M., Taylor, C.B. and Green, P.J. 1993. DST sequences, highly conserved among plant SAUR genes, target reporter transcripts for rapid decay in tobacco. Plant Cell 5: 701–714.

Oeller, P.W., Keller, J.A., Parks, J.E., Silbert, J.E. and Theologis, A. 1993. Structural characterization of the early indoleacetic acid-inducible genes *PS-IAA4/5* and *PS-IAA6*, of pea (*Pisum sativum* L). J. Mol. Biol. 233: 789–798.

Oono, Y., Chen, Q.G., Overvoorde, P.J., Kohler, C. and Theologis, A. 1998. *age* mutants of *Arabidopsis* exhibit altered auxin-regulated gene expression. Plant J. 10: 1649–1662.

Osterlund, M.T., Ang, L.-H. and Deng, X.-W. 1999. The role of *COP1* in repression of *Arabidopsis* photomorphogenic development. Trends Cell Biol. 9: 113–118.

Ouellet, F., Overvoorde, P.J. and Theologis, A. 2001. IAA17/AXR3. Biochemical insight into an auxin mutant phenotype. Plant Cell 13: 829–842.

Philippar, K., Fuchs, I., Luthen, H., Hoth, S., Bauer, C.S., Haga, K., Thiel, G., Ljung, K., Sandberg, G., Bottger, M., Becker, D. and Hedrich, R. 1999. Auxin-induced K(+) channel expression represents an essential step in coleoptile growth and gravitropism. Proc. Natl. Acad. Sci. USA 96: 12186–12191.

Reed, J.W., Elumalai, R.P. and Chory, J. 1998. Suppressors of an *Arabidopsis thaliana* phyB mutation identify genes that control light signaling and hypocotyl elongation. Genetics 148: 1295–1310.

Rogg, L.E., Lasswell, J. and Bartel, B. 2001. A gain-of-function mutation in *iaa28* suppresses lateral root development. Plant Cell 13: 465–480.

Roux, C. and Perrot-Rechenmann, C. 1997. Isolation by differential display and characterizationof a tobacco auxin-responsive cDNA Nt-gh3, related to *GH3*. FEBS Lett. 419: 131–136.

Sabatini, S., Beis, D., Wolkenfelt, H., Murfett, J., Guilfoyle, T., Malamy, J., Benfey, P., Leyser, O., Bechtold, N., Weisbeek, P. and Scheres, B. 1999. An auxin-dependent distal organizer of pattern and polarity in the *Arabidopsis* root. Cell 99: 463–472.

Sessions, A., Nemhauser, J.L., McColl, A., Roe, J.L., Feldman, K.A. and Zambryski, P.C. 1997. *ETTIN* patterns the *Arabidopsis* floral meristem and reproductive organs. Development 124: 4481–4491.

Singh, K., Dennis, E.S., Ellis, J.G., Llewellyn, D.J., Tokuhisa, J.G., Wahleithner, J.A. and Peacock, W.J. 1990. OCSBF-1, a maize ocs enhancer binding factor: isolation and expression during development. Plant Cell 1: 891–903.

Sitbon, F. and Perrot-Rechenmann, C. 1997. Expression of auxin-regulated genes. Physiol. Plant. 100: 443–455.

Swarup, R., Parry, G., Graham, N., Allen, T. and Bennett, M.S. 2002. Auxin cross-talk: integration of signalling pathways to control plant development. Plant Mol. Biol. 49: 409–424.

Theologis, A., Huynh, T.V. and Davis, R.W. 1985. Rapid induction of specific mRNAs by auxin in pea epicotyl tissue. J. Mol. Biol. 183: 53–68.

Tian, Q. and Reed, J.W. 1999. Control of auxin-regulated root development by the *Arabidopsis thaliana SHY2/IAA3* gene. Development 126: 711–721.

Timpte, C., Lincoln, C., Pickett, F.B., Turner, J. and Estelle, M. 1995. The *AXR1* and *AUX1* genes of *Arabidopsis* function in separate auxin-response pathways. Plant J. 8: 561–569.

Ulmasov, T., Liu, Z-B., Hagen, G. and Guilfoyle, T.J. 1995. Composite structure of auxin response elements. Plant Cell 7: 1611–1623.

Ulmasov, T., Hagen, G. and Guilfoyle, T.J. 1994. The ocs element in the soybean GH2/4 promoter is activated by both active and inactive auxin and salicylic acid analogs. Plant Mol. Biol. 26: 1055–1064.

Ulmasov, T Hagen, G. and Guilfoyle, T.J. 1997a. ARF1, a transcription factor that binds auxin response elements. Science 276: 1865–1868.

Ulmasov, T., Murfett, J., Hagen, G. and Guilfoyle, T.J. 1997b. Aux/IAA proteins repress expression of reporter genes containing natural and highly active synthetic auxin response elements. Plant Cell 9: 1963–1971.

Ulmasov, T., Hagen, G. and Guilfoyle, T.J. 1999a. Activation and repression of transcription by auxin response factors. Proc. Natl. Acad. Sci. USA 96: 5844–5849.

Ulmasov, T., Hagen, G. and Guilfoyle, T.J. 1999b. Dimerization and DNA binding of auxin response factors. Plant J. 19: 309–319.

Walker, J.C. and Key, J.L. 1982. Isolation of cloned cDNAs to auxin-responsive poly(A) RNAs of elongating soybean hypocotyl. Proc. Natl. Acad. Sci. USA 79: 7185–7189.

Worley, C.K., Zenser, N., Ramos, J., Rouse, D., Leyser, O., Theologis, A. and Callis, J. 2000. Degradation of Aux/IAA proteins is essential for normal auxin signalling. Plant J. 21: 553–562.

Wright, R., Hagen, G. and Guilfoyle, T. 1987. An auxin-induced polypeptide in dicot plants. Plant Mol. Biol. 9: 625–635.

Xie, Q., Frugis, G., Colgan, D. and Chua, N-H. 2000. *Arabidopsis* NAC1 transduces auxin signal downstream of TIR1 to promote lateral root development. Genes Dev. 14: 3024–3036.

Yamamoto, K.T., Mori, H. and Imaseki, H. 1992 cDNA cloning of indole-3-acetic acid-regulated genes:*Aux22* and *SAUR* from mung bean (*Vigna radiata*) hypocotyl tissue. Plant Cell Physiol. 33: 93–97.

Yang, T. and Poovaiah, B.W. 2000. Molecular and biochemical evidence for the involvement of calcium/calmodulin in auxin action. J. Biol. Chem. 275: 3137–3143.

Zhao Y., Christensen, S.K., Fankhauser, C., Cashman, J.R., Cohen, J.D., Weigel, D. and Chory, J. 2001. A role for flavin monooxygenase-like enzymes in auxin biosynthesis. Science 291: 306–309.

Plant Molecular Biology **49**: 387–400, 2002.
Perrot-Rechenmann and Hagen (Eds.), Auxin Molecular Biology.
© 2002 *Kluwer Academic Publishers.*

Genetics of Aux/IAA and ARF action in plant growth and development

E. Liscum[1],* and J.W. Reed[2]

[1]*Division of Biological Sciences, University of Missouri, 105 Tucker Hall, Columbia, MO 65211, USA (*author for correspondence);* [2]*Department of Biology, University of North Carolina at Chapel Hill, CB #3280, Coker Hall, Chapel Hill, NC 27599, USA*

Received 1 May 2002; accepted in revised form 13 July 2002

Key words: Arabidopsis mutants, ARF (auxin response factor), auxin, Aux/IAA, light, phytochrome, signal transduction

Abstract

Dramatic advances in our understanding of auxin signal-response pathways have been made in recent years. Much of this new knowledge has come through the study of mutants in *Arabidopsis thaliana*. Mutations have been identified in a wide variety of auxin-response components, including auxin transporters, protein kinases and phosphatases, components of a ubiquitin-proteosome pathway, and transcriptional regulators. This review focuses on mutations that affect auxin-modulated transcription factors, in particular those in the *Aux/IAA* and *AUXIN RESPONSE FACTOR* (*ARF*) genes. Mutants in members of these related gene families exhibit phenotypes that indicate both unique localized functions, as well as overlapping redundant functions, throughout plant development – from embryogenesis to flowering. Effects of specific mutations on Aux/IAA and ARF protein functions at the biochemical and physiological levels will be discussed. We will also discuss potential mechanisms for interactions between auxin and light response pathways that are suggested by these mutants.

Introduction

It has been known for decades that the plant hormone auxin (indole-3-acetic acid) can regulate gene expression (Key, 1969; Theologis, 1986). Several families of auxin-regulated genes (e.g., the *Aux/IAA*, *GH3*, and *SAUR* families) have been identified. Some of these genes are expressed in distinct spatial and temporal patterns, thus underscoring the diversity of auxin responses in different plant tissues and organs (Abel and Theologis, 1996; Sitbon and Perrot-Rechenmann, 1997; Guilfoyle *et al.*, 1998a). Many of these genes are 'primary-response genes', as their expression increases transiently within minutes of auxin application, independent of *de novo* protein synthesis (Abel and Theologis, 1996; Sitbon and Perrot-Rechenmann, 1997). A number of these primary-response genes can also be induced by the protein synthesis inhibitor cycloheximide, independently of auxin (Franco *et al.*, 1990; Abel *et al.*, 1995; Gil and Green, 1997), suggesting that the auxin-dependent induction of such

genes normally requires inactivation and/or removal of repressor proteins (Abel and Theologis, 1996). Consistent with this interpretation, recent studies have demonstrated the importance of ubiquitin-mediated protein turnover in normal auxin responses (del Pozo and Estelle, 1999; Gray and Estelle, 2000; Worley *et al.*, 2000; also see review by Dharmasiri and Estelle, this issue).

Two related families of proteins, the Aux/IAA proteins and the auxin response factors (ARFs) (see Figure 1), are key regulators of auxin-modulated gene expression (Guilfoyle *et al.*, 1998a, b; Walker and Estelle, 1998). The *Aux/IAA* genes are themselves auxin-induced (Abel *et al.*, 1995), while there is no evidence for auxin-dependent changes in transcription of the *ARF* genes (Ulmasov *et al.*, 1999a). The *Arabidopsis* genome may encode as many as 29 Aux/IAA (Table 1) and 23 ARF proteins (see review by Hagen and Guilfoyle, this issue). This review focuses on genetic studies of Aux/IAA and ARF actions in *Arabidopsis*. However, we will briefly review some

388

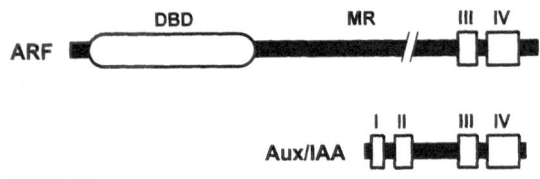

Figure 1. Domain properties of prototypical ARF and Aux/IAA proteins. While the sizes (see scale) of conserved domains – DNA-binding domain (DBD) in the ARFs, domains I and II in the Aux/IAAs, and shared carboxyl-terminal domains III/IV – are consistent from one protein to another, the divergent regions vary in size. The variable size of the ARF middle region (MR) is represented by the break.

features of the biochemistry of these proteins that are relevant to genetic studies of *aux/iaa* and *arf* mutants. For more detailed discussion of the biochemical properties and functions of the Aux/IAA and ARF proteins we refer the reader to the article by Hagen and Guilfoyle (this issue).

Overview of the biochemical properties of the Aux/IAA and ARF proteins

Both Aux/IAA and ARF proteins function as transcriptional regulators. While members of the Aux/IAA family are generally thought to act as repressors of auxin-induced gene expression (Abel *et al.*, 1994, 1995; Ulmasov *et al.*, 1997b), different ARF proteins can either activate or repress transcription (Ulmasov *et al.*, 1999b). Whether a given ARF acts as a transcription activator or repressor is determined by the sequence and corresponding structure of the middle region of the protein (Figure 1), which varies among ARFs (Ulmasov *et al.*, 1999b). For example, ARFs with Q-rich middle regions (e.g., ARF5 and ARF7) activate transcription in transient expression assays, while a P/S/T-rich middle region (e.g., ARF1) confers repressor activity (Ulmasov *et al.*, 1999b).

ARF proteins bind to auxin-responsive *cis*-acting promoter elements (AuxREs) using an amino-terminal DNA-binding domain (DBD) (Figure 1) (Ulmasov *et al.*, 1997a, 1999a). In contrast, despite the presence of a $\beta\alpha\alpha$-fold (within domain III, Figure 1) with similarity to the prokaryotic DNA-binding repressor proteins Arc and MetJ (Abel *et al.*, 1994), DNA-binding capacity of the Aux/IAA proteins has yet to be conclusively demonstrated (Abel and Theologis, 1996; Guilfoyle *et al.*, 1998a). It has therefore been hypothesized that the Aux/IAA proteins regulate transcription by modifying ARF activity (Guilfoyle *et al.*, 1998a, b).

Table 1. Aux/IAA genes of *Arabidopsis. Aux/IAA* genes listed here have been identified based upon PCR and hybridization analyses with DNA sequences encoding conserved domains of the proteins (Abel *et al.*, 1995; Ulmasov *et al.*, 1999a), yeast two-hybrid screens with other Aux/IAA proteins as baits (Kim *et al.*, 1997), and conceptual translation of the complete genome sequence (Arabidopsis Genome Initiative, 2000). Detailed information (e.g., cDNA accession number, protein properties, gene structure) about each gene/protein can be obtained at the Munich Information Center for Protein Sequences *Arabidopsis thaliana* Database (MATDB) (http://mips.gsf.de/proj/thal/db/index.html) using the MIPS protein code. The position of each gene within the *Arabidopsis* genome is given by chromsome number and approximate location on that chromosome in megabases (in parenthesis) as determined with The Arabidopsis Information Resource MapViewer (http://www.arabidopsis.org/servlets/mapper). Clones on which each *Aux/IAA* gene resides can also be found at MATDB. A similar table of the different ARF genes can be found in the article by Hagen and Guilfoyle (this issue).

Gene	MIPS protein code	Genome position
IAA1	At4g14560	4 (7.4)
IAA2	At3g23030	3 (9.2)
IAA3/SHY2	At1g04240	1 (1.8)
IAA4	At5g43700	5 (24.3)
IAA5	At1g15580	1 (6.0)
IAA6/SHY1	At1g52830	1 (19.7)
IAA7/AXR2	At3g23050	3 (9.2)
IAA8	At2g22670	2 (10.8)
IAA9	At5g65670	5 (26.4)
IAA10	At1g04100	1 (1.7)
IAA11	At4g28640	4 (19.0)
IAA12/BDL	At1g04550	1 (1.9)
IAA13	At2g33310	2 (15.2)
IAA14/SLR	At4g14550	4 (7.4)
IAA15	At1g80390	1 (32.6)
IAA16	At3g04730	3 (2.0)
IAA17/AXR3	At1g04250	1 (1.8)
IAA18	At1g51950	1 (22.7)
IAA19/MSG2	At3g15540	3 (6.2)
IAA20	At2g46990	2 (20.4)
IAA26/PAP1	At3g16500	3 (6.6)
IAA27/PAP2	At4g29080	4 (19.2)
IAA28	At5g25890	5 (9.1)
IAA29	At4g32280	4 (14.8)
IAA30	At3g62100	3 (23.7)
IAA31	At3g17600	3 (6.0)
IAA32	At2g01200	2 (0.9)
IAA33	At5g57420	5 (23.4)
IAA34	At1g15050	1 (5.8)

How might Aux/IAA proteins alter ARF activity? One clue comes from the two shared carboxyl-terminal protein-protein interaction motifs (domains III and IV) present in most Aux/IAA and ARF proteins (Figure 1). These domains have been shown to mediate both homo- and heterodimerization between members of the Aux/IAA and ARF families (Kim et al., 1997; Ulmasov et al., 1999a), and Aux/IAA proteins may therefore affect ARF function by dimerizing with them. Because ARF proteins appear to bind stably to DNA as ARF-ARF dimers (Ulmasov et al., 1999a), ARF-Aux/IAA heterodimerization may compromise ARF DNA-binding avidity. This could either decrease or increase target gene expression, depending on whether the Aux/IAA protein dimerized with an activating or a repressing ARF. Alternatively, ARF-Aux/IAA heterodimers might still bind DNA but have impaired ability to modulate transcription. Results from a number of transient expression studies are consistent with this hypothesis (Ulmasov et al., 1997a, b, 1999a, b).

Aux/IAA proteins also have two regions of conserved sequence in the amino-terminal portion (domains I and II, Figure 1) that are absent from ARF proteins (Guilfoyle et al., 1998a; Walker and Estelle, 1998). Although little is known about the function of domain I, domain II appears to confer instability to the Aux/IAA proteins (Worley et al., 2000; Ouellet et al., 2001; also see below). Aux/IAA proteins typically have in vivo half-lives of a matter of minutes (Abel et al., 1994), and thus have properties needed of factors that mediate rapid and transitory responses to signals. Moreover, given that auxin rapidly induces expression of the Aux/IAA genes (Abel and Theologis, 1996) and the encoded proteins can repress transcription (Abel et al., 1994, 1995; Ulmasov et al., 1997b), at least some Aux/IAAs activate negative feedback loops (Walker and Estelle, 1998). Thus, control over Aux/IAA protein turnover could have important regulatory functions in auxin responses.

A general working model for auxin regulation of gene expression

The available data on *Aux/IAA* and *ARF* genes and their corresponding proteins suggest a model for auxin regulation of gene expression and development. In this model, ARF proteins bind to AuxREs within promoters of auxin-responsive genes to activate or repress expression of the target gene. ARF proteins with Q-rich middle domains (e.g., ARF5, ARF6, ARF7, and

ARF8; Ulmasov et al., 1999b) will activate transcription, whereas others (e.g., ARF1; Ulmasov et al., 1997a, 1999b) may repress transcription, perhaps by interfering with Q-rich ARFs. ARF activity is also likely to be modulated through heterodimerization with Aux/IAA proteins (Kim et al., 1997; Ulmasov et al., 1999a), which act as transcriptional repressors (Ulmasov et al., 1997b). As the *Aux/IAA* gene promoters contain AuxREs that are themselves responsive to auxin-induced ARF activity, the Aux/IAA-ARF proteins are components of a potential (auto)feedback loop that may be very important in adaptation responses to auxin (Guilfoyle et al., 1998a, b; Leyser and Berleth, 1999). The rapid turnover of Aux/IAA proteins (Abel et al., 1994; Worley et al., 2000; Ouellet et al., 2001) also provides an opportunity for fast and efficient adaptation to endogenous or external signals. As such auxin, light (see below), or other signals may modulate Aux/IAA protein stability or activity, and hence ARF activity. Different cells may also exhibit differential responses to auxin because they express disparate sub-sets of ARFs, Aux/IAAs, and/or other tissue-specific factors that may interact with these regulators (Abel et al., 1995; Ulmasov et al., 1999a), providing yet another level of regulation. It therefore appears that the relative promoter binding affinities of activating and repressing ARFs, as well as the particular equilibrium of ARF-ARF, ARF-Aux/IAA, and Aux/IAA-Aux/IAA dimers in a given cell, may determine the degree of expression of various target genes, and thus morphological and developmental outcome. The *arf* and *aux/iaa* mutants discussed below represent important tools to address this model, and have already provided considerable support for it.

The *Arabidopsis* genome and evolution of the *Aux/IAA* and *ARF* gene families

Phylogenies of the *Arabidopsis* ARF and Aux/IAA protein sequences, based on conserved (alignable) amino acid sequence regions, are shown in Figure 2. Several interesting points arise from these phylogenies. For example, all of the ARF proteins that have Q-rich middle region activation domains (ARF5, 6, 7, 8, and 19) fall within the same clade, whereas most ARFs with other middle regions fall in one of several other monophyletic groups. Perhaps members of each of these other clades have distinct functional properties. The Aux/IAA sequences also fall into several

390

distinct clades, and closely related Aux/IAA proteins may share specific functional properties.

Comparison of these phylogenies with the genomic positions of the genes suggests that both local and large-scale genomic duplications contributed to the current set of genes. Seven out of eight members of a clade of very closely related *ARF*s (*12, 13, 14, 15, 20, 21, 22,* and *23*) are clustered in the genome, suggesting a quite recent local expansion. Could this local expansion have contributed to adaptation of *Arabidopsis* ecotypes to different local geographies? Many closely related pairs of *Aux/IAA* genes fall in regions of the genome that arose from large-scale duplications that predated the divergence of *Arabidopsis* from other plants (Vision *et al.,* 2000; D. Remington, personal communication). Perhaps these duplications have helped *Arabidopsis* to evolve distinct properties from other angiosperms. At a more macroevolutionary scale, an interesting unanswered question is the extent to which expansion of these gene families has facilitated evolution of new auxin responses. It may be very instructive to examine the complement of *ARF* and *Aux/IAA* genes in non-angiosperm plants that lack, or have divergent, auxin-regulated processes common to flowering vascular plants (see review by Cooke and Cohen, this issue). Of course, if we wish to understand the role of expansion of *ARF* and *Aux/IAA* gene families in evolution of plant development, it will be critical to understand their developmental functions in representative extant plants. Progress toward this goal in angiosperms is beginning to come from studies of mutations in these genes in *Arabidopsis.*

Biological functions of the Aux/IAA family

Identification of aux/iaa mutations

Screens for *Arabidopsis* mutants with altered auxin responses or morphological phenotypes have identified mutations in nine different *Aux/IAA* genes (Table 2): *IAA3/SHY2 (SHORT HYPOCOTYL* or *SUPPRESSOR OF HY2)* (Soh *et al.,* 1999; Tian and Reed, 1999), *IAA6/SHY1* (Kim *et al.,* 1996a, b; P. Nagpal and J.W. Reed, unpublished results), *IAA7/AXR2 (AUXIN RESISTANT 2)* (Nagpal *et al.,* 2000), *IAA12/BDL (BODENLOS)* (Hamann *et al.,* 1999; T. Hamann and G. Jürgens, personal communication), *IAA14/SLR (SOLITARY-ROOT)* (Fukaki and Tasaka, 1998; M. Tasaka, personal communication), *IAA17/AXR3* (Rouse *et al.,* 1998), *IAA18*

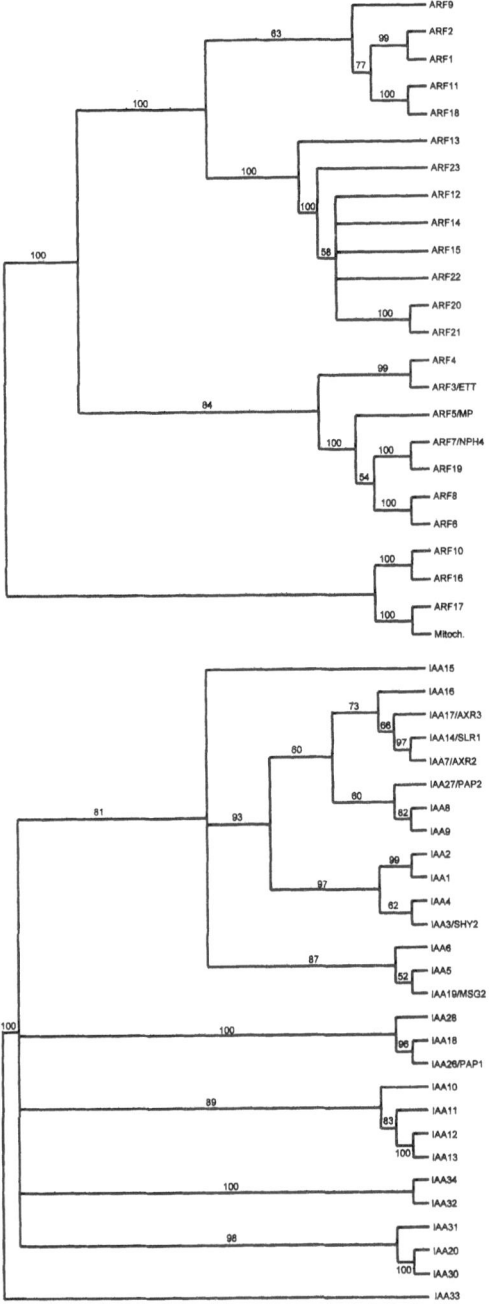

Figure 2. 'Majority rule' neighbor-joining bootstrap phylogenetic trees of ARF (left) and Aux/IAA (right) proteins from *Arabidopsis.* Amino acid sequences were aligned using Clustal with manual adjustments, and only conserved regions with unambiguous alignments were used for phylogenetic analyses. All branches with less than 50% bootstrap support have been collapsed. The 'Mitoch.' sequence shown in the left panel is an *Arabidopsis* mitochondrial sequence (GenBank accession number Y08502) that is also present (with a single residue change) within the region of chromosome 2 sequence that contains the mitochondrial genome duplication (Arabidopsis Genome Initiative, 2000). This sequence likely represents a pseudogene (see article by Hagen and Guilfoyle, this issue). (Phylogenies were generated by Dave Remington).

Table 2. Gain-of-function mutations in domain II of different *Aux/IAA* genes. Data are from the following sources: *shy2-2* (Reed *et al.*, 1998; Tian and Reed, 1999); *shy1-1* (Kim *et al.*, 1996a; P. Nagpal and J.W. Reed, unpublished results); *axr2-1* (Wilson *et al.*, 1990; Timpte *et al.*, 1992, 1994; Nagpal *et al.*, 2000); *bdl* (Hamann *et al.*, 1999; T. Hamann and G. Jürgens, unpublished results); *slr1-1* (Fukaki and Tasaka, 1998; M. Tasaka, personal communication); *axr3-1* (Leyser *et al.*, 1996; Rouse *et al.*, 1998); *iaa18-1* (P. Nagpal and J.W. Reed, unpublished results); *msg2* (Tatematsu *et al.*, 1999; E.L. Stowe-Evans and E. Liscum, unpublished results); and *iaa28-1* (Rogg *et al.*, 2001). Only the most dramatic phenotypes are listed. See the cited references for further details. For *shy1-1* and *iaa18-1*, the inference that the indicated mutation causes the phenotypes is based on genetic linkage only. The mutation in *IAA6* in the *shy1-1* mutant changes a non-conserved amino acid, and therefore requires additional proof.

Gene	Mutation	Most notable phenotype(s)
IAA3	*shy2-2*	short hypocotyl, leaves in dark, upcurled leaves
IAA6	*shy1-1*	short hypocotyl, upcurled leaves
IAA7	*axr2-1*	short hypocotyl, leaves in dark, wavy leaves, no root hairs, agravitropic root and inflorescence stems
IAA12	*Bdl*	no embryonic root, upcurled leaves
IAA14	*slr1-1*	no lateral roots, few root hairs, agravitropic root and hypocotyl
IAA17	*axr3-1*	short hypocotyl, leaves in dark, upcurled leaves, no root hairs, agravitropic root
IAA18	*iaa18-1*	long hypocotyl, fused cotyledons, short root, upcurled leaves
IAA19	*msg2*	agravitropic and aphototropic hypocotyl
IAA28	*iaa28-1*	few lateral roots, decreased shoot apical dominance

(P. Nagpal and J.W. Reed, unpublished results), *IAA19/MSG2(MASSUGU 2)* (Tatematsu et al., 1999), and *IAA28* (Rogg *et al.*, 2001). These mutations are believed to cause a gain of function of the corresponding genes, because (1) they are each dominant or semidominant for some phenotypes, (2) several of the mutant genes (*iaa3/shy2-2*, *iaa7/axr2-1*, *iaa12/bdl*, *iaa14/slr1-1*, *iaa17/axr3-1*, and *iaa28-1*) confer phenotypes when introduced into a wild-type background (Tian and Reed, 1999; Worley *et al.*, 2000; Rogg *et al.*, 2001; P. Nagpal and J.W.R., unpublished results; T. Hamann and G. Jürgens, personal communication; H. Fukaki and M. Tasaka, personal communication), and (3) intragenic suppressors of *iaa3/shy2-2*, *iaa7/axr2-1* and *iaa17/axr3-1* have molecular characteristics of loss-of-function mutations (Rouse *et al.*, 1998; Tian and Reed, 1999; Nagpal *et al.*, 2000).

Interestingly, each of these mutations changes a single amino acid in domain II of the corresponding protein. These mutations thus suggest that domain II plays a key role in the function of multiple Aux/IAA proteins, and characterizations of selected domain II mutations suggest that they stabilize the corresponding proteins. For example, fusion proteins between portions of the Ps-IAA6, At-IAA1, or At-IAA17 proteins containing domain II and GUS are degraded rapidly in plant cells, and engineered domain II mutations decrease this degradation (Worley *et al.*, 2000; S. Kepinski and O. Leyser, personal communication). Mutations in domain II also increase

steady-state levels of IAA3/SHY2 and IAA17/AXR3 (Colón-Carmona *et al.*, 2000; Ouellet *et al.*, 2001) and increase the half-life of IAA17/AXR3 by about sevenfold (Ouellet *et al.*, 2001). One attractive model to explain these results is that Aux/IAA proteins are substrates for the Skp1·cdc53/cullin·F-box[TIR1] (SCF[TIR1]) ubiquitin-ligase complex required for auxin responses (Gray and Estelle, 2000), and that auxin regulates either the activity of the SCF[TIR1] complex, or the recognition of Aux/IAA proteins by the complex (see Dharmasiri and Estelle, this issue).

Loss-of-function *aux/iaa* mutations have been isolated by screening for intragenic suppressors of gain-of-function alleles (*iaa3/shy2-2*, Tian and Reed, 1999; *iaa7/axr2-1*, Timpte *et al.*, 1994; and *iaa17/axr3-1*, Rouse *et al.*, 1998), or for insertions among T-DNA-insertional mutant populations (*iaa7/axr2*, Nagpal *et al.*, 2000; and *iaa19/msg2*, K. Yamamoto, personal communication). Several of the intragenic suppressors create premature stop codons or shift the reading frame, and are therefore probably null mutations. Others affect single amino acids in one of the other conserved domains. The T-DNA insertion mutation that has been described in *AXR2/IAA7* lacks the corresponding transcript and is probably a null allele (Nagpal *et al.*, 2000).

Developmental phenotypes of gain-of-function
aux/iaa mutants

The gain-of-function *aux/iaa* mutations each cause auxin-related developmental phenotypes. Interpreting these phenotypes is somewhat problematic because the stabilization caused by the domain II mutations may not mimic regulatory events actually occurring in wild-type plants. Thus, it is possible that the stabilized Aux/IAA proteins in the mutants are present at high enough concentrations to interact with ARF proteins that they would not normally contact. If this is true, then the phenotypes of the gain-of-function mutants may reflect the expression patterns of the mutated genes more than their 'real' function. Consistent with this notion, domain II mutations in five different genes (*IAA3/SHY2, IAA6/SHY1, IAA12/BDL, IAA17/AXR3,* and *IAA18*) from five different clades (Figure 2) cause leaves to curl up (Table 2). Although little detailed expression data are available, at least some of these phenotypes correlate with the expression domains of the *Aux/IAA* genes. For example, expression of a *uidA* gene from the *IAA3* promoter results in GUS protein activity in developing leaves at about the stage when they begin to curl (N. Uhlir, Q. Tian, and J.W. Reed, unpublished results).

Despite these caveats, the gain-of-function phenotypes do provide useful insight into the developmental processes that Aux/IAA proteins, as a class, can potentially regulate. Overall, the phenotypes shown in Table 2 suggest that Aux/IAA proteins mediate several distinct cellular processes at virtually all stages of development. First, they appear to regulate tissue patterning. For example, *iaa12/bdl-1* mutant embryos lack roots (Hamann *et al.*, 1999), and *iaa7/axr2-1* and *iaa17/axr3-1* cause altered gene expression patterns in the root meristem (Sabatini *et al.*, 1999). Similarly, *iaa18-1* mutant seedlings often have fused cotyledons (P. Nagpal and J.W. Reed, unpublished results), suggesting that Aux/IAA proteins can regulate phyllotactic pattern.

Second, several of the mutants have short hypocotyls, and several cause leaves to curl up (Table 2), suggesting that Aux/IAA proteins regulate cell enlargement in multiple growing tissues. This is also reflected in the short root hairs of *iaa7/axr2-1, iaa14/slr1-1,* and *iaa17/axr3-1* mutants, and in defects in differential growth responses required for tropic growth: *iaa7/axr2-1, iaa14/slr-1, iaa17/axr3-1* and *iaa19/msg2* mutant roots and/or shoots are agravitropic (Timpte *et al.*, 1994; Leyser *et al.*, 1996; H.

Fukaki and M. Tasaka, personal communication), and *iaa19/msg2* mutants also have defects in phototropism (E.L. Stowe-Evans and E. Liscum, unpublished results)

Third, multiple mutants have altered numbers of lateral roots and/or inflorescence branches, suggesting that Aux/IAAs regulate cell division in both lateral shoot meristems and lateral root initials in the pericycle. Consistent with the supposition that Aux/IAA proteins mediate auxin responses, most of the mutant phenotypes are in processes known to be influenced by the application of auxin (Davies, 1995). As discussed later, *aux/iaa* mutant phenotypes also suggest that light responses may be mediated in part through Aux/IAA proteins, pointing toward a previously under-appreciated connection between light and auxin signaling pathways.

The gain-of-function mutant phenotypes also suggest some of the complexity permitted by the large number of *Aux/IAA* genes in *Arabidopsis*. In at least three cases, mutations in different *Aux/IAA* genes have contrasting effects on the same auxin response. First, *iaa3/shy2-2, iaa6/shy1-1, iaa7/axr2-1,* and *iaa17/axr3-1* mutant seedlings have shorter hypocotyls than wild-type seedlings, whereas the *iaa18-1* mutant has a longer hypocotyl than the wild type (Timpte *et al.*, 1994; Kim *et al.*, 1996; Leyser *et al.*, 1996; Reed *et al.*, 1998; Nagpal *et al.*, 2000; P. Nagpal and J.W. Reed, unpublished results). Second, *iaa17/axr3-1* plants have increased shoot apical dominance (Leyser *et al.*, 1996), whereas *iaa14/slr-1* and *iaa28-1* mutants have decreased apical dominance in inflorescence shoots (Rogg *et al.*, 2001; H. Fukaki and M. Tasaka, personal communication). Third, *iaa3/shy2-2, iaa14/slr1-1* and *iaa28-1* mutants have fewer lateral roots than wild-type, but *iaa17/axr3-1* plants have more (Tian and Reed, 1999; Rogg *et al.*, 2001; H. Fukaki and M. Tasaka, personal communication). Thus, it appears that different *Aux/IAA* genes have opposing functions. It will be interesting to investigate whether these differences reflect expression in different cells or tissues, or whether they reflect functional differences among the corresponding proteins.

Developmental phenotypes of loss-of-function aux/iaa
mutants

In principle, phenotypes caused by loss-of-function mutations should provide a more accurate understanding of the normal function of the corresponding gene than would phenotypes caused by gain-of-function mutations. However, in contrast to their gain-of-function counterparts, all of the examined loss-of-function *aux/iaa* mutations have relatively minor effects on growth and development. For example, putative null *iaa3/shy2* and *iaa7/axr2* mutations cause subtly elongated hypocotyls, as compared to the strong *shy2-2* and *axr2-1* gain-of-function mutations that cause very short hypocotyls (Tian and Reed, 1999; Nagpal *et al.*, 2000). While *iaa3/shy2-2* gain-of-function mutant seedlings have fewer lateral roots and no 'root waving' growth, *shy2-22* and *shy2-24* loss-of-function mutant seedlings have more lateral roots and greater root waving than wild-type seedlings (Tian and Reed, 1999). In these cases, the loss-of-function phenotypes are the opposite of the gain-of-function phenotypes, suggesting that the latter indeed give useful information on the function of the wild-type gene. Loss-of-function *iaa17/axr3* mutants isolated as intragenic suppressors of *axr3-1* do not have obvious phenotypes, except insofar as they fail to revert all of the phenotypes caused by the parental gain-of-function *axr3-1* mutation (Rouse *et al.*, 1998).

The subtlety of these phenotypes suggests that other Aux/IAA proteins may compensate for absence of one of them in these mutants. This could occur if the genes are redundant, as seems likely considering the phylogeny (Figure 2), or if feedback regulatory mechanisms buffer the effects of loss of a single protein function. Efforts are underway in several laboratories to find insertion mutations in additional *Aux/IAA* genes (P. Nagpal and J.W. Reed, unpublished results; A. Theologis, personal communication; A.M. Jones, personal communication; K. Yamamoto, personal communication), and analyses of these mutants should reveal other functions of *Aux/IAA* genes.

Biological functions of the ARF family

In contrast to loss-of-function mutations in the *Aux/IAA* genes, loss-of-function mutations in ARF genes have quite profound phenotypes. Mutations in three different *ARF* genes (*ARF3*, *ARF5*, and *ARF7*) have been identified through morphological screens (Berleth and Jürgens, 1993; Liscum and Briggs, 1995; Sessions and Zambryski, 1995; Ruegger *et al.*, 1997; Watahiki and Yamamoto, 1997) (see Table 3). Mutations in these genes condition quite distinct phenotypes, indicating that ARF3, ARF5, and ARF7 each control separate morphological and develop-

mental processes and have little functional overlap. Mutations in several other *ARF* genes have been identified in reverse-genetic screens but do not confer such dramatic phenotypes (P. Nagpal and J.W. Reed, unpublished results; A. Theologis, personal communication), suggesting that other ARFs may act redundantly with each other or have less essential roles.

Isolation and characterization of arf3 mutants

Lesions in the *ARF3* gene are defined by the recessive floral patterning *ettin* (*ett*) mutations in *Arabidopsis* (Sessions *et al.*, 1997). The most dramatic phenotypes of the *arf3/ett* mutants are associated with alterations in the normal structure of the gynoecium (Alvarez, 1994; Sessions and Zambryski 1995; Sessions, 1997) (Table 3). Although to a lesser extent, the *arf3/ett* mutations affect the development of the other floral organs as well. In particular the number and size of stamens is reduced, while sepal and petal numbers are increased (Sessions, 1997). The phenotypic changes associated with *arf3/ett* mutants exhibit allele-specific severity that correlates well with the known molecular properties of the mutations. For example, with respect to gynoecium development the presumed *ett-1* null allele (Sessions *et al.*, 1997) loses all or nearly all valve tissue at the expense of expanded internode tissue at the base and stylar tissue at the apex (Sessions and Zambryski, 1995; Sessions, 1997). In contrast, the weak *ett-2* allele, which contains an R_{247} to K substitution within the DNA-binding domain (Sessions *et al.*, 1997), retains fertile valves that are only slightly basally reduced and the excess stylar tissue is restricted to the central region of the gynoecium (Sessions and Zambryski, 1995; Sessions, 1997). The *ett-3* and *ett-4* alleles that contain premature stop codons downstream of the DNA-binding domain are intermediate in their phenotypes, relative to the *ett-1* and *ett-2* alleles (Sessions and Zambryski, 1995; Sessions, 1997; Sessions *et al.*, 1997).

How is ARF3 functioning to regulate floral organ development? In a recent model, Nemhauser and colleagues (2000) proposed that ARF3/ETT regulates apical-basal boundaries within developing gynoecium tissues by positively regulating 'mid-level' auxin responses, or responses induced by concentrations of auxin found within the middle region of the developing gynoecium. As such weak *arf3/ett* alleles (e.g., *ett-2* or *ett-3*) are expected to destabilize both the apical-middle and basal-middle boundaries such that valves are only reduced apically and basally, whereas

Table 3. Loss-of-function mutations in *ARF* genes. Although other phenotypes may exist (see text), only ones common and obvious within an allelic series for each mutant gene are given. All mutations have been described previously: *ett* (Alvarez, 1994; Sessions and Zambryski, 1995); *mp* (Berleth and Jürgens, 1993); *nph4* (Liscum and Briggs, 1995), *tir5* (Ruegger *et al.*, 1997), *msg1* (Watahiki and Yamamoto, 1997).

Gene	Mutation(s)	Most notable phenotype(s)
ARF3	*Ett*	abnormal apical-basal gynoecium development
ARF5	*Mp*	severely reduced leaf vasculature (both numbers and connections)
ARF7	*nph4, tir5, msg1*	impaired differential growth responses in aerial tissues

strong alleles would be expected to so dramatically reduce these boundaries that little or no middle region would form, eliminating valve production at the cost of greatly expanded apical (stylar) and basal (internode/stem-like) tissues. Based on *arf3/ett* mutant phenotypes and the effects of the polar auxin transport inhibitor, *N*-1-naphthylphthalamic acid (NPA), Nemhauser et al. (2000) proposed that in addition to modulating 'mid-level' responses, ARF3/ETT normally represses 'high level' auxin responses in the apical region of developing gynoecium.

Although elegant in its basic simplicity, the aforementioned model is dependent upon two basic assumptions:. (1) that auxin is highest in concentration at the apical tip of the developing gynoecium, and (2) that a basipetal transport stream from the apex exists. At present nothing is currently known about auxin concentrations in different regions of floral apices (including developing gynoecia). Moreover, although polar basipetal transport of exogenously applied auxin has been directly demonstrated for excised floral apices (Okada *et al.*, 1991), a recent study of apical meristems from a *pin-formed 1* (*pin1-1*) inflorescence mutant of *Arabidopsis* that lacks a putative polar auxin transporter (Gälweiler *et al.*, 1998) and NPA-generated 'pins' from tomato vegetative stems suggest that auxin may be transported acropetally into the meristem from subtending tissues (Reinhardt *et al.*, 2000). More sensitive methods for measuring auxin gradients within plant tissues (Uggla *et al.*, 1996; Tuominen *et al.*, 1997; Uggla *et al.*, 1998) will need to be explored to address fully the potential limitations of the aforementioned model.

Unfortunately, any model proposed to explain the *arf3/ett* mutant phenotypes suffers in mechanistic terms since we know so little about the regulatory properties of the ARF3/ETT protein. For example, while full-length ARF3 can bind to a palindromic

AuxRE multimer *in vitro*, this interaction is weak and unstable (Ulmasov *et al.*, 1999a). ARF3 lacks the carboxyl-terminal domains III and IV (Sessions *et al.*, 1997; Ulmasov *et al.*, 1999a) that appears necessary for homo- and heterodimerization and stable DNA binding of all ARFs except ARF1 (Ulmasov *et al.*, 1999a). Thus, if ARF3 acts as a stable DNA-binding protein *in planta* it likely does so through some novel mechanism, such as interaction with an unidentified 'coupling' protein (Ulmasov *et al.*, 1999a). Such a hypothesis seems in line with the findings that *arf7* mutants containing C-terminal truncations that remove the protein-protein interaction domain exhibit weaker phenotypes than null mutants (see below). However, unlike ARF7, ARF3 does not appear to function as either a transcriptional repressor or enhancer when examined in a transient expression assay in carrot protoplasts (Ulmasov *et al.*, 1999b). It is therefore unclear at present if ARF3/ETT acts as a transcriptional regulator, or whether it may have an unrelated auxin-dependent function.

Isolation and characterization of arf5 mutants

Historically, the first identified mutations in an *ARF* gene came with the isolation of the *monopteros* (*mp*) mutants of *Arabidopsis* (Berleth and Jürgens, 1993), which were later shown to be alleles of the *ARF5* locus (Hardke and Berleth, 1998) (Table 3). Although originally found in screens for mutants with embryo patterning defects (e.g., strong *arf5/mp* alleles completely fail to form a proper embryo axis) (Berleth and Jürgens, 1993), both strong and weak *arf5/mp* alleles also exhibit post-embryonic defects. The most common phenotype of *arf5/mp* mutants is the disruption of vascular tissues, both with respect to patterning and differentiation (Berleth and Jürgens, 1993; Przemeck *et al.*, 1996; Hardke and Berleth, 1998; Deyholos *et al.*, 2000). In particular, cotyledons and true leaves of *arf5/mp* mutants have dramatically reduced numbers of veins, including a general lack of marginal veins, and fewer connective branches between major viens (Przemeck *et al.*, 1996; Mattsson *et al.*, 1999). Less severe vascular defects have been observed in *arf5/mp* mutant stem tissues with xylem strands often being improperly aligned and disconnected (Przemeck *et al.*, 1996).

Although null *arf5/mp* alleles are presumed lethal, strong and weak alleles have been identified and the strength of their phenotypes are correlated with the position of the lesion within the predicted mutant ARF5/MP protein (Hardtke and Berleth, 1998). In particular, all six *arf5/mp* alleles that have been sequenced contain premature stop codons downstream of the DBD, and allele strength decreases as the mutation position moves more carboxyl-terminally. ARF5 function therefore appears to be tightly coupled to the presence of the carboxyl-terminal dimerization region (domains III and IV, Figure 1), suggesting that dimerization is critical for normal function. Consistent with this hypothesis is the knowledge that ARF5 binds strongly to AuxREs as a dimer (Ulmasov *et al.*, 1999a) and can act as a strong transcriptional activator in transient expression assays (Ulmasov *et al.*, 1999b).

Auxin has long been proposed to regulate vascular development, with polar transport being of particular importance (Sachs, 1981, 1991; Nelson and Dengler, 1997). Studies in *Arabidopsis* with inhibitors of polar auxin transport and mutants disrupted specifically in components of polar auxin transport systems (e.g., *pin1* mutants) provide support for this hypothesis (Mattsson *et al.*, 1999; Sieburth, 1999; Tsiantis *et al.*, 1999). The *arf5/mp* mutants also exhibit impaired polar auxin transport (Przemeck *et al.*, 1996). However, it remains possible that a defect in auxin transport may not cause the venation problems, but rather may secondarily arise from the altered vascular organization (Mattsson *et al.*, 1999). Thus, while the extent to which auxin transport may directly regulate vascular development is still debatable (Mattsson *et al.*, 1999; Koizumi *et al.*, 2000), it is clear that ARF5/MP, an auxin-responsive transcriptional activator (Ulmasov *et al.*, 1999b), likely regulates the expression of genes necessary for the patterning and differentiation of vascular tissues.

Isolation and characterization of arf7 mutants

Mutations in *ARF7* have been isolated in three independent and distinct genetic screens in *Arabidop-*

sis. Three *arf7* alleles were identified as the *nph4* *(non-phototropic hypocotyl 4)* mutants (representing the *nph4-1* to *nph4-3* alleles) that failed to exhibit hypocotyl phototropism in response to long-term low fluence rate unilateral blue light (Liscum and Briggs, 1995). Two additional *arf7* alleles were identified in a screen for seedlings resistant to NPA, and were originally designated *tir5-1* and *tir5-2* (for *transport inhibitor-resistant 5*) to reflect this phenotype (Ruegger *et al.*, 1997), but later designated *nph4-4* and *nph4-5* (Stowe-Evans *et al.*, 1998). Finally, the nine *msg1 (massugu 1)* mutants identified in a screen for seedlings that did not exhibit curvature in response to auxin applied externally to one side of the hypocotyl have been shown to represent *arf7* alleles (Watahiki and Yamamoto, 1997; Harper *et al.*, 2000). These latter mutants have been re-designated *nph4-101* to *nph4-109* (Harper *et al.*, 2000). In addition to the phenotypes for which they were identified, the *arf7/nph4* mutants have been shown to exhibit impaired hypocotyl gravitropism, altered apical hook maintenance, and epinastic or hyponastic leaves (depending on the allele) (Liscum and Briggs, 1996; Watahiki and Yamamoto, 1997; Stowe-Evans *et al.*, 1998; Watahiki *et al.*, 1999; Harper *et al.*, 2000). Because all of the documented growth defects of *arf7/nph4* mutants are associated with differential growth it appears that ARF7/NPH4 responds to a gradient of auxin concentration within a given tissue to modulate localized cell elongation processes (Harper *et al.*, 2000). At least for organs responding to tropic stimuli there is considerable evidence for the establishment of a lateral auxin gradient capable of promoting differential growth (Went and Thimann, 1937; Kaufman *et al.*, 1995; Lomax *et al.*, 1995).

Like the *arf3/ett* and *arf5/mp* mutants, *arf7/nph4* mutants form an allelic series that varies in phenotypic strength (Watahiki and Yamamoto, 1997; Harper *et al.*, 2000). Recovery of alleles containing premature stop codons within the middle region activation domain indicates that the C-terminal dimerization region (domains III and IV) of ARF7 is required for its normal function (Harper *et al.*, 2000), similar to what has been hypothesized from the phenotypes of like *arf5/mp* mutants (Hardke and Berleth, 1998). However, the finding that such *arf7/nph4* alleles are phenotypically weaker than the apparent null allele, *arf7/nph4-1* (Harper *et al.*, 2000), suggests that some activity is retained by truncated ARF7/NPH4 protein. Although the DNA-binding properties of ARF7/NPH4 have not been reported it is likely to form strong stable interactions with AuxREs as a homodimer with itself or another ARF since this seems to be the case for all ARF proteins tested to date (Ulmasov et al., 1999a). How then can we explain the apparent activity of ARF7/NPH4 proteins lacking the C-terminal region? Two obvious possibilities exist. First, a truncated ARF7/NPH4 protein might still dimerize with itself or another ARF through an undefined amino-terminal/middle region protein-protein interaction motif. Alternatively, an N-terminal or middle region motif might allow interaction with an unrelated coupling protein to promote interaction with AuxRE sequences. Both of these scenarios are dependent upon the amino-terminal or middle region of an ARF protein being able to support protein-protein interactions. Some evidence does exist to support this thesis. For example, truncated ARF1 lacking domains III and IV has been shown to bind palindromic AuxREs as stably as full-length protein that prefers such palindromic sequences (two binding sites) over AuxRE half-sites (one binding site) (Ulmasov et al., 1997a). With respect to ARF7/NPH4 itself, support for protein-protein interaction within the amino terminus comes from the nph4-3 allele that is predicted to make a protein only 192 amino acids long that actually lacks the DNA-binding motif (Harper et al., 2000). This mutant exhibits a gravitropic defect that is stronger than the nph4-1 null allele, suggesting that mutant ARF7/NPH4-3 protein may act as a dominant-negative by interacting with and sequestering another ARF or coupling factor, thus preventing auxin-dependent transcription (Harper et al., 2000).

Consistent with ARF7's function as a transcriptional activator in transient expression assays (Ulmasov et al., 1999b), the arf7/nph4-1 null allele exhibits dramatically reduced levels of mRNA for a number of genes that are normally responsive to exogenously applied auxin (Stowe-Evans et al., 1998). Reduced expression of the auxin-responsive DR5::GUS reporter gene in the basal region of a strong arf5/mp mutant seedling has also been reported (Sabatini et al., 1999). However, because the developmental defects of the arf5/mp mutant are so severe, it is unclear whether the reduced expression of the reporter gene results from altered direct interaction between ARF5/MP and the synthetic promoter, or as a secondary consequence (e.g., altered expression of some other auxin-responsive transcription factor) of the patterning and morphology defects conditioned by the arf5/mp mutation. No changes in auxin-dependent gene expression have been reported for arf3/ett mu-

tants. While it is not known whether any of the genes exhibiting reduced expression in the arf7/nph4 background (Stowe-Evans et al., 1998) represent targets for ARF7/NPH4 regulation in response to endogenous fluxes of auxin in wild-type plants, the findings are significant because they indicate that ARF7 protein can act as a transcriptional activator in planta and that at least normal differential growth responses are likely to require auxin-induced gene expression.

Aux/IAA and ARF proteins may link light perception and auxin response pathways

Several of the mutants discussed in previous sections exhibit altered light responses, suggesting that light may act in part by altering auxin responses. For example, gain-of-function mutations in IAA3/SHY2 and IAA6/SHY1 were isolated as suppressors of the morphological phenotypes of mutants deficient in the activity of the phytochrome red/far-red photoreceptors (Kim et al., 1996a; Reed et al., 1998). In addition, iaa3/shy2, iaa7/axr2 and iaa17/axr3 gain-of-function mutations all promote leaf formation in darkness (Kim et al., 1996a, b; Soh et al., 1999; Nagpal et al., 2000), a response normally triggered by light (Kendrick and Kronenberg, 1994). Loss-of-function alleles of ARF7/NPH4 also exhibit phenotypes that provide a connection between phytochromes and auxin. In particular, arf7/nph4 null mutants lack a phytochrome B (phyB)-dependent hypocotyl curvature response (Stowe-Evans et al., 1998). Moreover, arf7/nph4 mutants recover a phototropic response when both the phototropic receptor phototropin 1 (phot1, previously called nph1; Briggs et al., 2001; Sakai et al., 2001) and phyA are activated, despite being completely aphototropic when only phot1 is active (Liscum and Briggs, 1996; Stowe-Evans et al., 2001). As discussed above, loss of differential growth responses, such as basal phot1-dependent phototropism (Liscum and Briggs, 1996; Stowe-Evans et al., 2001), in the arf7/nph4 mutants indicates that auxin-induced ARF7/NPH4-dependent gene expression is an integral component of such responses. The simplest hypothesis to explain why phyA activation can bypass the need for ARF7/NPH4 is that another ARF may respond to phyA signaling by activating the expression of genes that ARF7/NPH4 would otherwise activate (Stowe-Evans et al., 2001).

What is the mechanistic connection between phytochromes and Aux/IAA-ARF proteins? Two lines of evidence suggest that phytochromes may modulate Aux/IAA protein activity directly. First, two IAA proteins, IAA26/PAP1 (PHYTOCHROME-ASSOCIATED PROTEIN 1) and IAA27/PAP2, were identified in a yeast two-hybrid screen using phyA as a bait (Choi *et al.*, 1999; Soh *et al.*, 1999). Second, phyA can phosphorylate several Aux/IAA proteins *in vitro* (Colón-Carmona *et al.*, 2000). The lack of light-dependence for this *in vitro* phosphorylation (Colón-Carmona *et al.*, 2000) may seem confounding, however phytochrome migration from the cytoplasm to the nucleus in response to light is a major regulatory event in the modulation of light responses (Nagy and Schäfer, 2000; Fankhauser, 2001). Moreover, recent studies of phytochrome action have revealed that phytochromes indeed interact directly with at least one other nuclear transcription factor, PIF3, to regulate its activity (Martinez-Garcia *et al.*, 2000). Thus, *in planta* phytochrome is activated by light while in the cytoplasm and then translocates to the nucleus where it may interact with and phosphorylate substrates such as Aux/IAA proteins. The findings that at least seven different Aux/IAAs interact with phyA, either *in vitro* (Colón-Carmona *et al.*, 2000) or in a yeast two-hybrid system (Soh *et al.*, 1999), suggest that this phytochrome activity may be quite general.

Although it is not yet clear if Aux/IAAs are phosphorylated by phytochromes *in vivo* (Colón-Carmona *et al.*, 2000), such an activity, if it exists, might regulate the stability of Aux/IAA proteins, which could in turn modulate the amount of ARF-dependent changes in transcription. Would phosphorylation of an Aux/IAA protein stabilize or destabilize that protein? As mentioned earlier, the Aux/IAA proteins might be degraded through the SCF^{TIR1} ubiquitin-ligase complex (Gray and Estelle, 2000). Interestingly, in yeast and flies, most proteins destined for degradation by the SCF complex are first phosphorylated (Maniatis, 1999). To the extent that Aux/IAA proteins repress ARF activity (Ulmasov *et al.*, 1997b), a phytochrome-dependent destabilization of Aux/IAAs would lead to increased ARF activity. This scenario is consistent with the model above that phyA potentiates phot1-induced phototropism in the *arf7/nph4*-null mutant by activating another ARF protein (Liscum and Briggs, 1996; Stowe-Evans *et al.*, 2001). Interestingly, mutations in *AXR1*, which encodes one of the two components (the other being ECR1) of a heterodimeric ubiquitin-E1-like enzyme that functions upstream of the SCF^{TIR1} complex (Walker and Estelle, 1998; del Pozo and Estelle, 1999; Gray and

Estelle, 2000), eliminate this effect in the *nph4* mutant background (E.L. Stowe-Evans and E. Liscum, unpublished results). This suggests that AXR1/ECR1-stimulated protein turnover (possibly of Aux/IAA proteins) is essential for this hypothesized redundant ARF activation. The recent identification of an F-box protein (EID1) involved in phyA signaling further suggests a role for a proteolytic pathway in the processing of phyA-derived/modified signals (Büche *et al.*, 2000; Dieterle *et al.*, 2001).

An alternative (contrasting) hypothesis would be required to explain why some gain-of-function mutations in *Aux/IAA* genes cause ectopic 'light' responses such as hypocotyl growth inhibition and leaf formation in darkness (Kim *et al.*, 1998; Soh *et al.*, 1999; Nagpal *et al.*, 2000). In this case, phytochrome phosphorylation of Aux/IAA proteins might promote increased, rather than decreased, stability. The stabilized Aux/IAA proteins might then dimerize with and inhibit other ARF or Aux/IAA proteins that might otherwise repress light response outputs. This alternative could also be used to explain the phyA-mediated recovery of phototropism in the *arf7/nph4* mutants, if the targets of the stabilized Aux/IAA proteins themselves repress (through heterodimerization) the ARF that can substitute for ARF7. It is clear that additional *in planta* biochemical studies will be required to address more directly how light affects Aux/IAA and ARF stability and function.

Conclusions and future prospects

Observed spatial patterns of auxin-mediated gene expression, cell division, and cell elongation, as well as experimental manipulations of auxin concentrations, suggest that auxins act as morphogens to pattern various tissues and responses, as classically defined in metazoan systems (Neumann and Cohen, 1997). As many of the *aux/iaa* and *arf* mutants discussed above confer resistance to exogenous auxin, it seems likely that Aux/IAA and ARF proteins may be central to 'reading' auxin concentration gradients, and then translating this information into gradients of gene expression that cause morphological and developmental patterning outputs. For example, the *arf3/ett*, *arf5/mp*, and *arf7/nph4* phenotypes are consistent with a failure to 'sense' a gradient of auxin along the length of the developing carpel, in developing vasculature, and laterally across growing stems, respectively (Went and Thimann, 1937; Lomax *et al.* 1995; Kaufman

398

et al., 1995; Mattsson *et al.*, 1999; Sieburth, 1999; Nemhauser *et al.*, 2000). Similarly, altered root patterning in *iaa7/axr2* and *iaa17/axr3* mutants suggests that Aux/IAA proteins help to pattern the root meristem (Sabatini *et al.*, 1999). Moreover, cotyledon fusion in *iaa18-1* seedlings suggests that the IAA18 protein may be required for the response to a positional gradient in a ring around the meristem (P. Nagpal and J.W. Reed, unpublished results). Aux/IAAs or ARFs might conceivably act as auxin receptors (an untested possibility that deserves attention), or their activities and/or stability may be regulated quantitatively by specific concentrations of auxin perceived at the start of an unknown signal transduction system. On the other hand, the possibility that altered polar auxin transport in *arf5/mp* mutant stems may result from altered developmental patterning of vascular tissues (Mattsson *et al.*, 1999) suggests that Aux/IAA and ARF protein activities may be required to establish the auxin gradient rather than respond to it. These alternative visions may come together in the intercellular feedback loops embodied in the canalization hypothesis for vascular patterning (Sachs, 1991), whereby auxin itself regulates auxin transport.

While it is most probable that the Aux/IAA and ARF proteins are involved in 'sensing' auxin gradients by activating/repressing gene expression in localized regions, many questions still remain. How is auxin perceived? Could Aux/IAA and/or ARF proteins be nuclear auxin-binding proteins in a fashion reminiscent of the steroid hormone receptors in metazoans (Freedman 1998; Karin 1998)? If not receptors themselves, how does auxin regulate the transcriptional activities of the Aux/IAA and ARF proteins? Is proteolysis of Aux/IAAs key to this regulation, or just one of many players? Once 'activated' by auxin, what are the target genes for regulation by Aux/IAA and ARF proteins that lead to the observed changes in patterning and morphology? How do other sensory-response pathways (e.g., phytochrome pathways) modulate Aux/IAA-ARF-dependent transcription? Are there direct connections as proposed here for phyA and Aux/IAAs, or are such modulatory activities indirect through changes in auxin levels, for example? Many of the answers to these questions are now within reach with the recent completion of the *Arabidopsis* genome sequencing project (Arabidopsis Genome Initiative, 2000). We are now in a position to identify loss- and gain-of-function alleles of all of the *Aux/IAA* and *ARF* genes, as well as the genes encoding putative auxin transporters and enzymes involved in auxin

metabolism (synthesis, conjugation, and catabolism). Moreover, studies of auxin induced/repressed gene expression can now be done on a whole genome level to identify targets for regulation by ARF and Aux/IAA proteins under various conditions. These are exciting times for auxinologists as we are limited mainly by our ability to come up with clever and interesting questions to ask, and then sort through all of the data it is possible to generate with the technologies at hand!

Acknowledgements

We would like to thank Dr Dave Remington for sharing data from his phylogenetic studies of the Aux/IAA and ARF proteins. We also thank Drs Tom Guilfoyle and Gretchen Hagen for critical reading of the manuscript. Unpublished works cited herein from the Liscum laboratory have been supported by grants from NSF (MCB-9723124 and MCB-0077312).

References

Abel, S. and Theologis, A. 1996. Early genes and auxin action. Plant Physiol. 111: 9–17.

Abel, S., Oeller, P.W. and Theologis, A. 1994. Early auxin-induced genes encode short-lived nuclear proteins. Proc. Natl. Acad. Sci. USA 91: 326–330.

Abel, S., Nguyen, M.D. and Theologis, A. 1995. The *PS-IAA4/5*-like family of early auxin-inducible mRNAs in *Arabidopsis thaliana*. J. Mol. Biol. 251: 533–549.

Alvarez, J. 1994. The *ETTIN* gene. In: J., Bowman (Ed.) Arabidopsis: An Atlas of Morphology, Springer-Verlag, New York, pp. 268–269.

Arabidopsis Genome Initiative. 2000. Analysis of the genome sequence of the flowering plant *Arabidopsis thaliana*. Nature 408: 796–815.

Berleth, T. and Jürgens, G. 1993. The role of the *MONOPTEROS* gene in organizing the basal body region of the *Arabidopsis* embryo. Development 118: 575–587.

Briggs, W.R., Beck, C.F., Cashmore, A.R., Christie, J.M., Hughes, J., Jarillo, J.A., Kagawa, T., Kanegae, H., Liscum, E., Nagatani, A., Okada, K., Salomon, M., Rüdiger, W., Sakai, T., Takano, M., Wada, M. and Watson, J.C. 2001. The phototropin family of photoreceptors. Plant Cell 13: 993–997.

Büche, C., Poppe, C., Schäfer, E. and Kretsch, T. 2000. *eid1*: a new *Arabidopsis* mutant hypersensitive in phytochrome A-dependent high-irradiance responses. Plant Cell 12: 547–558.

Choi, G., Yi, H., Lee, J., Kwon, Y.-K., Soh, M.S., Shin, B., Luka, Z., Hahn, T.-R. and Song, P.-S. 1999. Phytochrome signaling is mediated through nucleoside diphosphate kinase 2. Nature 401: 610–613.

Colón-Carmona, A., Chen, D.L., Yeh, K.-C. and Abel, S. 2000. Aux/IAA proteins are phosphorylated by phytochrome *in vitro*. Plant Physiol. 124: 1728–1738.

Cooke, T.J., Poli, D.B., Sztein, A.E. and Cohen, J.D. 2002. Evolutionary patterns in auxin action. Plant Mol. Biol. 49: 319–338.

Davies, P.J. 1995. Plant Hormones, Kluwer Academic Publishers, Dordrecht, Netherlands.

del Pozo, J.C. and Estelle, M. 1999. Function of the ubiquitin-proteosome pathway in auxin responses. Trends Plant Sci. 4: 107–112.

Deyholos, M.K., Cordner, G., Beebe, D. and Sieburth, L.E. 2000. The SCARFACE gene is required for cotyledon and leaf vein patterning. Development 127: 3205–3213.

Dharmasiri, S. and Estelle, M. 2002. The role of regulated protein degradation in auxin response. Plant Mol. Biol. 49: 401–408.

Dieterle, M., Zhou, Y.-C., Schäfer, E., Funk, M. and Kretsch, T. 2001. EID1, an F-box protein involved in phytochrome A-specific light signaling. Genes Dev. 15: 939–944.

Fankhauser, C. 2001. The phytochromes, a family of red/far-red absorbing photoreceptors. J. Biol. Chem. 276: 11453–11456.

Franco, A.R., Gee, M.A. and Guilfoyle, T. 1991. Induction and superinduction of auxin-responsive mRNAs with auxin and protein synthesis inhibitors. J. Biol. Chem. 265: 15845–15849.

Freedman, L.P. 1998. Molecular Biology of Steroid and Nuclear Hormone Receptors. Birkhäuser, Switzerland.

Fukaki, H. and Tasaka, M. 1998. SLR, a novel genetic locus involved in auxin signaling in Arabidopsis thaliana. 9th International Conference on Arabidopsis Research, Madison, WI.

Gälweiler, L., Changhui, G., Müller, A., Wisman, E., Mendgen, K., Yephremov, A. and Palme, K. 1998. Regulation of polar auxin transport by AtPIN1 in Arabidopsis vascular tissue. Science 282: 2226–2230.

Gil, P. and Green, P.J. 1997. Regulatory activity exerted by the SAUR-AC1 promoter region in transgenic plants. Plant Mol. Biol. 34: 803–808.

Gray, W.M. and Estelle, M. 2000. Function of the ubiquitin-proteosome pathway in auxin response. Trends Biochem. Sci. 25: 133–138.

Guilfoyle, T., Hagen, G., Ulmasov, T. and Murfett, J. 1998a. How does auxin turn on genes? Plant Physiol. 118: 341–347.

Guilfoyle, T., Ulmasov, T. and Hagen, G. 1998b. The ARF family of transcription factors and their role in plant hormone-responsive transcription. Cell Mol. Life Sci. 54: 619–627.

Hagen, G. and Guilfoyle, T. 2002. Auxin-responsive gene expression: genes, promoters and regulatory factors. Plant Mol. Biol. 49: 373–385.

Hamann, T., Mayer, U. and Jürgens, G. 1999. The auxin-insensitive bodenlos mutation affects primary root formation and apical-basal patterning in the Arabidopsis embryo. Development 126: 1387–1395.

Hardtke, C.S. and Berleth, T. 1998. The Arabidopsis gene MONOPTEROS encodes a transcription factor mediating embryo axis formation and vascular development. EMBO J. 17: 1405–1411.

Harper, R.M., Stowe-Evans, E.L., Luesse, D.R., Muto, H., Tatematsu, K., Watahiki, M.K., Yamamoto, K. and Liscum, E. 2000. The NPH4 locus encodes the auxin response factor ARF7, a conditional regulator of differential growth in aerial Arabidopsis tissue. Plant Cell 12: 757–770.

Karin, M. 1998. New twists in gene regulation by glucocorticoid receptor: Is DNA binding dispensable? Cell 93: 487–490.

Kaufman, P.B., Wu, L.-L., Brock, T.G. and Kim, D. 1995. Hormones and orientation of growth. In: P.J. Davies (Ed.) Plant Hormones, Kluwer Academic Publishers, Dordrecht, Netherlands, pp. 547–571.

Kendrick, R.E. and Kronenberg, G.H.M. 1994. Photomorphogenesis in Plants. Kluwer Academic Publishers, Dordrecht, Netherlands.

Key, J.L. 1969. Hormones and nucleic acid metabolism. Annu. Rev. Plant Physiol. 20: 449–474.

Kim, B.C., Soh, M.S., Kang, B.J., Furuya, M. and Nam, H.G. 1996a. Two dominant photomorphogenic mutations of Arabidopsis thaliana identified as suppressor mutations of hy2. Plant J. 9: 441–456.

Kim, B.C., Soh, M.O., Hong, S.H., Furuya, M. and Nam, H.G. 1996b. Photomorphogenic development of the Arabidopsis shy2-1D mutation and its interaction with phytochromes in darkness. Plant J. 15: 61–68.

Kim, J., Harter, K. and Theologis, A. 1997. Protein-protein interactions among the Aux/IAA proteins. Proc. Natl. Acad. Sci. USA 94: 11786–11791.

Koizumi, K., Sugiyama, M. and Fukuda, H. 2000. A series of novel mutants of Arabidopsis thaliana that are defective in the formation of continuous vascular network: calling the auxin signal flow eanalization hypothesis into question. Development 127: 3197–3204.

Leyser, O. and Berleth, T. 1999. A molecular basis for auxin action. Cell Mol. Dev. Biol. 10: 131–137.

Leyser, H.M.O., Pickett, F.B., Dharmasiri, S. and Estelle, M. 1996. Mutations in the AXR3 gene of Arabidopsis result in altered auxin response including ectopic expression from the SAUR-AC1 promoter. Plant J. 10: 403–413.

Liscum, E. and Briggs, W.R. 1995. Mutations in the NPH1 locus of Arabidopsis disrupt the perception of phototropic stimuli. Plant Cell 7: 473–485.

Liscum, E. and Briggs, W.R. 1996. Mutations of Arabidopsis in potential transduction and response components of the phototropic signaling pathway. Plant Physiol. 112: 291–296.

Lomax, T.L., Muday, G.K. and Rubery, P.H. 1995. Auxin transport. In: P.J. Davies (Ed.) Plant Hormones, Kluwer Academic Publishers, Dordrecht, Netherlands, pp. 509–530.

Maniatis, T. 1999. A ubiquitin ligase complex essential for the NF-B, Wnt/Wingless, and Hedgehog signaling pathways. Genes Dev. 13: 505–510.

Martinez-Garcia, J.F., Huq, E. and Quail, P.H. 2000. Direct targeting of light signals to a promoter element-bound transcription factor. Science 288: 859–863.

Mattsson, J., Sung, Z.R. and Berleth, T. 1999. Responses of plant vascular systems to auxin transport inhibition. Development 126: 2979–2991.

Nagpal, P., Walker, L., Young, J., Sonawala, A., Timpte, C., Estelle, M. and Reed, J.W. 2000. AXR2 encodes a member of the Aux/IAA protein family. Plant Physiol. 123: 563–573.

Nagy, F. and Schäfer, E. 2000. Nuclear and cytosolic events of light-induced, phytochrome-regulated signaling in higher plants EMBO J. 19: 157–163.

Nelson, T. and Dengler, N. 1997. Leaf vascular pattern formation. Plant Cell 9: 1121–1135.

Nemhauser, J.L., Feldman, L.J. and Zambryski, P.C. 2000. Auxin and ETTIN in Arabidopsis gynoecium morphogenesis. Development 127: 3877–3888.

Neumann, C. and Cohen, S. 1997. Morphogens and pattern formation. BioEssays 19: 721–729.

Okada, K., Ueda, J., Komaki, M.K., Bell, C.J. and Shimura, Y. 1991. Requirement of the auxin polar transport system in early stages of Arabidopsis floral bud formation. Plant Cell 3: 677–684.

Ouellet, F., Overvoorde, P.J. and Theologis, A. 2001. IAA17/AXR3: biochemical insight into an auxin mutant phenotype. Plant Cell 13: 829–841.

Przemeck, G.K.H., Mattsson, J., Hardtke, C.S., Sung, Z.R. and Berleth, T. 1996. Studies on the role of the Arabidopsis gene

400

MONOPTEROS in vascular development and plant cell axialization. Planta 200: 229–237.

Reed, J.W., Elumalai, R.P. and Chory, J. 1998. Suppressors of an *Arabidopsis thaliana phyB* mutation identify genes that control light signalling and hypocotyl elongation. Genetics 148: 1295–1310.

Reinhardt, D., Mandel, T. and Kuhlemeier, C. 2000. Auxin regulates the initiation and radial position of plant lateral organs. Plant Cell 12: 507–518.

Rogg, L.E., Lasswell, J. and Bartel, B. 2001. A gain-of-function mutation in *IAA28* suppresses lateral root development. Plant Cell 13: 465–480.

Rouse, D., Mackay, P., Stirnberg, P., Estelle, M. and Leyser, O. 1998. Changes in auxin response from mutations in an *AUX/IAA* gene. Science 279: 1371–1373.

Ruegger, M., Dewey, E., Hobbie, L., Brown, D., Bernasconi, P., Turner, J., Muday, G. and Estelle, M. 1997. Reduced naphthylphthalamic acid bidning in the *tir3* mutant of *Arabidopsis* is associated with a reduction in polar auxin transport and diverse morphological defects. Plant Cell 9: 745–757.

Sabatini, S., Beis, D., Wolkenfelt, H., Murfett, J., Guilfoyle, T., Malamy, J., Benfey, P., Leyser, O., Bechtold, N., Weisbeek, P and Scheres, B. 1999. An auxin-dependent distal organizer of pattern and polarity in the *Arabidopsis* root. Cell 99: 463–472.

Sachs, T. 1981. The control of the patterned differentiation of vascular tissues. Adv. Bot. Res. 9: 152–262.

Sachs, T. 1991. Cell polarity and tissue patterning in plants. Development (Suppl. 1): 83–93.

Sakai, T., Kagawa, T., Kasahara, M., Swartz, T.E., Christie, J.M., Briggs, W.R., Wada, M. and Okada, K. 2001. *Arabidopsis* nph1 and npl1: Blue light receptors that mediate both phototropism and chloroplast relocation. Proc. Natl. Acad. Sci. USA, in press.

Sessions, R.A. 1997. *Arabidopsis* (Brassicaceae) flower development and gynoecium patterning in wild-type and *ettin* mutants. Am. J. Bot. 84: 1179–1191.

Sessions, R.A. and Zambryski, P.C. 1995. *Arabidopsis* gynoecium structure in the wild-type and *ettin* mutants. Development 121: 1519–1532.

Sessions, A., Nemhauser, J.L., McColl, A., Roe, J.L., Feldmann, K.A. and Zambryski, P.C. 1997. *ETTIN* patterns the *Arabidopsis* floral meristem and reproductive organs. Development 124: 4481–4491.

Sieburth, L. 1999. Auxin is required for leaf vein pattern in Arabidopsis. Plant Physiol. 121: 1179–1190.

Sitbon, F. and Perrot-Rechenmann, C. 1997. Expression of auxin-regulated genes. Physiol. Plant. 100: 443–455.

Soh, M.S., Hong, S.H., Kim, B.C., Vizir, I., Park, D.H., Choi, G., Hong, M.Y., Chung, Y.-Y., Furuya, M. and Nam, H.G. 1999. Regulation of both light- and auxin-mediated development by the *Arabidopsis IAA3/SHY2* gene. J. Plant Biol. 42: 239–246.

Stowe-Evans, E.L., Harper, R.M., Motchoulski, A.V. and Liscum, E. 1998. NPH4, a conditional modulator of auxin-dependent differential growth responses in *Arabidopsis*. Plant Physiol. 118: 1265–1275.

Stowe-Evans, E.L., Luesse, D.R. and Liscum, E. 2001. The enhancement of phototropin-induced phototropic curvature in Arabidopsis occurs via a photoreversible phytochrome A-dependent modulation of auxin responsiveness. Plant Physiol. 126: 826–834.

Tatematsu, K., Watahiki, M.K. and Yamamoto, K.T. 1999. Evidences for a dominant mutation of *IAA19* that disrupts hypocotyl

growth curvature responses and alters auxin sensitivity. 10th International Conference on Arabidopsis Research, Melbourne, Australia.

Theologis, A. 1986. Rapid gene regulation by auxin. Annu. Rev. Plant Physiol. 37: 407–438.

Tian, Q. and Reed, J.W. 1999. Control of auxin-regulated root development by the *Arabidopsis thaliana SHY2/IAA3* gene. Development 126: 711–721.

Timpte, C.S., Wilson, A.K. and Estelle, M. 1992. Effects of the *axr2* mutation of *Arabidopsis* on cell shape in hypocotyl and inflorescence. Planta 188: 271–278.

Timpte, C., Wilson, A. and Estelle, M. 1994. The *axr2-1* mutation of *Arabidopsis thaliana* is a gain-of-function mutation that disrupts an early step in auxin response. Genetics 138: 1239–1249.

Tsiantis, M., Brown, M.I.N., Skibinski, G. and Langdale, J.A. 1999. Disruption of auxin transport is associated with aberrant leaf development in maize. Plant Physiol. 121: 1163–1168.

Tuominen, H., Puech, L., Fink, S. and Sundberg, B. 1997. A radial concentration gradient in indole-3-acetic acid is related to secondary xylem development in hybrid aspen. Plant Physiol. 115: 577–585.

Ulmasov, T., Hagen, G. and Guilfoyle, T.J. 1997a. ARF1, a transcription factor that binds to auxin response elements. Science 276: 1865–1868.

Ulmasov, T., Murfett, J., Hagen, G. and Guilfoyle, T.J. 1997b. Aux/IAA proteins repress expression of reporter genes containing natural and highly active synthetic auxin response elements. Plant Cell 9: 1963–1971.

Ulmasov, T., Hagen, G. and Guilfoyle, T.J. 1999a. Dimerization and DNA binding of auxin response factors. Plant J. 19: 309–319.

Ulmasov, T., Hagen, G. and Guilfoyle, T.J. 1999b. Activation and repression of transcription by auxin-response factors. Proc. Natl. Acad. Sci. USA 96: 5844–5849.

Uggla, C., Mellerowicz, E.J. and Sundberg, B. 1998. Indole-3-acetic acid control cambial growth in scots pine by positional signaling. Plant Physiol. 117: 113–121.

Uggla, C., Moritz, T., Sandberg, G. and Sundberg, B. 1996. Auxin as a positional signal in pattern formation in plants. Proc. Natl. Acad. Sci. USA 93: 9282–9286.

Vision, T.J., Brown, D.G. and Tanksley, S.D. 2000. The origins of genomic duplications in *Arabidopsis*. Science 290: 2114–2117.

Walker, L. and Estelle, M. 1998. Molecular mechanisms of auxin action. Curr. Opin. Plant Biol. 1: 434–439.

Watahiki, M.K. and Yamamoto, K. 1997. The *massagu1* mutation of *Arabidopsis* identified with failure of auxin-induced growth curvature of hypocotyl confers auxin insensitivity to hypocotyl and leaf. Plant Physiol. 115: 419–426.

Watahiki, M.K., Tatematsu, K., Fujihira, K., Yamamoto, M. and Yamamoto, K. 1999. The *MSG1* and *AXR1* genes of *Arabidopsis* are likely to act independently in growth-curvature responses of hypocotyl. Planta 207: 362–369.

Went, F.W. and Thimann, K.V. 1937. Phytohormones, Macmillan, New York.

Wilson, A.K., Pickett, F.B., Turner, J.C. and Estelle, M. 1990. A dominant mutation in *Arabidopsis* confers resistance to auxin, ethylene and abscisic acid. Mol. Gen. Genet. 222: 377–383.

Worley, C.K., Zenser, N., Ramos, J., Rouse, D., Leyser, O., Theologis, A. and Callis, J. 2000. Degradation of Aux/IAA proteins is essential for normal auxin signaling. Plant J. 21: 553–562.

Plant Molecular Biology **49**: 401–409, 2002.
Perrot-Rechenmann and Hagen (Eds.), Auxin Molecular Biology.
© 2002 *Kluwer Academic Publishers.*

The role of regulated protein degradation in auxin response

Sunethra Dharmasiri and Mark Estelle
Institute for Cellular and Molecular Biology, Section of Molecular Cell and Developmental Biology, University of Texas at Austin, Austin, TX 7871, USA

Received 1 June 2001; accepted is revised form 9 July 2001

Key words: auxin, cullin, Nedd8, RUB, SCF, ubiquitin

Abstract

Auxin-regulated gene expression is mediated by two families of transcription factors. The ARF proteins bind to a conserved DNA sequence called the AuxRE and activate transcription. The Aux/IAA proteins repress ARF function, presumably by forming dimers with ARF proteins. Recent genetic studies in *Arabidopsis* indicate that auxin regulates this system by promoting the ubiquitin-mediated degradation of the Aux/IAA proteins, thus permitting ARF function. Mutations in components of SCF^{TIR1}, a ubiquitin protein ligase (E3) result in stabilization of Aux/IAA proteins and decreased auxin response. Further, recent biochemical experiments indicate that the Aux/IAA proteins bind SCF^{TIR1} in an auxin-dependent manner.

Introduction

Auxin plays a crucial role in diverse aspects of plant growth and development including tropic responses, apical dominance in the shoot, lateral root formation and differentiation of the vascular system. To understand the molecular basis of auxin action, two major approaches have been used. The first approach is to identify mutants that lack normal auxin responses, and to determine the genetic basis for these defects (Hobbie *et al.*, 1994). The second is to take a direct molecular approach to identify genes and proteins that are regulated by the auxin signal (Abel and Theologis, 1994; Guilfoyle *et al.*, 1998). Both of these approaches have provided significant insights into auxin signaling. One of the most exciting developments of the past few years, at least from our perspective, is the convergence of the two approaches. In this review we will focus on recent results that indicate auxin response depends on the regulated degradation of a large family of transcriptional regulators, the Aux/IAA proteins.

Ubiquitin proteasome pathway

Ubiquitin is a small conserved protein that is covalently attached to diverse proteins, usually targeting them for degradation by 26S proteasome (Hershko and Ciechanover, 1998). The involvement of regulated protein degradation by the ubiquitin-proteasome pathway has been demonstrated in a wide variety of cellular processes (Hershko and Ciechanover, 1998). The ubiquitin conjugation pathway begins by the ATP-dependent activation of ubiquitin by a ubiquitin-activating (E1) enzyme. All E1 enzymes have a conserved cysteine residue that forms a thiol ester linkage with the C-terminal glycine of ubiquitin. Usually, E1 enzymes are encoded by a single gene or a small family of related genes (Hershko and Ciechanover, 1998). The activated ubiquitin molecule is then transferred onto the second component of the pathway, a ubiquitin-conjugating (E2) enzyme, again forming a thiol ester linkage between the terminal glycine of ubiquitin and a conserved cysteine on the E2 enzyme. The E2 enzymes are encoded by a large gene family. In yeast, there are 11 related E2 enzymes while in *Arabidopsis* there are at least 36 *Arabidopsis* E2 isoforms belonging to 12 groups (Hershko and Ciechanover, 1998; R. Vierstra, personal communication). The third step in the ubiquitination reaction is the transfer of ubiquitin onto the substrate protein. This step is assisted by an E3 ubiquitin ligase that confers substrate specificity to the pathway. The ubiquitinated protein is then degraded by the 26S proteasome, a large complex that comprises over 30 subunits, arranged in two subcomplexes of 19S and 20S (Hershko and Ciechanover, 1998; Vierstra and Callis, 1999). The 19S subcomplex is responsible for recognizing the tagged protein, while the lumen of the 20S subcomplex degrades the protein.

Compared to the E1 and E2 enzymes, E3 ligases are very diverse and complex. So far, four types of E3 ligases have been characterized. These are; HECT (homology to E6-AP carboxyl terminus) domain containing, Ubr1p (ubiquitin amino-end recognizing protein 1), APC/C (anaphase-promoting complex/cyclosome), and SCF (Skp1-cullin-F-box/RING-H2) E3 ligases. All four types have been identified in plants (Callis and Vierstra, 2000).

SCF E3 ligases

The SCF family of E3 ligases has been well characterized in plants, animals and yeast, both structurally and functionally (Patton et al., 1998). The name of the complex comes from the first three subunits identified in yeast, Skp1, Cdc53 (or cullin) and the F-box protein. Recently, a fourth subunit, a RING-H2 motif containing protein (Rbx1/Hrt1/Roc1) was recognized as an essential component of the complex (Tyers and Willems, 1999). The architecture of the SCF complex derived from protein interaction studies and structural analyses, suggests a common protein core consisting of Cullin, Rbx1 and Skp1. This core recruits different F-box proteins to form functionally distinct SCFs (Patton et al., 1998; Tyers and Willems, 1999). There is only a single Skp1 protein identified so far in man, but there are 18 Skp1-related genes present in the *Arabidopsis* genome (called ASKs) (Crosby, personal communication). Cullins also belong to gene families. For example, there are at least six cullins in man and five in *Arabidopsis*. The F-box proteins function in substrate recognition and are the most diverse and specific components of the complex. Genome sequencing projects in *Arabidopsis* and *Caenorhabditis elegans* have identified over 300 F-box-containing proteins in each organism. Some of these F-box proteins may function in non-SCF complexes (Kipreos and Pagano, 2000; Galan et al., 2001). Nonetheless, the large number of F-box-containing protein encoding genes underscores the potential importance of SCF complexes and regulated protein degradation in various cellular processes.

Among the best-known SCF complexes in yeast is SCF^{Cdc4}. This E3 facilitates degradation of Sic1 (an inhibitor of Cdc28/cyclinB), and promotes G1/S transition of the cell cycle. Yeast SCF^{Grr1} promotes degradation of the G1 cyclins, Cln1 and Cln2, and SCF^{Met30} is involved in sulfur metabolism (Patton et al., 1998). In mammalian cells, SCF^{Skp2} regulates the destruction of $P27^{kip1}$, an inhibitor of cyclin-dependent kinase Cdk2 (Carrano et al., 1999). In *Arabidopsis*, F-box proteins regulate jasmonic acid response (Coi1), floral development (UFO), circadian rhythm (ZTL and FKF1) (Xie et al., 1998; Samach et al., 1999; Nelson et al., 2000; Somers et al., 2000), and phytochrome-dependent light signaling (Dieterle et al., 2001). The SCF^{TIR1} complex is required for auxin response, and will be discussed in detail below (Gray et al., 1999).

The ubiquitin-like proteins

In addition to ubiquitin, several families of ubiquitin like proteins (Ubls) have been described (Vierstra and Callis, 1999; Hochstrasser, 2000; Yeh et al., 2000). As early as 1987, UCRP/ISG15, a ubiquitin-like protein modifier, was identified in animals followed by the SUMO/Sentrin family and the NEDD8/RUB family (Yeh et al., 2000). Ubls are activated and conjugated to the substrate in a mechanism similar to that of ubiquitin. However, a single Ubl polypeptide is attached at the site of conjugation rather than a chain as for ubiquitin. In addition, Ubls do not target the modified substrate for degradation. Rather, the modification appears to have a number of functions depending on the substrate. In some cases the modification is important for cellular localization while in others, the modified protein is stabilized. Members of both the SUMO and RUB families have been identified in plants (Vierstra and Callis, 1999).

Auxin-response mutants in *Arabidopsis*

By utilizing the effects of applied auxin on root elongation, a large number of auxin response mutants have been identified in *Arabidopsis* (Hobbie et al., 1994). Two general classes of mutants have been recovered. One class is composed of the recessive loss-of-function mutations *axr1*, *axr4*, and *tir1*. The *axr1* mutants were the first to be characterized in this group (Lincoln et al., 1990). *Axr1* plants display a number of severe auxin-related growth defects including reduced apical dominance, reduced cell elongation and reduced tropic responses. The *tir1* (*transport inhibitor response 1*) mutants were initially isolated in a screen for resistance to auxin transport inhibitors, and later identified as defective in auxin response rather than auxin transport (Ruegger *et al.*, 1998). Mature *tir1* plants are similar to wild-type plants in appearance. However, many auxin responses, including auxin inhibition of root growth, lateral root formation and auxin dependent hypocotyl elongation in seedlings are diminished in these mutants. The *axr4* mutants also show defects in many aspects of auxin response (Hobbie and Estelle, 1995). The *tir1* and *axr4* mutations both act synergistically with *axr1* suggesting that these genes act in the same or overlapping pathways (Hobbie and Estelle, 1995; Ruegger *et al.*, 1998).

The second class of mutations confer a dominant gain-of-function. These mutants include *axr2, axr3,* and *iaa28* (Rouse *et al.*, 1998; Nagpal *et al.*, 2000; Rogg *et al.*, 2001). The related mutants *shy2* and *msg2* also belong to this class (Tian and Reed, 1999). In general, these mutations confer reduced auxin response and result in a variety of growth defects.

The TIR1 protein, an SCF connection

The TIR1 protein encodes a 594 amino acid protein that contains an F-box domain and 16 degenerate leucine rich repeats (LRR) (Ruegger *et al.*, 1998). The F-box motif is located at the N-terminus, and is required for the assembly of the F-box protein onto the core of the SCF complex (cullin, Rbx1 and Skp1) through binding to Skp1 (Patton *et al.*, 1998). In a yeast two-hybrid screen, TIR1 interacted with two highly related genes, ASK1 (*Arabidopsis* Skp1-like1) and ASK2 (Gray *et al.*, 1999). Additionally, extracts from transgenic plants carrying c-myc-TIR1 co-immunoprecipitated both ASK1 and ASK2, further confirming the *in vivo* interactions between TIR1 and ASK proteins (Gray *et al.*, 1999; Gray and Estelle, 2000). The two-hybrid interaction between TIR1 and ASK1/ASK2 was completely abolished when the F-box motif was mutated. Interestingly, this point mutation did not interfere with co-immunoprecipitation of ASK proteins with TIR1 suggesting that other regions on the protein contribute to stability of the complex. Recent experiments have linked ASK1 with auxin response. Knockout *ask1-1* mutants are male sterile and show auxin-resistant root elongation (Gray *et al.*,

1999; Yang *et al.*, 1999). The *Arabidopsis* genome encodes at least 18 ASK proteins (Crosby, personal communication). Because the yeast two-hybrid analysis recovered only two of these, it is likely that individual ASK proteins are specialized for specific SCF complexes. There are 6 known cullin proteins in *Arabidopsis*. So far only AtCul1 is known to function in SCF^{TIR1} (Gray *et al.*, 1999).

Leucine-rich repeats are found in many F-box proteins and are thought to interact with substrates, conferring specificity to the SCF complex (Patton *et al.*, 1998). Since the *tir1* mutants are loss-of-function mutants, at least some TIR1 substrates must be negative regulators of auxin response. The levels of TIR1 appear to be important for auxin response, since plants that are heterozygous for *tir1* have an intermediate level of auxin resistance. In fact, over-expression of TIR1 confers increased auxin response (Gray *et al.*, 1999).

Auxin response and the RUB conjugation pathway

Recessive mutations in the *AXR1* gene confer an array of auxin-related phenotypes, placing AXR1 in a central position in auxin response (Lincoln *et al.*, 1990; Hobbie *et al.*, 1994). Molecular cloning of the *AXR1* gene demonstrated that the protein is related to the N-terminus of ubiquitin activating (E1) enzyme (Leyser *et al.*, 1993). AXR1-related proteins were later identified in other organisms including man and yeasts (Hochstrasser, 2000). These enzymes were shown to activate various ubiquitin-like proteins with the co-operation of a second subunit related to the C-terminus of E1. The second subunit carries the active-site cysteine.

The *AXR1* orthologue in budding yeast is *ENR2* (E1 amino terminus-related 2). Genetic studies indicated that *ENR2* is essential for modification of Cdc53 with Rub1. Although not essential for viability, this modification is required for normal function of SCF^{CDC4} in the degradation of the CDK inhibitor Sic1 (Lammer *et al.*, 1998). These results demonstrated for the first time that members of the cullin family are modified by RUB1, and that this modification is important for SCF function. In *Arabidopsis*, AXR1 functions together with ECR1 (E1 C-terminus-related) to activate RUB1 and form a thiol ester linkage between RUB1 and a conserved cysteine on ECR1 *in vitro* (del Pozo *et al.*, 1998). Mutation of this cysteine (C215A) completely abolishes the activity of the enzyme. The *RCE1* (RUB1-conjugating enzyme1) gene was subsequently identified because of its similarity to yeast Ubc12. In the presence of AXR1, ECR1 and ATP, RCE1 can form a thiol ester linkage with RUB1 confirming that this protein is a RUB E2 (del Pozo and Estelle, 1999).

Based on the yeast studies, we proposed that the *Arabidopsis* cullin AtCul1 may be modified by RUB1. This has now been confirmed *in vitro* and *in vivo* (del Pozo and Estelle, 1999; del Pozo and Estelle, 2002). In addition, our recent studies confirm that RUB1 modification of CUL1 is dependent upon both AXR1 and ECR1 (del Pozo *et al.*, 2002). These results are consistent with our earlier genetic studies demonstrating an interaction between the *axr1* and *tir1* mutations (Ruegger *et al.*, 1998). It now appears very likely that the primary function of AXR1 in auxin response is through RUB1 modification of CUL1 (Figure 1).

It is noteworthy that there are at least 3 RUB/NEDD8 family proteins in *Arabidopsis* (Rao-Naik *et al.*, 1998). RUB1 and RUB2 differ from each other by a single amino acid, while RUB3 is about 78% identical to the other two (Rao-Naik *et al.*, 1998). Whether RUB2 and RUB3 proteins can modify cullin proteins is currently unknown, and worth investigating. The presence of another AXR1-like protein (AXL1), another RCE1-like protein (RCE2), and four additional cullins (AtCul2-5) in the *Arabidopsis* proteome may suggest combinatorial regulation of additional SCF complexes employing different F-box proteins and potential targets in a tissue-specific manner. In fact, double mutants between *axr1* and *axl1* display severe auxin-related morphological defects suggesting a role for AXL1 in the same or a redundant pathway in auxin response (N. Dharmasiri and M. Estelle, unpublished results). There are also 3 other F-box proteins (LRF1, LRF2, and LRF3) with significant sequence similarity to TIR1 along the entire length of the protein encoded in the *Arabidopsis* genome (Dharmasiri *et al.*, unpublished results). These proteins may also function in auxin response, a possibility that is consistent with the relatively weak phenotype of the *tir1* mutants.

Regulation of SCF^{TIR1} complex

Why is RUB1 modification of CUL1 important for function of SCF^{TIR1}? At this point the answer to this question is not clear. In one study the modification appeared to be important for centrosomal localization of CUL1 (Freed *et al.*, 1999). The RUB conjugation pathway is present in the nucleus of both plant and animal cells and it is possible that the modification helps to retain the SCF in the nucleus or target it to a specific structure within the nucleus (del Pozo *et al.*, 1998; Yeh *et al.*, 2000). A number of other studies indicate that the modification increases SCF activity *in vitro* (Morimoto *et al.*, 2000; Podust *et al.*, 2000; Read *et al.*, 2000; Wu *et al.*, 2000; Tanaka *et al.*, 2001). At present there is no information on how this might happen. Some of the obvious possibilities have been excluded. For example, the modification does not appear to affect assembly or stability of the SCF (Yeh *et al.*, 2000). Neither does it seem to influence the interaction between the SCF and substrates.

Recently some important clues to this puzzle have emerged from genetic studies in *Arabidopsis* and the yeast *Schizosaccharomyces pombe* (Lyapina *et al.*, 2001; Schwechheimer *et al.*, 2001). Two groups have shown in these species, that the COP9 signalosome (CSN) possess an activity that removes RUB/NEDD8 from cullin. The CSN is a large multisubunit complex related to the 19S lid subcomplex of the proteasome (Wei and Deng, 1999). The complex was first identified in *Arabidopsis* as a negative regulator of photomorphogenesis (Wei and Deng, 1999). The CSN is found in plants, animals, fission yeast, but not budding yeast. In *Arabidopsis*, the complete absence of the complex results in early seedling lethality. However, Schwechheimer *et al.* (2001) generated transgenic lines with reduced levels of the CSN5 protein, a subunit of the complex. These plants had an increased level of modified CUL1 compared to wild type suggesting that the CSN is responsible for removing RUB from the cullin. Strong support for this idea comes from the companion paper in which purified mammalian CSN was shown to remove Nedd8 from the fission yeast cullin Pcu1 *in vitro*. Strikingly, the *CSN5* line is also resistant to auxin and has a morphology that is quite similar to that of *axr1* plants. Furthermore, the rate of degradation of PSIAA6, a potential target of SCF complex (see below), was reduced in the *CSN5* transgenic lines. Thus an increase in RUB-modified cullin has the same effect on auxin response as a decrease in the amount of modified cullin (*axr1* plants). A very similar story emerges from a series of experiments involving the RBX1 protein. This subunit of the SCF is known to be essential for RUB-modification of the cullin (Kamura *et al.*, 1999). In *Arabidopsis* plants that over-express RBX1, there is a dramatic increase in the relative amount of modified cullin (Gray

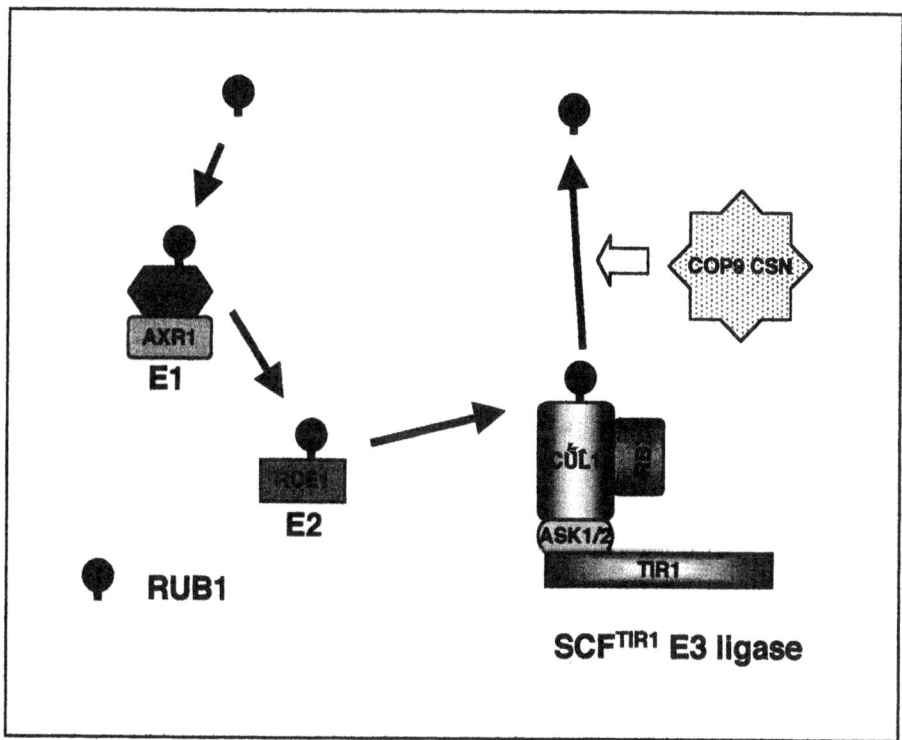

Figure 1. Model for the RUB1 conjugation / deconjugation pathway. RUB1 is activated by AXR1-ECR1 heterodimer (E1), and conjugated to CUL1 via the action of RUB1 conjugating enzyme, RCE1 (E2). The COP9 signalosome (CSN) regulates the deconjugation of RUB1 from CUL1. Activity of SCF^TIR1 E3 ligase complex appears to be regulated by RUB1 conjugation and deconjugation of CUL1 for normal auxin response.

and Estelle, unpublished). This change is associated with a reduction in auxin response and a similar suite of growth defects as is observed in the *axr1* and the *CSN5* transgenic lines. Taken together, these results suggest that cullin modification and RUB-removal are regulated and dynamic processes. Perhaps RUB conjugation and removal are each required at specific moments in the SCF catalytic cycle.

Targets of the SCF^TIR1 complex

Considering the prominent role of SCF^TIR1 in auxin response, one of the most important and interesting questions is the identity of its substrates. One group of likely candidates are the AUX/IAA proteins (Abel *et al.*, 1994; Gray and Estelle, 2000; Worley *et al.*, 2000). These proteins are short-lived nuclear localized proteins that have been implicated in auxin response. There are 24 *Aux/IAA* genes in the *Arabidopsis* genome, many of which are auxin regulated. The Aux/IAA proteins all have a characteristic structure with four conserved domains (I–IV). Domains III

and IV are involved in homo- and hetero-dimerization within the family or with members of the ARF protein family (auxin-response factor)(Abel and Theologis, 1996; Guilfoyle *et al.*, 1998). The ARF proteins share domains III and IV with the Aux/IAA proteins and also have a B1 DNA binding domain that is required for binding to a DNA sequence called the auxin-response element (AuxRe) (Guilfoyle *et al.*, 1998). ARF proteins can act as either transcriptional activators or repressors (Ulmasov *et al.*, 1999). In at least some cases, Aux/IAA proteins inhibit ARF-mediated transcriptional regulation (Ulmasov *et al.*, 1997). Although not yet proven, these results suggest that the Aux/IAA proteins inhibit ARF function through formation of an Aux/IAA-ARF heterodimer (Guilfoyle *et al.*, 1998). Since the ARFs can have either positive or negative effects on transcription, this implies that the AUX/IAA proteins can act as either positive or negative regulators of auxin response.

A growing number of *Aux/IAA* genes have been identified through genetic studies including *AXR2/IAA7, AXR3/IAA17, SHY2/IAA3,* and *IAA28* (Morimoto *et al.*, 2000; Read *et al.*, 2000; Rogg

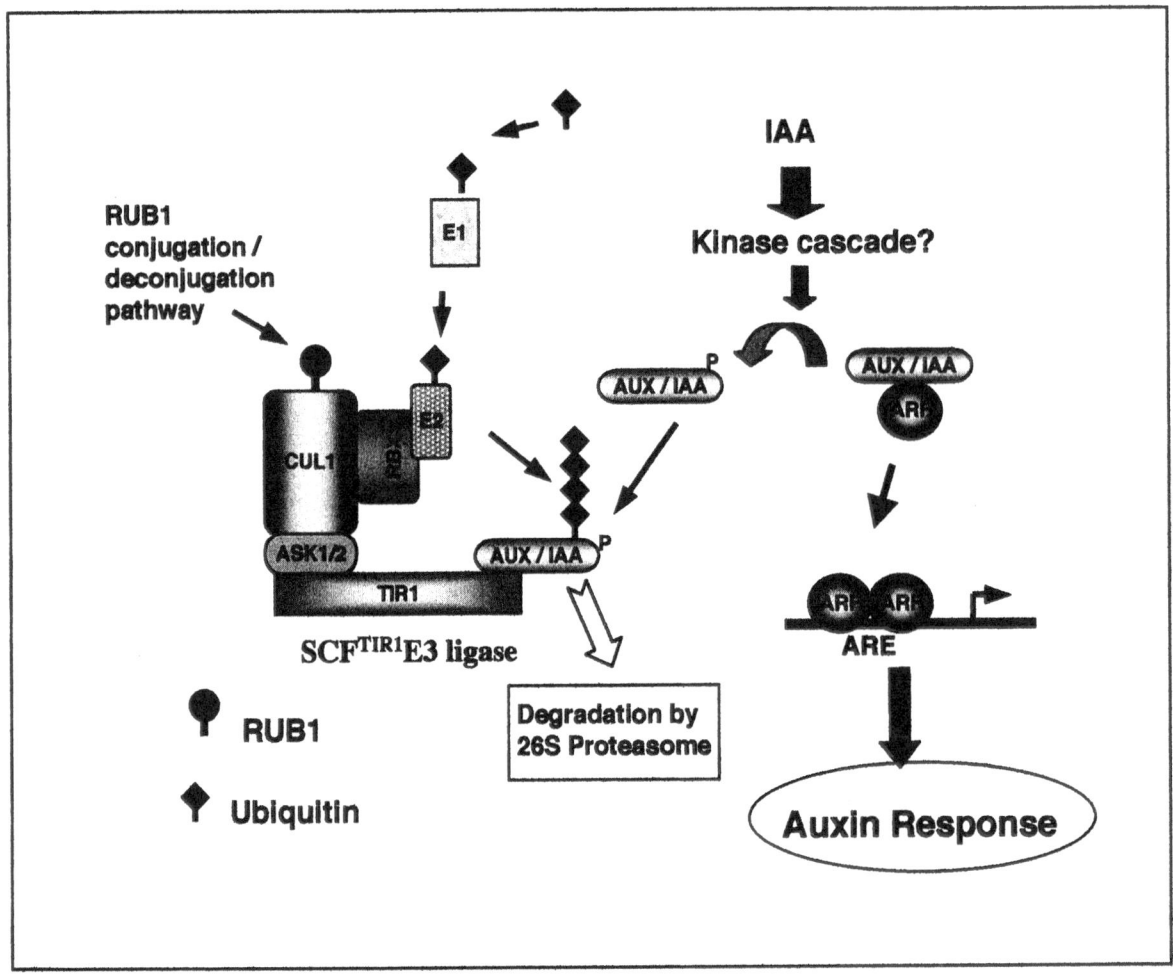

Figure 2. Model for the function of SCF^{TIR1} complex during auxin response. In this model we suggest that under un-induced conditions, AUX/IAA proteins remain bound to ARFs (auxin-response factors) and prevent their function. Auxin may induce the ubiquitin proteasome pathway-mediated degradation of AUX/IAA proteins, releasing ARFs. Dimerized ARFs then bind to AuxRE (auxin-response elements) in auxin-response gene promoters. It is not yet clear if an auxin-induced kinase cascade is directly involved in phosphorylation of AUX/IAA proteins or a factor associated with AUX/IAAs.

et al., 2001; Tanaka *et al.*, 2001) (see Liscum and Reed, 2002). Gain-of-function mutations in each of these genes resulted in defects in auxin response and auxin-regulated growth processes. Molecular analysis of the mutant genes revealed that in every case, the phenotype was caused by an amino acid substitution within domain II of the protein. In fact, subsequent studies showed that domain II includes an instability determinant and the gain-of-function mutations result in stabilization of the affected protein (Worley *et al.*, 2000).

All of these results suggest that the Aux/IAA proteins might be substrates for SCF^{TIR1}. This has now been confirmed using genetic and biochemical ap-

proaches. An AXR3-GUS fusion protein is unstable in a wild-type line and further destabilized by auxin (Gray et al., 2001). However, the fusion is more stable in both the *tir1* and *axr1* mutants suggesting that SCF^{TIR1} is required for normal degradation of the protein. Similar results were obtained with an AXR2-GUS fusion protein.

Most strikingly, recent studies show that both recombinant AXR2 and AXR3 interact with SCF^{TIR1} when added to plant extracts (Gray *et al.*, 2001). Further, this interaction was stimulated by auxin pretreatment of the seedlings used to prepare the extract. This result suggests that the interaction between SCF^{TIR1} and the Aux/IAA proteins is auxin regulated. Finally, the domain II mutant proteins corresponding to the gain-of-function forms of AXR2 and AXR3 did not interact with the SCF indicating that stabilization of mutant protein is caused by the failure to interact with the SCF (Gray et al., 2001).

Auxin regulation of Aux/IAA degradation

A model describing our current view of SCF^{TIR1} regulation of auxin response is outlined in Figure 2. The results described above largely confirm many aspects of this model. However, a major outstanding question concerns auxin regulation of the pathway. The genes in the RUB conjugation pathway (*AXR1, ECR1*, and *RCE1*) are not rapidly induced by auxin treatment and auxin does not appear to affect the level of RUB-CUL1 (del Pozo et al., 2002). Similarly, expression of *TIR1* is not rapidly auxin-regulated (Gray et al., 1999). Instead, all of these genes are expressed in meristems, organ primordia and other growing cells suggesting that auxin regulation of the pathway occurs by some other mechanism (Gray et al., 1999; del Pozo et al., unpublished results). As mentioned above, one possibility is through regulation of the interaction between SCF^{TIR1} and the Aux/IAA proteins. In most cases, SCF-substrate recognition requires phosphorylation of the substrate (Hershko and Ciechanover, 1998). Auxin may act to phosphorylate either the Aux/IAA proteins or an unknown adapter protein, thereby promoting interaction with the SCF. It is interesting to note that phytochrome A has recently been shown to phosphorylate Aux/IAA proteins *in vitro* (Colon-Carmona et al., 2000). At this point, the physiological significance of this activity is unclear, but it is certainly possible to imagine that PHYA acts to regulate degradation of these proteins. This would be consistent with growing evidence for a close relationship between light and hormone signaling.

Conclusions and future prospects

This review and the others in this issue demonstrate how far we have progressed in the past several years. At the same time, our recent progress has led us to a fuller understanding of the complexity of the auxin regulatory system. The recent completion of the *Arabidopsis* genome clearly illustrates the challenge. According to the latest annotation information there are 24 *Aux/IAA* genes, 23 *ARF* genes, and 3 genes closely related to *TIR1*. The other well-known auxin-regulated genes, *GH3* and *SAUR,* are also represented by very large families in the *Arabidopsis* genome. There is also extraordinary complexity in the ubiquitin pathway. In addition to the *CUL* and *ASK* gene families, there are over 350 different F-box proteins in the proteome and similar numbers of other families of E3 ubiquitin ligases. To understand this complexity we will need to utilize all of the resources of the post-genomic era including forward and reverse genetic procedures, expression profiling, metabolic profiling and advanced proteomics. Of equal importance, an unprecedented level of cooperation and collaboration will be required for this effort. This volume was inspired by a meeting devoted to auxin biology that took place in Corsica in May, 2000. Judging by the collegial and convivial atmosphere of that meeting, the auxin community is up to this challenge.

Acknowledgements

Research in our lab is supported by grants from the NIH (43644), NSF (PGR-0077769), and the US Department of Energy (DE-FG02-98-ER20313).

References

Abel, S., Oeller, P.W. and Theologis, A. 1994. Early auxin-induced genes encode short-lived nuclear proteins. Proc. Natl. Acad. Sci. USA 91: 326–330.

Abel, S. and Theologis, A. 1996. Early genes and auxin action. Plant Physiol. 111: 9–17.

Callis, J. and Vierstra, R.D. 2000. Protein degradation in signaling. Curr. Opin. Plant Biol. 3: 381–386.

Carrano, A.C., Eytan, E., Hershko, A. and Pagano, M. 1999. SKP2 is required for ubiquitin-mediated degradation of the CDK inhibitor p27. Nature Cell Biol. 1: 193–199.

Colon-Carmona, A., Chen, D.L., Yeh, K.C. and Abel, S. 2000. Aux/IAA proteins are phosphorylated by phytochrome *in vitro*. Plant Physiol. 124: 1728–1738.

del Pozo, J.C. and Estelle, M. 1999. The *Arabidopsis* cullin AtCUL1 is modified by the ubiquitin-related protein RUB1. Proc. Natl. Acad. Sci. USA 96: 15342 15347.

del Pozo, J.C., Timpte, C., Tan, S., Callis, J. and Estelle, M. 1998. The ubiquitin-related protein RUB1 and auxin response in *Arabidopsis*. Science 280: 1760–1763.

del Pozo, J.C., Dharmasiri, S., Hellman, H., Walker, L., Gray, W.M. and Estelle, M., 2002. AXR1-ECR1-dependent conjugation of RUB1 to the Arabidopsis Cullin AtCul1 is required for auxin response. Plant Cell (in press).

Dieterle, M., Zhou, Y.C., Schafer, E., Funk, M. and Kretsch, T. 2001. EID1, an F-box protein involved in phytochrome A-specific light signaling. Genes Dev. 15: 939–944.

Freed, E., Lacey, K.R., Huie, P., Lyapina, S.A., Deshaies, R.J., Stearns, T. and Jackson, P.K. 1999. Components of an SCF

408

ubiquitin ligase localize to the centrosome and regulate the centrosome duplication cycle. Genes Dev. 13: 2242–2257.

Galan, J.M., Wiederkehr, A., Seol, J.H., Haguenauer-Tsapis, R., Deshaies, R.J., Riezman, H. and Peter, M. 2001. Skp1p and the F-box protein Rcy1p form a non-SCF complex involved in recycling of the SNARE Snc1p in yeast. Mol. Cell Biol. 21: 3105–3117.

Gray, W.M. and Estelle, I. 2000. Function of the ubiquitin-proteasome pathway in auxin response. Trends Biochem. Sci. 25: 133–138.

Gray, W.M., del Pozo, J.C., Walker, L., Hobbie, L., Risseeuw, E., Banks, T., Crosby, W.L., Yang, M., Ma, H. and Estelle, M. 1999. Identification of an SCF ubiquitin-ligase complex required for auxin response in *Arabidopsis thaliana*. Genes Dev. 13: 1678–1691.

Gray, W.M., Kepinski, S., Rouse, D., Leyser, O., and Estelle, M., 2001. Auxin regulates SCF^{TIR1} - dependent degradation of AUX/IAA proteins. Nature 414: 271–276.

Guilfoyle, T., Hagen, G., Ulmasov, T. and Murfett, J. 1998. How does auxin turn on genes? Plant Physiol. 118: 341–347.

Hershko, A. and Ciechanover, A. 1998. The ubiquitin system. Annu. Rev. Biochem. 67: 425–479.

Hobbie, L. and Estelle, M. 1995. The *axr4* auxin-resistant mutants of *Arabidopsis thaliana* define a gene important for root gravitropism and lateral root initiation. Plant J. 7: 211–220.

Hobbie, L., Timpte, C. and Estelle, M. 1994. Molecular genetics of auxin and cytokinin. Plant Mol. Biol. 26: 1499–1519.

Hochstrasser, M. 2000. Evolution and function of ubiquitin-like protein-conjugation systems. Nature Cell Biol. 2: E153–E157.

Kamura, T., Conrad, M.N., Yan, Q., Conaway, R.C. and Conaway, J.W. 1999. The Rbx1 subunit of SCF and VHL E3 ubiquitin ligase activates Rub1 modification of cullins Cdc53 and Cul2. Genes Dev. 13: 2928–2933.

Kipreos, E.T. and Pagano, M. 2000. The F-box protein family. Genome Biol. 1 (2000).

Lammer, D., Mathias, N., Laplaza, J.M., Jiang, W., Liu, Y., Callis, J., Goebl, M. and Estelle, M. 1998. Modification of yeast Cdc53p by the ubiquitin-related protein rub1p affects function of the SCFCdc4 complex. Genes Dev. 12: 914–926.

Leyser, H.M., Lincoln, C.A., Timpte, C., Lammer, D., Turner, J. and Estelle, M. 1993. *Arabidopsis* auxin-resistance gene *AXR1* encodes a protein related to ubiquitin-activating enzyme E1. Nature 364: 161–164.

Lincoln, C., Britton, J.H. and Estelle, M. 1990. Growth and development of the axr1 mutants of Arabidopsis. Plant Cell 2: 1071–1080.

Liscum, E. and Reed, J.W. 2002. Genetics of Aux/IAA and ARF action in plant growth and development. Plant Mol. Biol. 49: 387–400.

Lyapina, S., Cope, G., Shevchenko, A., Serino, G., Tsuge, T., Zhou, C., Wolf, D.A., Wei, N. and Deshaies, R.J. 2001. Promotion of NEDD8-CUL1 conjugate cleavage by COP9 signalosome. Science 292: 1382–1385.

Morimoto, M., Nishida, T., Honda, R. and Yasuda, H. 2000. Modification of cullin-1 by ubiquitin-like protein Nedd8 enhances the activity of SCF(skp2) toward p27(kip1). Biochem. Biophys. Res. Comm. 270: 1093–1096.

Nagpal, P., Walker, L.M., Young, J.C., Sonawala, A., Timpte, C., Estelle, M. and Reed, J.W. 2000. AXR2 encodes a member of the Aux/IAA protein family [In Process Citation]. Plant Physiol. 123: 563–574.

Nelson, D.C., Lasswell, J., Rogg, L.E., Cohen, M.A. and Bartel, B. 2000. FKF1, a clock-controlled gene that regulates the transition to flowering in *Arabidopsis*. Cell 101: 331–340.

Patton, E.E., Willems, A.R. and Tyers, M. 1998. Combinatorial control in ubiquitin-dependent proteolysis: don't Skp the F-box hypothesis. Trends Genet. 14: 236–243.

Podust, V.N., Brownell, J.E., Gladysheva, T.B., Luo, R.S., Wang, C., Coggins, M.B., Pierce, J.W., Lightcap, E.S. and Chau, V. 2000. A Nedd8 conjugation pathway is essential for proteolytic targeting of p27Kip1 by ubiquitination. Proc. Natl. Acad. Sci. USA 97: 4579–4584.

Rao-Naik, C., delaCruz, W., Laplaza, J.M., Tan, S., Callis, J. and Fisher, A.J. 1998. The rub family of ubiquitin-like proteins. Crystal structure of *Arabidopsis* rub1 and expression of multiple rubs in *Arabidopsis*. J. Biol. Chem. 273: 34976–34982.

Read, M.A., Brownell, J.E., Gladysheva, T.B., Hottelet, M., Parent, L.A., Coggins, M.B., Pierce, J.W., Podust, V.N., Luo, R.S., Chau, V. and Palombella, V.J. 2000. Nedd8 modification of cul-1 activates SCF(β(TrCP))-dependent ubiquitination of IκBα. Mol. Cell Biol. 20: 2326–2333.

Rogg, L.E., Lasswell, J. and Bartel, B. 2001. A gain-of-function mutation in iaa28 suppresses lateral root development. Plant Cell 13: 465–480.

Rouse, D., Mackay, P., Stirnberg, P., Estelle, M. and Leyser, O. 1998. Changes in auxin response from mutations in an AUX/IAA gene. Science 279: 1371–1373.

Ruegger, M., Dewey, E., Gray, W.M., Hobbie, L., Turner, J. and Estelle, M. 1998. The TIR1 protein of *Arabidopsis* functions in auxin response and is related to human SKP2 and yeast grr1p. Genes Dev. 12: 198–207.

Samach, A., Klenz, J.E., Kohalmi, S.E., Risseeuw, E., Haughn, G.W. and Crosby, W.L. 1999. The *UNUSUAL FLORAL ORGANS* gene of *Arabidopsis thaliana* is an F-box protein required for normal patterning and growth in the floral meristem. Plant J. 20: 433–445.

Schwechheimer, C., Serino, G., Callis, J., Crosby, W.L., Lyapina, S., Deshaies, R.J., Gray, W.M., Estelle, M. and Deng, X.W. 2001. Interactions of the COP9 signalosome with the E3 ubiquitin ligase SCFTIR1 in mediating auxin response. Science 292: 1379–1382.

Somers, D.E., Schultz, T.F., Milnamow, M. and Kay, S.A. 2000. ZEITLUPE encodes a novel clock-associated PAS protein from *Arabidopsis*. Cell 101: 319–329.

Tanaka, K., Kawakami, T., Tateishi, K., Yashiroda, H. and Chiba, T. 2001. Control of IκBα proteolysis by the ubiquitin-proteasome pathway. Biochimie 83: 351–356.

Tian, Q. and Reed, J.W. 1999. Control of auxin-regulated root development by the *Arabidopsis thaliana SHY2/IAA3* gene. Development 126: 711–721.

Tyers, M. and Willems, A.R. 1999. One ring to rule a superfamily of E3 ubiquitin ligases [comment]. Science 284: 601, 603, 604.

Ulmasov, T., Murfett, J., Hagen, G. and Guilfoyle, T.J. 1997. Aux/IAA proteins repress expression of reporter genes containing natural and highly active synthetic auxin response elements. Plant Cell 9: 1963–1971.

Ulmasov, T., Hagen, G. and Guilfoyle, T.J. 1999. Activation and repression of transcription by auxin-response factors. Proc. Natl. Acad. Sci. USA 96: 5844–5849.

Vierstra, R.D. and Callis, J. 1999. Polypeptide tags, ubiquitous modifiers for plant protein regulation. Plant Mol. Biol. 41: 435–442.

Wei, N. and Deng, X.W. 1999. Making sense of the COP9 signalosome. A regulatory protein complex conserved from *Arabidopsis* to human. Trends Genet. 15: 98–103.

Worley, C.K., Zenser, N., Ramos, J., Rouse, D., Leyser, O., Theologis, A. and Callis, J. 2000. Degradation of Aux/IAA proteins is essential for normal auxin signalling. Plant J. 21: 553–562.

Wu, K., Chen, A. and Pan, Z.Q. 2000. Conjugation of Nedd8 to CUL1 enhances the ability of the ROC1-CUL1 complex to promote ubiquitin polymerization. J. Biol. Chem. 275: 32317–32324.

Xie, D.X., Feys, B.F., James, S., Nieto-Rostro, M. and Turner, J.G. 1998. COI1: an *Arabidopsis* gene required for jasmonate-regulated defense and fertility. Science 280: 1091–1094.

Yang, M., Hu, Y., Lodhi, M., McCombie, W.R. and Ma, H. 1999. The *Arabidopsis SKP1-LIKE1* gene is essential for male meiosis and may control homologue separation. Proc. Natl. Acad. Sci. USA 96: 11416–11421.

Yeh, E.T., Gong, L. and Kamitani, T. 2000. Ubiquitin-like proteins: new wines in new bottles. Gene 248: 1–14.

Plant Molecular Biology **49**: 411–426, 2002.
Perrot-Rechenmann and Hagen (Eds.), Auxin Molecular Biology.
© 2002 *Kluwer Academic Publishers.*

411

Auxin cross-talk: integration of signalling pathways to control plant development

Ranjan Swarup, Geraint Parry, Neil Graham, Trudie Allen and Malcolm Bennett*
*School of Biosciences, Sutton Bonington, University of Nottingham, Leicestershire, LE12 5RD, UK (*author for correspondence; e-mail Malcolm.bennett@nottingham.ac.uk)*

Received 11 July 2001; accepted in revised form 3 September 2001

Key words: auxin, cross-talk, light, phytohormones, signal transduction

Abstract

Plants sense and respond to endogenous signals and environmental cues to ensure optimal growth and development. Plant cells must integrate the myriad transduction events into a comprehensive network of signalling pathways and responses. The phytohormone auxin occupies a central place within this transduction network, frequently acting in conjunction with other signals, to co-ordinately regulate cellular processes such as division, elongation and differentiation. As a non-cell autonomous signal, auxin also interacts with other signalling pathways to regulate inter-cellular developmental processes. As part of this especially themed edition of Plant Molecular Biology, we will review examples of 'cross-talk' between auxin and other signalling pathways. Given the current state of knowledge, we have deliberately focused our efforts reviewing auxin interactions with other phytohormone and light signalling pathways. We conclude by discussing how new genomic approaches and the *Arabidopsis* genome sequence are likely to impact this area of research in the future.

Abbreviations: ABA, abscisic acid; ACC, 1-aminocyclopropane-1-carboxylic acid; BL, brassinolide; BR, brassinosteroid; 2,4-D, 2,4-dichlorophenoxyacetic acid; FR, far-red; GA, gibberellic acid; IAA, indole-3-acetic acid; JA, jasmonic acid; 1-NAA, 1-naphthaleneacetic acid; NPA, 1-naphthylphthalamic acid; PAT, polar auxin transport; R, red; TIBA, 2,3,5-triiodobenzoic acid; TILLING, Targeting Induced Local Lesions in Genomes

Introduction

Over sixty-five years have elapsed since Went and Thimann (1937) published their classic book entitled 'Phytohormones' which described how the only hormone then identified, auxin, regulates plant development. Since then, a whole series of plant signalling molecules have been identified (Letham *et al.*, 1978; Kende and Zeevaaart, 1997) which interact with auxin to regulate developmental processes such as cell division, elongation and differentiation. For example, auxin and cytokinin synergistically regulate tobacco pith cell proliferation (Skoog and Tsui, 1948; Miller *et al.*, 1956); in conjunction with GA (gibberellic acid), auxin promotes stem elongation in pea (Ross *et al.*, 2000), whilst auxin and ethylene co-ordinately regulate *Arabidopsis* root hair differen-

tiation (Masucci and Schiefelbein, 1994). As a non-cell autonomous signalling mechanism, polar auxin transport is a frequent target for other signalling pathways to regulate plant development (Figure 1). For example, ethylene-induced hook formation in etiolated *Arabidopsis* hypocotyl tissue is dependent on auxin transport activity (Raz and Ecker, 1999), whilst blue-light-induced hypocotyl phototropic curvature requires auxin transport to facilitate hormone redistribution and hence cause differential growth (Harper *et al.*, 2000).

Auxin cross-talk with other signalling molecules was originally studied by observing the effects of exogenous application of combinations of hormones on plant proliferation or morphology (Skoog and Miller, 1957; Wickson and Thimann, 1958). Despite the utility of the spray-and-pray strategy to identify in-

412

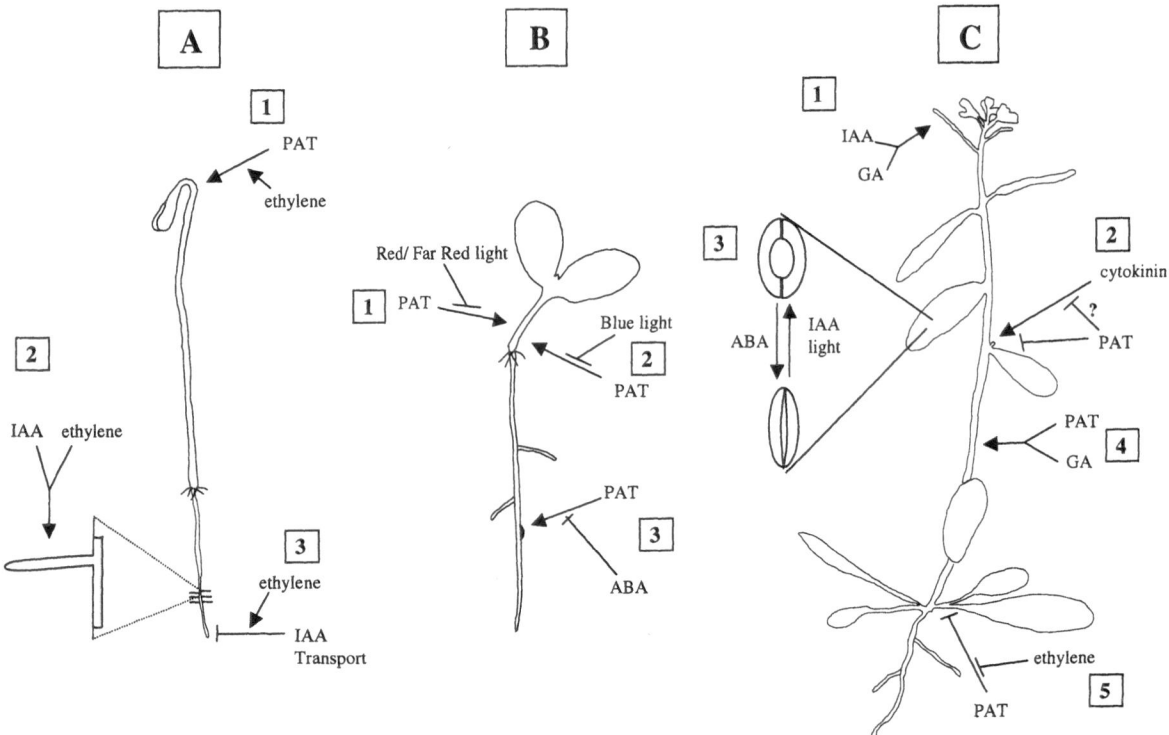

Figure 1. Auxin interacts with other signals to regulate plant development. A. Etiolated seedling stage; 1, ethylene promotes apical hook formation via polar auxin transport (PAT; Raz and Ecker, 1999); 2, root hair differentiation and elongation is regulated by ethylene and auxin (Masucci and Schiefelbein, 1994; Pitts *et al.*, 1998); 3, root growth inhibition by ethylene requires auxin transport (Rahman *et al.*, 2001). B. Light-grown seedling stage; 1, PAT represents a target for red and far-red light to inhibit hypocotyl elongation (Jensen *et al.*, 1998); 2, blue-light-induced phototropic curvature requires PAT-mediated auxin redistribution (Briggs and Huala, 1999); 3, lateral root initiation requires PAT (Casimiro *et al.*, 2001) but is repressed by ABA (H. Zhang, personal communication). C. Mature plant stage. 1, both auxin and GA signals promote parthenocarpic fruit development (Vivian-Smith and Koltunov, 1999); 2, PAT is required for apical dominance, whilst cytokinin promotes lateral bud formation (Wickson and Thimann, 1958); 3, ABA and auxin/light interact antagonistically to regulate stomatal aperture (Eckert and Kaldenhoff, 2000); 4, stem elongation requires both auxin and GA to positively regulate each others biosynthesis (Ross *et al.*, 2000); 5, ethylene-promoted leaf abscission correlates with reduced polar auxin transport (Beyer and Morgan, 1971).

teractions between classes of hormone, this approach was not always reproducible due to variations in, for example, hormone uptake and transport. To overcome these limitations, transgenic plants were created that ectopically express *Agrobacterium* auxin and cytokinin biosynthesis genes (Klee *et al.*, 1987; Medford *et al.*, 1989), thereby allowing researchers to study the developmental effects of directly manipulating endogenous hormone levels. During the last decade, our understanding of hormone 'cross-talk' at the signal transduction level has been greatly facilitated by the application of genetic approaches in the model plant *Arabidopsis thaliana* (McCourt, 1999). Mutant screens originally designed to isolate lesions within auxin signal transduction components (Maher and Martindale, 1980; Estelle and Somerville, 1987) largely resulted in the identification of mutations with altered responses to multiple hormones (Pickett *et al.*,

1990; Wilson *et al.*, 1990; Leyser *et al.*, 1996). The pleiotropic signalling defects exhibited by many of the auxin-response mutants highlighted the prevalence of 'cross-talk' between auxin and other signalling pathways (summarised in Table 1). At the level of the individual gene, the altered hormone 'signature' provided insight into the integrative function of the wild-type protein. For example, mutations in the auxin influx and efflux carrier genes *AUX1* and *AtPIN2* alter both auxin- and ethylene-sensitive root elongation (Pickett *et al.*, 1990; Müller *et al.*, 1998; Table 1), suggesting that ethylene requires auxin transport to inhibit root growth. More recently, the application of targeted genetic screens designed to identify second site enhancer or suppresser mutations have uncovered hitherto unexpected interactions between auxin and other signalling pathways (McCourt, 1999). For example, the identification of the *shy2* suppresser of the phytochrome mutant *hy2* led Tian and Reed (1999) to propose that the auxin transcriptional regulator SHY2/IAA3 represents a link between light and auxin signalling.

This review aims to describe pertinent examples of auxin cross-talk that have been revealed using the experimental approaches highlighted above. Given the current state of knowledge, we have deliberately focused our efforts on reviewing auxin interactions with other phytohormone and light signalling pathways.

Auxin and cytokinin

Classic experiments by Skoog and colleagues originally demonstrated that auxin and cytokinin synergistically regulated tobacco pith cell proliferation (Skoog and Tsui, 1948; Miller *et al.*, 1956). Recent studies conclude that auxin and cytokinin control tobacco cell proliferation by regulating the expression of interacting cell cycle components (Figure 2). In common with other eukaryotes, the cell cycle in plants is controlled by serine-threonine kinases whose function is controlled by their association with regulatory cyclin subunits (Figure 2). The expression of proteins related to the class of *cdc2* class of cyclin-dependent kinases in tobacco pith explants is up-regulated by auxin (John *et al.*, 1993). Although the expression of these genes is elevated by auxin, the catalytic activity of the cdc2 kinase is only increased when tobacco explants are also treated with cytokinin (John *et al.*, 1993). The D-type cyclin *cycD3* transcript has been observed to increase in *Arabidopsis* mutants containing elevated cytokinin levels and could be induced by the application of exogenous cytokinin (Nogue *et al.*, 2000). Significantly, cell division can be induced and maintained in the absence of exogenous cytokinin in transgenic plants over-expressing *CycD3* (Riou-Khamlichi *et al.*, 1999), thus demonstrating that cytokinin regulates shoot cell proliferation primarily via *CycD3*. Therefore, auxin and cytokinin appear to co-ordinately regulate shoot cell proliferation by controlling the expression of the interacting cell cycle regulatory components, *cdc2* and *cycD3* (Figure 2). In contrast to their synergistic effects on shoot cells, auxin and cytokinin have antagonistic effects on lateral root formation. Immunological assays have revealed that auxin and cytokinin have antagonistic effects on root *cdc2* expression; whereas auxin increased the abundance of immunologically detectable cdc2-like proteins in extracts from pea roots, cytokinin reduced the levels (John *et al.*, 1993).

It is well established that exposing callus cultures to a high auxin-to-cytokinin ratio results in root formation, whilst a low ratio of these hormones promotes shoot development (Skoog and Miller, 1957). The ratio hypothesis has been extended to the intact plant to explain other hormone-regulated developmental processes such as apical dominance (Chatfield *et al.*, 2000; Kotov and Kotova, 2000; Emery *et al.*, 1998). Classical experiments have demonstrated that auxin is required for apical dominance, whilst cytokinin promotes lateral bud formation (Wickson and Thimann, 1958; Figure 1C). The altered apical dominance phenotype of *Arabidopsis* auxin and cytokinin mutants supports this model. For example, *axr3* mutant plants exhibit an increased response to auxin and develop an enhanced apical dominance phenotype (Cline *et al.*, 2001; Leyser *et al.*, 1996). Conversely, *axr1* plants are deficient in their auxin response and feature increased lateral bud outgrowth (Lincoln *et al.*, 1990; Stirnberg *et al.*, 1999). Similarly, the cytokinin over-producer *amp1* exhibits increased lateral branching (Chaudhury *et al.*, 1993). Bangerth (1994) reported that decapitation of the apical bud in *Phaseolus vulgaris* reduced polar auxin transport and increased cytokinin transport from the roots by 4000%. This observation led Bangerth (1994) to propose the control of root cytokinin levels by auxin as a mechanism for the regulation of lateral bud outgrowth. However, Chatfield *et al.* (2000) have shown, using an excised lateral bud assay system, that apically applied auxin arrests lateral bud outgrowth in the absence of roots and was therefore independent of root-derived cytokinin. Nevertheless, it remains possible that auxin application arrests bud outgrowth by modifying local levels of cytokinin.

Auxin and cytokinin frequently control plant development by regulating each other's abundance (Eklof *et al.*, 2000 and references therein; Figure 2). For example, Eklof *et al.* (1997) observed that transformed *Nicotiana tabacum* plants expressing the *ipt* gene had lower levels of IAA due to reduced rates of IAA synthesis and turnover. Conversely, overproduction of IAA when using the *Agrobacterium iaaM/iaaH* genes resulted in a reduced pool size of cytokinins (Eklof *et al.*, 1997, 2000). One mechanism by which cytokinin might increase auxin levels was demonstrated by Yip and Yang (1986) who observed that cytokinin inhibited the formation of the inactive conjugate IAA-aspartate in mungbean hypocotyl tissues. Cytokinin levels can be reduced in two ways, through oxidative breakdown and by glucosylation. The enzyme responsible for cytokinin oxidative break-

Table 1. *Arabidopsis* auxin mutants with documented response defects to other hormones and other external stimuli.

Mutant	Signalling defect							References
	auxin	ethylene	cytokinin	ABA	GA	light	other	
Atpin2/eir1	✖	✖	–	–	–	–	✖	Müller *et al.* (1998), Luschnig *et al.* (1998)
aux1	✖	✖	✖	–	–	-	✖	Bennett *et al.* (1996), Marchant *et al.* (1999)
axr1	✖	✖	✖	–	–	✖	✖	Lincoln *et al.* (1990), Stirnberg *et al.* (1999)
axr2	✖	✖	✖	✖	–	✖	✖	Nagpal *et al.* (2000), Timpte *et al.* (1994)
axr3	✖	✖	✖	✖	–	–	✖	Leyser *et al.* (1996)
axr4	✖	–	–	–	–	–	✖	Hobbie and Estelle (1995)
dfl1	✖	–	–	–	–	✖	–	Nakazawa *et al.* (2001)
iaa28	✖	✖	✖	–	–	–	–	Rogg *et al.* (2001)
nph4, msg1	✖	✖	–	–	–	✖	–	Harper *et al.* (2000), Watahiki and Yamamoto (1997)
sax1	✖	✖	–	✖	✖	–	✖	Ephritikhine *et al.* (1999a)
shy2	✖	✖	✖	✖	–	✖	✖	Tian and Reed (1999)
tir1	✖	–	–	–	–	–	✖	Ruegger *et al.* (1998)
tir3/big	✖	–	–	–	✖	✖	✖	Gil *et al.* (2001), Ruegger *et al.* (1997), Sponsel *et al.* (1997)

down is cytokinin oxidase (Hare and van Staden, 1994) and its activity is increased by the action of auxin (Figure 2). This has been shown by studies of cytokinin oxidase purified from maize and tobacco pith explants (Palni *et al.*, 1988) and by transgenic tobacco tissues expressing the *ipt* gene (Zhang *et al.*, 1995). The experiments performed on tobacco pith explants demonstrated that the breakdown of radiolabelled zeatin riboside was increased by 1-NAA (1-naphthaleneacetic acid). This observation was confirmed using tobacco tissues expressing the *ipt* gene, where 1-NAA was shown to increase the activity of purified cytokinin oxidase *in vitro*, but did not increase the levels of the enzyme in callus. These experiments indicate that auxin influences the activity of cytokinin oxidase directly (Figure 2). Additional observations suggest that auxin and auxin conjugates may play a role in regulating the metabolism of cytokinin conjugates. For example, β-glucosidases that catalyse the hydrolysis of the cytokinin-*O*-glucoside bond, releasing active cytokinin, are inhibited by esters of IAA-glucose (Brzobohaty *et al.*, 1993; Brzobohaty *et al.*, 1994).

Whilst we have described how auxin and cytokinin co-ordinate plant development by regulating the expression of cell cycle components (Hemerly *et al.*, 1993; Soni *et al.*, 1995) or each other's abundance (Ecklof *et al.*, 2000), little is known about how auxin and cytokinin interact at the signal transduction level. Evidence that this is indeed occurring comes from the cytokinin-response phenotype of auxin mutants such as *axr1* (Hobbie *et al.*, 1994) and *axr2* (Wilson *et al.*, 1990) (Table 1). However, a clearer understanding of how the wild-type AXR1 and AXR2/IAA17 proteins interact with the cytokinin-response pathway awaits further characterisation of several recently discovered components of the cytokinin-signalling pathway (Inoue *et al.*, 2001; reviewed by Schmulling, 2001; Figure 2).

Auxin and ethylene

Auxin and ethylene co-ordinately regulate several developmental programs in plants (Figure 1). For example, in *Arabidopsis* auxin and ethylene have been described to regulate apical hook formation (Lehman *et al.*, 1996; Raz and Ecker, 1999), root hair differentiation (Masucci and Schiefelbein, 1994), root hair elongation (Pitts *et al.*, 1998), root growth (Rahman *et al.*, 2001) and hypocotyl phototropism (Harper *et al.*, 2000) (Figure 3), whilst in cotton, ethylene-promoted leaf abscission is correlated with reduced polar auxin transport (Beyer and Morgan, 1971).

Auxin and ethylene have been described to interact at the level of ethylene biosynthesis (Abel *et al.*, 1995; Woeste *et al.*, 1999). Ethylene biosynthesis requires the conversion of the precursor *S*-adenosylmethionine (SAM) by ACC synthase and ACC oxidase (reviewed by Yang and Hoffman, 1984; Figure 3). Auxin has been described to induce ethylene biosynthesis by up-regulating the expression of ACC synthase in *Arabidopsis* (Abel *et al.*, 1995), tomato (Abel and Theol-

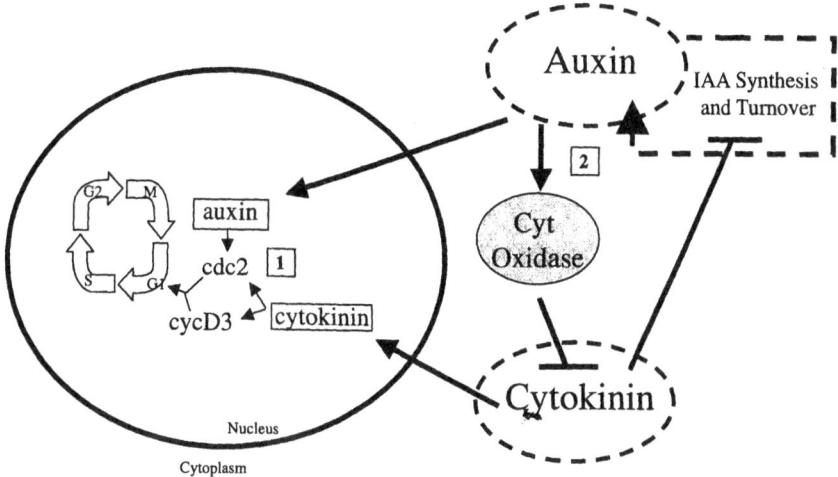

Figure 2. Auxin-cytokinin cross-talk in plant cells. Cytokinin and auxin regulate plant cell proliferation by (1) controlling the expression of the cell cycle components cdc2 and cycD3 (John *et al.*, 1993; Nogue *et al.*, 2000), whilst the size of cellular pools of auxin and cytokinin are controlled via (2) hormone regulated enzymes such as cytokinin oxidase (Cyt oxidase; Zhang *et al.*, 1995).

ogis, 1996) and lupin (Beckman *et al.*, 2000), whereas both ACC synthase and ACC oxidase expression are enhanced in mung bean (Yu *et al.*, 1998). In *Arabidopsis*, one of the six ACC synthase genes (Liang *et al.*, 1992), termed *ACS4*, has been demonstrated to be up-regulated by auxin (Abel *et al.*, 1995). The *ACS4* promoter contains a number of auxin-response elements (Woeste *et al.*, 1999). The ability of auxin to stimulate ACS expression has also been shown by studying auxin-insensitive mutants of *Arabidopsis* and tomato (Abel and Theologis, 1996). Given that auxin stimulates the production of ethylene, it is often unclear whether developmental effects attributed to auxin are due to auxin, ethylene or synergistic interactions between both classes of phytohormones. Romano *et al.* (1993) elegantly addressed this question by engineering transgenic plants that over-produced auxin with the *iaaM* gene, whilst simultaneously expressing the ethylene synthesis inhibiting ACC deaminase transgene. The authors clearly demonstrated that apical dominance and leaf epinasty were primarily controlled by auxin, whilst ethylene was partially responsible for the inhibition of stem elongation.

The induction of apical hook formation in *Arabidopsis* represents one of the best described examples of auxin-ethylene cross-talk in plants (Lehman *et al.*, 1996; Raz and Ecker, 1999). After germination, hypocotyl tissues proximal to the shoot apical meristem transiently form an apical hook (Raz and Ecker, 1999; Figure 1A). The ethylene-dependent induction of differential growth that promotes apical hook formation is well established experimentally. Wild-type

Arabidopsis seedlings germinated in the presence of ethylene exhibit exaggerated hook curvature (Guzman and Ecker, 1990). Similarly, the *Arabidopsis* ethylene over-producing mutant *eto* (Guzman and Ecker, 1990) and the constitutive ethylene-response mutation *ctr1* (Kieber *et al.*, 1993) feature exaggerated apical hooks in the absence of exogenous ethylene. In contrast, ethylene-insensitive mutants have a hookless phenotype (Guzman and Ecker, 1990). Auxin has also been demonstrated to regulate hook development. Wild-type *Arabidopsis* seedlings germinated in the presence of auxin or the auxin transport inhibitor NPA (1-naphthylphthalamic acid) exhibit no apical hook (Lehman *et al.*, 1996). Likewise, the *Arabidopsis* auxin-over-producing mutant alleles of *SUPER ROOT1, sur1/alf1/hls3* (Boerjan *et al.*, 1995; Celenza *et al.*, 1995; Lehman *et al.*, 1996) or auxin transport mutant *aux1* (Roman *et al.*, 1995) disrupt hook formation.

Auxin appears to act downstream of ethylene to control hook development since the auxin transport inhibitor NPA is able to disrupt apical hook formation in *eto and ctr1* plants (Lehman *et al.*, 1996; Figure 1A). The auxin-up-regulated genes *AtAux2-11* and *SAUR-AC1* are non-symmetrically expressed in wild-type hook tissues (Lehman *et al.*, 1996), suggesting that differential growth results from either an asymmetric pattern of auxin sensitivity or auxin redistribution in hook tissues. This pattern of *AtAux2-11* and *SAUR-AC1* hook expression, together with ethylene-induced hook formation, is disrupted by mutations in the *hookless1* (*HLS1*) gene (Lehman *et al.*, 1996). The

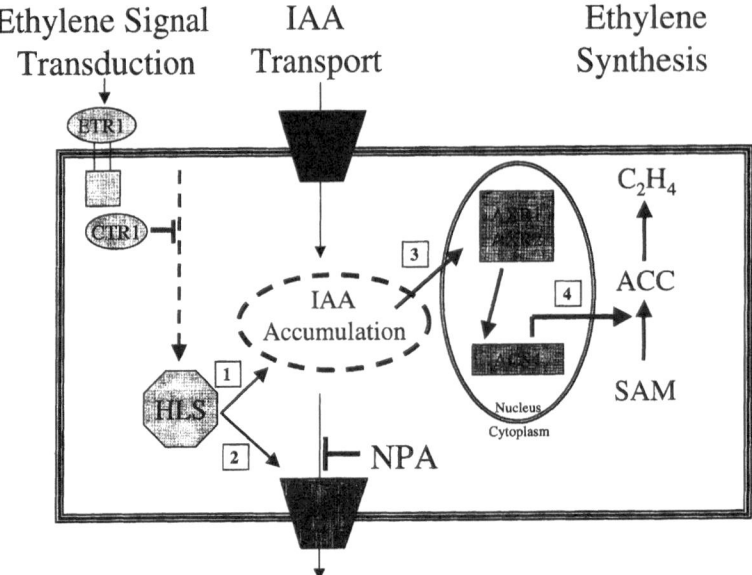

Figure 3. Auxin-ethylene cross-talk in plant cells. Ethylene up-regulates *HOOKLESS1* gene expression to facilitate apical hook formation by regulating either (1) IAA levels (Lehman *et al.*, 1996) or (2) auxin transport activity (Kieber, 1997). Auxin up-regulates ethylene biosynthesis via (3) AXR1- and AXR2-dependent expression of the (4) *ACC synthase 4* (*ACS4*) gene (Abel *et al.*, 1995).

HLS1 protein has homology to *N*-acetyltransferases, prompting Lehman *et al.* (1996) to propose that HLS1 may acetylate an IAA-related metabolite. Alternatively, Kieber (1997) suggests that HLS1 may acetylate an auxin transport component, a mechanism used by eukaryotes to block protein degradation (Figure 3). Ethylene appears to regulate hook formation via *HLS1* since *HSL1* expression is up-regulated by ethylene and HLS1 over-expressing plants exhibit an exaggerated hook phenotype. Nevertheless, the uniform pattern of *HLS1* expression across the apical hook (Lehman *et al.*, 1996) suggests that hook formation does not arise from differential *HSL1* gene expression. Raz and Ecker (1999) have reported that the ACC oxidase gene, *AtACO2*, is differentially expressed in outer versus inner hook tissues. Hence, the characteristic pattern of differential hook growth may result from asymmetric ethylene biosynthesis.

Auxin and ABA

Abscisic acid (ABA) and auxin have been observed to interact antagonistically to regulate stomatal aperture (Eckert and Kaldenhoff, 2000; Figure 1C). Auxin causes a reduction in turgor within guard cells which concomitantly serves to open the stomatal pore. Conversely, ABA serves to increase turgor within guard cells thus closing the stomatal pore and reducing wa-

ter loss via transpiration. The antagonistic nature of this interaction requires the precise co-ordination of ion channel activity within guard cells. These channels allow the flow of ions that decrease (auxin) or increase (ABA) the cytosolic pH and therefore effect the turgor of the guard cells (Grabov and Blatt, 1998; all authors, J. Exp. Bot. special issue March 1998). The effect of ABA on guard cell signalling has been extensively studied at the electrophysiological and genetic level. For example, the ABA-response mutant *abi1* is defective in the control of guard cell K^+ channels, causing a wilty phenotype (Armstrong *et al.*, 1995). The plethora of ABA-related information contrasts the scarcity of knowledge regarding auxin. However, auxin has been shown to affect inward and outward rectifying K^+ channels in a dose-dependent manner (Blatt and Thiel, 1994). Moreover, the guard cell auxin-signalling cascade has been linked to other hormonal and environmental signals. For example, increased red-light intensities stimulate stomatal opening which concomitantly increase transpiration rate and CO_2 uptake (Eckert and Kaldenhoff, 2000). Mutant *axr1-3* seedlings exhibited a faster (but not greater) stomatal response in the presence of both red and blue light. This behaviour contrasts the speed of the response observed in *aba3-2* plants, further clarifying the antagonistic relationship of auxin and

ABA in the control of stomatal aperture (Eckert and Kaldenhoff, 2000).

Aside from interactions of auxin and ABA at the level of guard cell aperture, genetic evidence from *Arabidopsis* indicates these two hormones may interact to influence root growth and seed germination. Dominant mutations in the auxin-response gene *AXR2/IAA7* confer an ABA-insensitive phenotype to roots (Wilson *et al.*, 1990; Timpte *et al.*, 1994; Nagpal *et al.*, 2000; Table 1) and both *axr1* and *axr2* have weak ABA-insensitive phenotype as measured by seed germination (P. McCourt, personal communication). Furthermore, *abi3* mutants that were originally identified as highly insensitive to ABA are insensitive to NPA, an auxin transport inhibitor at the level of lateral root growth. Casimiro *et al.* (2001) have recently demonstrated that auxin transport is required for lateral root initiation since NPA is able to block the initial pericycle divisions. Recent studies have observed that ABA inhibits lateral root development (H.Zhang, personal communication). Interestingly, ABA inhibition coincides with a suppression of auxin response and the down-regulation of the putative auxin influx and efflux carriers, AUX1 and EIR1/PIN2/AGR1, in the lateral root primordia. Such observations suggest that ABA acts by modulating the response and/or transport of auxin in developing lateral root primordia.

Auxin and GA

Auxin and GA have been described to co-ordinately regulate several developmental programs in plants (Figure 1) including pea stem elongation (Ross and O'Neill, 2001) and parthenocarpy (Sastry and Muir, 1963; Ikeda *et al.*, 1999; Vivian-Smith and Koltunov, 1999). GA is best known for promoting elongation growth. Many *Arabidopsis* mutants exhibiting altered hypocotyl length are altered in either the biosynthesis or signal transduction of GA. For example, the GA biosynthesis mutant *ga1* has a shortened hypocotyl, whilst the constitutive GA-response mutant *spy* has a longer hypocotyl (reviewed by Richards *et al.*, 2001). Auxins are also known to promote *Arabidopsis* hypocotyl growth (Smalle *et al.*, 1997). Transgenic *Arabidopsis* plants in which endogenous IAA levels have been artificially lowered by expressing the *iaaL* gene have reduced hypocotyl length (Romano *et al.*, 1991). Conversely, transgenic plants expressing the *iaaM* gene have elevated auxin levels and longer hypocotyls (Romano *et al.*, 1991). Similarly, the *Ara-*

bidopsis auxin over-producing mutant *sur1* has longer hypocotyls than the wild type (Boerjan *et al.*, 1995), whilst the auxin-response mutants *axr1* and *axr3* have shorter hypocotyls (Lincoln *et al.*, 1990; Leyser *et al.*, 1993, 1996; Rouse *et al.*, 1998). Despite dual control of a common growth process, current evidence suggests that auxin and GA independently regulate *Arabidopsis* hypocotyl elongation (Collett *et al.*, 2000). For example, the hypocotyl of the auxin-response mutant *axr1* shows a normal GA response (Table 1). Likewise, the *spy* mutant hypocotyl exhibits a wild-type auxin response (Collett *et al.*, 2000).

Several laboratories have reported that mutations in the *Arabidopsis TIR3* gene disrupt polar auxin transport and GA responses (Table 1), causing characteristic auxin and GA phenotypic defects such as reduced lateral root number and impaired elongation of the stem, pedicels and siliques tissues, respectively (Ruegger *et al.*, 1997; Sponsel *et al.*, 1997; Gil *et al.*, 2001; R. Hooley, personal communication). The *10.3* mutant allele of TIR3 was originally identified as a GA-responsive semi-dwarf (Sponsel *et al.*, 1997). Sponsel *et al.* (1997) reported that the *10.3 tir3* allele exhibited an abnormal stem elongation response to C20-GA, suggesting that TIR3 influences C20 oxidation of GAs. Recent observations in pea suggest that shoot apically derived IAA is required to maintain normal levels of bio-active GA in elongating stem tissues (Ross *et al.*, 2000; Ross and O'Neill, 2001; Figure 1C). Ross *et al.* (2000) observed that decapitated pea plants had almost no capacity to synthesise GA1 from GA20. The enzyme that catalyses the conversion of GA20 to GA1 is encoded by the *LE* gene (*PsGA3Ox1*) (Lester *et al.*, 1997). IAA has been demonstrated to regulate LE expression (Ross *et al.*, 2000). Decapitated pea plants have reduced levels of *LE* mRNA. However, apical application of IAA was sufficient to restore GA1 synthesis, but GA1 biosynthesis was reduced when auxin transport inhibitors were applied to just beneath the apical bud (Ross, 1998). Collectively, these results suggest that polar auxin transport is required for normal GA1 biosynthesis in pea stem tissue (Figure 1C). The observation that the *le* mutant had a 30% reduction in IAA (Law and Davies, 1990) led Ross *et al.* (2000) to further conclude that IAA and GA positively regulate each other's biosynthesis.

Auxin and brassinosteroids

Auxin and brassinosteroids (BR) have been reported to co-ordinately regulate several plant developmental processes. For example, studies in monocotyledons have shown that auxin and BR positively interact to control lamina joint bending (Yamamuro et al., 2000). Similarly, Kim et al. (2000) have observed that gravitropic curvature is increased in maize (Zea mays) by either brassinolide (BL) alone and in the presence of IAA and BL. The BL-mediated gravitropic response can be abolished in the presence of the auxin transport inhibitor TIBA (2,3,5-triiodobenzoic acid) showing that BL is acting via an auxin-mediated process. Auxin and BR have recently been shown to synergistically and antagonistically regulate expression of the ACS family of ethylene biosynthesis genes in mung bean (Vigna radiata) hypocotyl tissues (Yi et al., 1999). IAA induces the up-regulation of three ACS genes (pVR-ACS1, 6 and 7) in etiolated hypocotyls whilst 24-epibrassinolide (BR) induces pVR-ACS6 and 7 but abolishes the up-regulation of pVR-ACS1 induced by IAA and kinetin. However, neither auxin nor BR could induce pVR-ACS7 in light-grown seedlings showing that the relationship between auxin, ethylene and BR is further complicated by light (Yi et al., 1999). A molecular link has recently been described between the auxin-response pathway and BR biosynthesis upon characterisation of the sax1 mutant (Ephritikhine et al., 1999b). The sax1 mutant plant is partially restored by treatment with exogenous brassinosteroid; sax1 root growth is hypersensitive to both ABA and auxin, whereas hypocotyl elongation is resistant to both ethylene and GA (Ephritikhine et al., 1999a). The pleiotropic hormone phenotype of the sax1 mutant (Table 1) suggests that BR antagonistically and synergistically regulates hormonal growth responses in Arabidopsis roots and shoots, respectively.

Auxin and jasmonic acid

The majority of papers that have researched the interactions between jasmonic acid (JA) and auxin have concluded that the relationship between the two hormones is antagonistic. Ueda et al. (1994) and Irving et al. (1999) have shown that JA inhibits auxin-regulated elongation in etiolated oat coleoptiles. Tang et al. (2001) have recently reported that two vacuolar glycoprotein acid phosphatases (VspA and VspB) from soybean are differentially regulated by JA and auxin during early stages of seedling growth. A promoter dissection study of the VspA and VspB sequences has revealed specific motifs that positively respond to JA whilst some are negatively regulated by auxin. In contrast to these findings, Wang et al. (1999) have shown that upon germination, auxin and JA act synergistically to up-regulate two soybean vacuolar lipoxygenases (LOX4 and LOX5). Our understanding of the integration of auxin and JA signalling pathways has been greatly facilitated after the isolation of the JA-response gene COI1 (Xie et al., 1998). The close homology of COI1 to TIR1 and its ability to interact with other components of the plant SCF complex (Del Pozo and Estelle, 1999), raises the intriguing possibility that the JA response, like auxin, is regulated by ubiquitin-mediated protein degradation (Gray and Estelle, 2000). Further research is needed to determine whether any integration of auxin- and JA-signalling pathways is achieved through degradation of common target regulatory proteins.

Auxin and light

Plants respond to environmental cues to ensure optimal growth and development. Auxin mediates plant responses to environmental signals such as gravity (reviewed by Chen et al., this issue), temperature stress (reviewed by Delong et al., this issue) and light (this section). Darwin originally proposed that light and a transmissible signal (later discovered to be auxin) interact to cause phototropic curvature in plants (Darwin, 1880). Recent physiological and genetic experiments have revealed that auxin is closely involved with many other light-regulated developmental processes (Table 1) including hypocotyl elongation, shade avoidance and photomorphogenesis (Figure 1). This section will review our current understanding of the interaction between auxin and light-signalling pathways that regulate these developmental processes.

Auxin and light-dependent growth responses

Classic experiments using maize coleoptiles have demonstrated that a unilateral blue-light stimulus causes auxin redistribution towards the shaded side of the illuminated organ, resulting in a differential growth response termed phototropic curvature (reviewed by Briggs and Huala, 1999; Figure 1A). The recent adoption of a genetic approach in Arabidopsis has significantly advanced our understanding of how phototropic and auxin-signalling pathways interact. Phototropic defects have been described for mutations disrupting several auxin transport and signal transduction components. For example, reverse genetic studies on the auxin efflux carrier gene *At-PIN3* have uncovered a phototropic defect (J. Friml and K. Palme, personal communication). Localisation studies have revealed that *At*PIN3 is expressed in the endodermal layer of hypocotyl tissue, where it is preferentially targeted to the plasma membrane facing the cortical cell layer (J. Friml and K. Palme, personal communication). This pattern of expression is consistent with *At*PIN3 facilitating the asymmetric redistribution of auxin after a phototropic stimulus. Lesions in later steps of the *Arabidopsis* hypocotyl phototropic signal transduction pathway have also been described. The *nph4* mutant was originally identified by its reduced phototropic response (Liscum and Briggs, 1996). The hypocotyl growth of *NPH4* mutant alleles *nph4/msg1/tir5* is resistant to the auxins IAA, 2,4-D (2,4-dichlorophenoxyacetic acid) and 1-NAA (Ruegger *et al.*, 1997; Watahiki and Yamamoto, 1997; Harper *et al.*, 2000), suggesting that the NPH4 protein plays a central role in auxin-mediated differential growth (Table 1). The *NPH4* gene has been cloned and found to encode the auxin response factor, ARF7 (Harper *et al.*, 2000; see review by Liscum and Reed, this issue; Figure 4). Consistent with a transcriptional activator function for NPH4/ARF7, the expression of the auxin-inducible genes SAUR-AC1, IAA12, GH3, IAA4 and IAA6 was severely impaired in *nph4* seedlings (Harper *et al.*, 2000). In summary, mutational studies in *Arabidopsis* have revealed how several auxin-signalling components participate within the phototropic response pathway to control differential growth (Figure 4).

In addition to mediating the light-regulated differential growth response associated with phototropism, auxin appears to influence hypocotyl elongation by light. For example, when compared to wild-type seedlings, auxin-over-producing *sur1* mutant and 35S::*iaaM* transgenic seedlings have longer hypocotyls when grown in the light, yet exhibit normal hypocotyl length in the dark (Boerjan *et al.*, 1995; Romano *et al.*, 1995). Conversely, auxin-underproducing *iaaL* plants exhibit a shorter hypocotyl, yet have normal hypocotyl length in the dark (Romano *et al.*, 1995). Likewise, the auxin transport inhibitor NPA is a potent inhibitor of light-grown hypocotyl elongation whereas in the dark it has no significant effect on hypocotyl growth (Tamimi and Firn, 1985;

Jensen *et al.*, 1998). Steindler *et al.* (1999) reported that the auxin transport inhibitor NPA significantly reduces the hypocotyl elongation of wild-type seedlings under a low R/FR ratio. Plants grown under low R/FR ratio conditions (mimicking shade) exhibit the shade avoidance response, a phytochrome-dependent process leading to a stimulation of elongation growth (Smith and Whitelam, 1997). Many lines of evidence suggest that ATHB-2, a homeodomain-leucine zipper protein, controls the shade avoidance response in *Arabidopsis* (Steindler *et al.*, 1999; Morelli and Ruperti, 2000). The ATHB-2 gene is rapidly and strongly up-regulated under low R/FR ratio conditions that induce the shade avoidance response (Carbelli *et al.*, 1996; Steindler *et al.*, 1997). Functionally, transgenic plants over-expressing the ATHB-2 gene exhibit typical shade avoidance phenotypes including longer hypocotyls, smaller cotyledons and reduced numbers of lateral roots (Steindler *et al.*, 1999). Using NPA, the authors demonstrated that ATHB-2-induced hypocotyl elongation was dependent on auxin transport. In summary, polar auxin transport represents a target for several light-signalling pathways to regulate hypocotyl elongation growth in *Arabidopsis* (Figure 4).

Auxin and photomorphogenesis

Recent genetic and biochemical experiments indicate that auxin is closely associated with photomorphogenesis (Table 1; Walton and Ray, 1981; Iino and Karr, 1982; Jones *et al.*, 1991; Behringer and Davies, 1992; Boerjan *et al.*, 1995; Kraepiel *et al.*, 1995; Gil *et al.*, 2001). Several lines of evidence suggest that members of the *GH3* family of auxin-inducible genes are involved in phytochrome signalling. *FIN219* is a member of the *GH3* family that is involved in *phyA* signalling. *Fin219* was originally isolated as a suppressor mutation of the constitutive photomorphogenic mutant *cop1-6* (Hseih *et al.*, 2000), suggesting that FIN219 normally regulates COP1 activity negatively. FIN219 has been demonstrated to interact with the phytochrome A signalling component FHY1 (Hseih *et al.*, 2000; Figure 4). Like the *phyA* mutant, fin219 mutant seedlings exhibit a long hypocotyl in FR light and have lost the ability to induce *CHS* and *CAB* genes in FR light. Transgenic lines over-expressing FIN219 are hyper-photomorphic in FR light. However, the absence of any auxin-related mutant phenotype suggests that FIN219 is not an auxin-signalling component. Nevertheless, since FIN219 is auxin-inducible it remains possible that auxin modulates the expres-

420

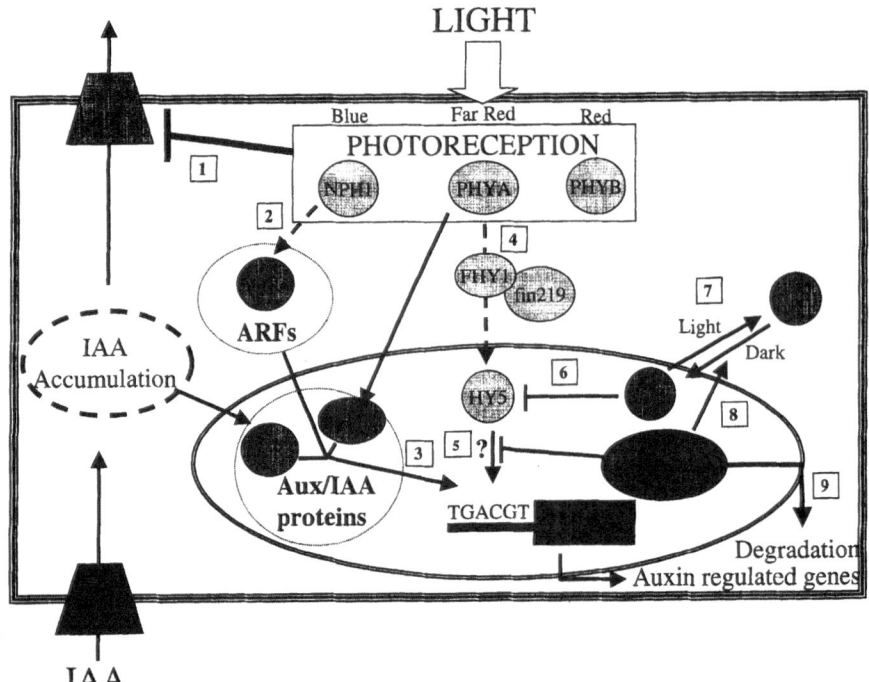

LIGHT

Figure 4. Auxin-light cross-talk in plant cells. 1. Several light signalling pathways regulate hypocotyl elongation via an auxin transport-dependent mechanism. 2. The photoreceptor NPH1 regulates hypocotyl phototropic curvature via the auxin-response factor, ARF7/NPH4 (Harper *et al.*, 2000). 3. Light is proposed to modulate auxin-regulated gene expression via the PhyA-dependent phosphorylation of IAA proteins such as SHY2/IAA3 (Tian and Reed, 1999; Colon-Carmona *et al.*, 2000). 4. The auxin-inducible GH3 family member, FIN219, interacts with the PHYA signal transduction component FHY1 to regulate photomorphogenesis via the HY5 transcription factor (Hseih *et al.*, 2000). 5. Oyama *et al.* (1997) have proposed that HY5 regulates auxin-responsive gene expression. 6. The photomorphogenic repressor COP1 regulates HY5 abundance. 7. COP1 nuclear localisation occurs in the dark by a (8) COP9 signalosome-dependent mechanism (Osterlund *et al.*, 2000). 9. The COP9 signalosome is required for the efficient degradation of HY5 and the IAA protein, PSIAA6 (Schwechheimer *et al.*, 2001).

sion level of FIN219 to co-ordinate phytochrome signalling through COP1. *DFL1* represents another auxin-responsive *GH3* gene homologue that also regulates several light responses (Nakazawa *et al.*, 2001). For example, the dominant *dfl1-D* mutant exhibits a short hypocotyl phenotype in the light (Nakazawa *et al.*, 2001). In contrast to *fin219*, the dominant *dfl1-D* mutant exhibits an auxin-resistant hypocotyl phenotype in the light (Nakazawa *et al.*, 2001; Table 1). Collectively, these observations suggest that light regulates shoot cell elongation via an IAA-dependent mechanism (Nakazawa *et al.*, 2001).

Genetic screens designed to isolate second site mutations that suppress the light-insensitive phenotype of the phytochrome mutant *hy2* resulted in the identification of the *shy2* locus (Tian and Reed, 1999; Table 1). The *shy2* gain-of-function mutant exhibits a constitutive photomorphogenic phenotype, forming leaves in the dark. The *SHY2* gene has been cloned and found to encode *IAA3*, a member of the auxin-inducible *IAA* gene family (Tian and Reed, 1999). The

IAA proteins act as transcription regulators that modulate auxin-regulated gene expression (for a review, see Hagen and Guilfoyle, this issue; Figure 4). Since *shy2* acts as a suppresser of *hy2*, Tian and Reed (1999) have proposed that SHY2/IAA3 represents a link between light and auxin-signalling pathways. Recombinant IAA proteins from *Arabidopsis* (SHY2/IAA3, AXR3/IAA17, IAA1 and IAA9) and pea (Ps IAA4) have recently been demonstrated to be phosphorylated by oat phytochrome A *in vitro* (Colon-Carmona *et al.*, 2000). If established *in vivo*, the phosphorylation of IAA proteins by phytochrome would explain how light is able to fine-tune an auxin response. The results of Colon-Carmona *et al.* (2000) conclude that the N-terminal region of Ps-IAA4 is a potent substrate for oat PHYA. Interestingly, this region of the IAA protein is Ser/Thr-rich and contains functionally identified signals for nuclear localisation (Abel and Theologis, 1995) and protein degradation (Worley *et al.*, 2000). It is therefore possible that some PHYA-mediated responses are mediated by changes in the phosphorylation status of IAA proteins, thereby fine-tuning the expression of auxin-regulated genes.

The auxin- and light-signalling pathways also appear to interact via another class of transcription regulator. HY5, a bZIP transcription factor that binds to the promoters of light-induced genes, acts as a positive regulator of photomorphogenesis (Oyama et al., 1997; Chattopadhyay et al., 1998). The hy5 mutant exhibits a light-insensitive hypocotyl elongation phenotype, increased lateral root formation and elongation, a lack of lateral root gravitropism, reduced secondary thickening in the root and in the hypocotyl and a deficiency in chlorophyll accumulation. The phenotypic similarity between hy5 and the auxin-insensitive diageotropica (dgt) mutant (Zobel, 1974), led Oyama et al. (1997) to propose that DGT and HY5 regulate common signalling pathways in tomato and Arabidopsis. The primary sequence of HY5 is very similar to the soybean transcription factor STF1A. STF1A preferentially binds to DNA containing the TGACGT core sequence, the same motif that confers auxin inducibility on the soybean GH3 promoter (Liu et al., 1994; Figure 4). The basic region of STF1A has been shown to interact with the target DNA. Interestingly, the basic region of HY5 is completely identical to the basic region of STF1A raising the possibility that HY5 protein may bind the same sequence of DNA as STF1A. Collectively, these lines of evidence suggest that HY5 may regulate the expression of auxin-induced genes (Figure 4).

The photomorphogenic repressor protein, COP1, has recently been demonstrated to regulate HY5 abundance by targeting the transcription factor for degradation via the proteasome (Osterlund et al., 2000; Figure 4). The COP1 protein contains several motifs that appear to be important for this function. These domains include an N-terminal ring finger motif, which has been shown to be essential for the function of many ubiquitin ligases (Deshaies, 1999) and a C terminal WD40 repeat unit, which serves as a substrate-recruiting module (Torii et al., 1998). The C-terminal domain of COP1 has been demonstrated to interact with HY5 (Osterlund et al., 2000) and can repress photomorphogenesis (Oyama et al., 1997). The subcellular localisation of COP1 appears critical for its function (Osterlund et al., 2000; Figure 4). In the light, COP1 is localised to the cytoplasm. However, in the dark, COP1 is nuclear-localised, thereby preventing HY5 from activating gene expression. The nuclear accumulation of COP1 is dependent upon several genes including DET1, COP8, 9, 10 and 11 and FUS 5, 6, 11

and 12 (von Arnim et al., 1997). Most of these genes encode subunits of COP9 signalosome, a nuclear protein complex (Wei and Deng, 1999). All COP9 signalosome subunits have distant relatives in the 19S cap of the 26S proteasome (Wei and Deng, 1999) suggesting that the function of the COP9 signalosome is related to proteolysis (Figure 4). Schwechheimer et al. (2001) have recently shown that COP9 signalosome interacts with the E3 ubiquitin ligase SCF (TIR1) in vivo and plants with reduced COP9 signalosome have reduced auxin responses similar to tir1 mutants (Table 1). In addition, Schwechheimer et al. (2001) reported that the COP9 signalosome is required for efficient degradation of PSIAA6, a candidate substrate of SCF (TIR1). Hence, auxin- and light-signalling pathways appear to employ related regulatory mechanisms to control the turnover of common regulatory proteins such as PSIAA6 and HY5.

Future developments

The past decade has seen impressive strides in dissecting the molecular basis of auxin cross-talk in plants. Advances in our understanding of the integration of plant signalling processes at the transduction level have relied, rely and will continue to rely heavily on the application of genetic approaches in the model plant Arabidopsis thaliana (McCourt, 1999; see Introduction). Such studies have helped, in the first instance, to identify important components of auxin and other signalling pathways. An integrative signalling function has often been revealed through the pleiotropic hormone response phenotype of the null mutation (Wilson et al., 1990), or by subsequent second-site mutation screens (Tian and Reed, 1999). In the near future, reporter-based screens (Ishitani et al., 1997; Kieber et al., 2001) and activation tagging approaches (Kakimoto, 1996; McCourt, 1999) represent additional strategies likely to reap dividends. At the protein level, novel interactions between newly discovered components from nominally discrete signalling pathways will be detected through the application of two-hybrid/proteomic-based approaches or the use of high-throughput protein chip-based technologies (Williams and Cole, 2001). Nevertheless, microarray-based expression analysis represents the genomic technology most likely to have an immediate impact in this area of research. The ability to transcript profile the entire Arabidopsis genome opens up unprecedented opportunities to study cross-talk at the level of gene expression (Kuhn, 2001; Celis *et al.*, 2000; Gil *et al.*, 2001). However, great care must be taken in experimental design to ensure that meaningful results are obtained. For example, the researcher must ensure that comparisons are made between materials at equivalent developmental stages when expression profiling a hormone mutant versus wild type. Equally importantly, validation of initial expression profiling results must be obtained with either independent alleles or related hormone mutants.

Transcript profiling primary response mutants from distinct transduction pathways is likely to uncover common target genes, whose 'potential' dual signalling function can be tested by a reverse-genetic approach. Large-scale reverse genetic approaches will be needed to address the signalling function(s) of members within families of related transduction components, given the high level of gene duplication revealed by the *Arabidopsis* genome sequence (*Arabidopsis* Genome Initiative, 2000). For example, genetic approaches have revealed distinct, but biochemically related, signalling functions for TIR1 and COI1 in the auxin- and JA-response pathways, respectively (reviewed by Del Pozo and Estelle, 1999; Figure 4). Reverse-genetic studies will determine whether other F-box sequences have related functions in either auxin, JA, both or other response pathways. Whilst knock-out mutants are initially useful to define the null phenotype of a particular sequence, this approach cannot be used to define discrete domains that are likely to exist within a transduction component that performs an integrative signalling function. In the absence of homologous recombination in plants, the availability of an allelic series of missense mutations has considerable merit. For example, an allelic series of mutations in the COP1 sequence helped define several functionally distinct domains/motifs/residues (McNellis *et al.*, 1994). The recent development of the TILLING (Targeting Induced Local Lesions in Genomes)-based reverse-genetic approach (McCallum *et al.*, 2000a, b; Colbert *et al.*, 2001) holds particular promise for researchers attempting to generate an allelic series in the absence of a simple genetic screen. TILLING allows researchers to rapidly pinpoint an allelic series of point mutations within the sequence of interest, thereby aiding the identification of discrete signalling domains that will be crucial for future studies on the integration of transduction pathways.

In summary, given the rich molecular resources available, *Arabidopsis* will continue to represent the model experimental system to study hormone cross-

talk. Nevertheless, we must not overlook the rich diversity of signalling mechanisms that have evolved in other plant species and endeavour to adopt a comparative research approach.

Acknowledgements

We would like to thank Jiri Friml, Peter McCourt, Klaus Palme and Hanma Zhang for providing information prior to publication. We would also like to acknowledge funding from the Plant and Microbial Science Committee of the BBSRC and EC framework IV and V.

References

Abel, S. and Theologis, A. 1995. A polymorphic bipartite motif signals nuclear targeting of early auxin-inducible proteins related to Ps-IAA4 from pea (*Pisum sativum*). Plant J. 8: 87–96.

Abel, S. and Theologis, A. 1996. Early gene and auxin action. Plant Physiol. 111: 9–17.

Abel, S., Nguyen, M.D., Chow, W. and Theologis, A. 1995. *ACS4*, a primary indoleacetic acid-responsive gene encoding 1-aminocyclopropane-1-carboxylate synthase in *Arabidopsis thaliana*. Structural characterisation, expression in *Escherichia coli*, and expression characteristics in response to auxin. J. Biol. Chem. 270: 19093–19099.

Arabidopsis Genome Initiative. 2000. Analysis of the genome sequence of the flowering plant *Arabidopsis thaliana*. Nature 408: 796–815.

Armstrong, F., Leung, J., Grabov, A., Brearley, J., Giraudat, J. and Blatt M. 1995. Sensitivity to abscisic-acid of guard-cell K^+ channels is suppressed by *abi1-1*, a mutant *Arabidopsis* gene encoding a putative protein phosphatase. Proc. Natl. Acad. Sci. USA 92: 9520–9524.

Bangerth, F. 1994. Response of cytokinin concentration in the xylem exudate of bean (*Phaseolus vulgaris* L.) plants to decapitation and auxin treatment, and relationship to apical dominance. Planta 194: 439–442.

Beckman, E.P., Saibo, N.J.M., Di Cataldo, A., Regalado, A.P., Ricardo, C.P. and Rodrigues-Pousada, C. 2000. Differential expression of four gene encoding 1-aminocyclopropane-1-carboxylate synthase in *Lupinus albus* during germination and in response to indole-3-acetic acid and wounding. Planta 211: 663–672.

Behringer, F.J. and Davies, P.J. 1992. Indole-3-acetic-acid levels after phytochrome-mediated changes in the stem elongation rate of dark-grown and light-grown *Pisum* seedlings. Planta 188: 85–92.

Bennett, M.J., Marchant, A., Green, H.G., May, S.T., Ward, S.P., Millner, P.A., Walker, A.R., Schultz, B. and Feldmann, K.A. 1996. *Arabidopsis AUX1* gene: A permease-like regulator of root gravitropism. Science 273: 948–950.

Beyer, E.M. and Morgan, P.W. 1971. Abscission: the role of ethylene modification of auxin transport. Plant Physiol. 48: 208–212.

Blatt, M.R. and Thiel G. 1994. K^+ channels of Stomatal guard cells: Bimodal control of the K^+ inward-rectifier evoked by auxin. Plant J. 5: 55–68.

Boerjan, W., Cervera, M.T., Delarue, M., Beeckman, T., Dewitte, W., Bellini, C., Caboche, M., Vanonckelen, H., Van

Montagu, M. and Inzé, D. 1995. Superroot, a recessive mutation in *Arabidopsis*, confers auxin overproduction. Plant Cell 7: 1405–1419.

Briggs, W.R. and Huala, E. 1999. Blue-light photoreceptors in higher plants. Annu. Rev. Cell Dev. Biol. 15: 33–62.

Brzobohaty, B., Moore, I., Kristoffersen, P., Bako, L., Campos, N., Schell, J. and Palme K. 1993. Release of active cytokinin by a β-glucosidase localised to the maize root meristem. Science 262: 1051–1054.

Brzobohaty, B., Moore, I. and Palme K. 1994. Cytokinin metabolism: implications for regulation of plant growth and development. Plant Mol. Biol. 26: 1483–1497.

Carabelli, M., Morelli, G., Whitelam, G. and Ruberti, I. 1996. Twilight-zone and canopy shade induction of the Athb-2 homeobox gene in green plants. Proc. Natl. Acad. Sci. USA 93: 3530–3535.

Casimiro, I., Marchant, A., Bhalerao, R., Beeckman, T., Dhooge, S., Inzé, D., Sandberg, G., Casero, P. and Bennett, M.J. 2001. Auxin transport promotes *Arabidopsis* lateral root initiation. Plant Cell 13: 843–852.

Celis, J.E., Kruhoffer, M., Gromova, I., Frederiksen, C., Ostergaard, M., Thykjaer, T., Gromov, P., Yu, J.S., Palsdottir, H., Magnusson, N. and Orntoft, T.F. 2000. Gene expression profiling: monitoring transcription and translation products using DNA microarrays and proteomics. FEBS Lett. 480: 2–16.

Celenza, J., Grisafi, P. and Fink, G. 1995. A pathway for lateral root formation in *Arabidopsis thaliana*. Genes Dev. 9: 2131–2142.

Chatfield, S.P., Stirnberg, P., Forde, B.G. and Leyser, O. 2000. The hormonal regulation of axillary bud growth in *Arabidopsis*. Plant J. 24: 159–169.

Chattopadhyay, S., Ang, L.H., Puente, P., Deng, X.W. and Wei, N. 1998. *Arabidopsis* bZIP protein HY5 directly interacts with light-responsive promoters in mediating light control of gene expression. Plant Cell 10: 673–683.

Chaudhury, A.M., Letham, S., Craig, S. and Dennis, E.S. 1993. AMP1: a mutant with high cytokinin levels and altered embryonic pattern, faster vegetative growth, constitutive photomorphogenesis and precocious flowering. Plant J. 4: 907–916.

Chen, R., Rosen, E., Guan, C., Boonsirichai, K. and Masson, P.H. 2002. Complex physiological and molecular processes underlying root gravitropism. Plant Mol. Biol. 49: 305–317.

Cline, M.G., Chatfield, S.P. and Leyser, O. 2001. NAA restores apical dominance in the *axr3-1* mutant of *Arabidopsis thaliana*. Ann. Bot. 87: 61–65.

Colbert, T., Till, B.J., Tompa, R., Reynolds, S., Steine, M.N., Yeung, A.T., McCallum, C.M., Comai, L. and Henikoff, S. 2001. High throughput screening for induced point mutations. Plant Physiol. 126: 480–484.

Collett, C.E., Harberd, N.P. and Leyser, O. 2000. Hormonal interactions in the control of *Arabidopsis* hypocotyl elongation. Plant Physiol. 124: 553–561.

Colon-Carmona, A, Chen, D.L., Yeh, K.C. and Abel, S. 2000. Aux/IAA proteins are phosphorylated by phytochrome *in vitro*. Plant Physiol. 124: 1728–1738.

Darwin, C. 1880. The Power of Movement in Plants. John Murray, London.

DeLong, A., Mockaitis, K. and Christensen, S. 2002. Protein phosphorylation in the delivery of and response to auxin signals. Plant Mol. Biol. 49: 285–303.

Del Pozo, J.C, and Estelle, M. 1999. Function of the ubiquitin-proteosome pathway in auxin response. Trends Plant Sci. 4: 107–112.

Deshaies, R.J. 1999. SCF and cullin/ring H2-based ubiquitin ligases. Annu. Rev. Cell Dev. Biol. 15: 435–467.

Eckert, M. and Kaldenhoff, R. 2000. Light-induced stomatal movement of selected *Arabidopsis thaliana* mutants. J. Exp. Bot. 51: 1435–1442.

Eklof, S., Astot, C., Blackwell, J., Moritz, T., Olsson, O. and Sandberg, G. 1997. Auxin-cytokinin interactions in wild-type and transgenic tobacco. Plant Cell Physiol. 38: 225–235.

Eklof, S., Astot, C., Stibon, F., Moritz, T., Olsson, O. and Sandberg, G. 2000. Transgenic tobacco plants co-expressing *Agrobacterium iaa* and *ipt* genes wild-type hormone levels but display both auxin- and cytokinin-overproducing phenotypes. Plant J. 23: 279–284.

Emery, R.J.N., Longnecker, N.E.and Atkins, C.A. 1998. Branch development in *Lupinus angustifolius* L. II. Relationship with endogenous ABA, IAA and cytokinins in axillary and main stem buds. J. Exp. Bot. 49: 555–562.

Ephritikhine, G., Fellner, M., Vannini, C., Lapous, D. and Barbier-Brygoe H. 1999a. The *sax1* dwarf mutant of *Arabidopsis thaliana* shows altered sensitivity of growth responses to abscisic acid, auxin, gibberellins and ethylene and is partially rescued by exogenous brassinosteroid. Plant J. 18: 303–314.

Ephritikhine, G., Pagant, S., Fujioka, S., Takatsuto, S., Lapous, D., Caboche, M., Kendrick, R.E. and Barbier-Brygoo, H. 1999b. The *sax1* mutation defines a new locus involved in the brassinosteroid biosynthesis pathway in *Arabidopsis thaliana*. Plant J. 18: 315–320.

Estelle, M.A. and Somerville, C. 1987. Auxin-resistant mutants of *Arabidopsis thaliana* with an altered morphology. Mol. Gen. Genet. 206: 200–206.

Gil, P., Dewey, E., Friml, J., Zhao, Y., Snowden, K.C., Putterill, J., Palme K., Estelle, M. and Chory, J. 2001. BIG: a calossin-like protein required for polar auxin transport in *Arabidopsis*. Genes Dev. 15: 1985–1997.

Grabov, A. and Blatt, M.R. 1998. Co-ordination of signalling elements in guard cell ion channel control. J. Exp. Bot. 49: 351–360.

Gray, W.M and Estelle, M. 2000. Function of the ubiquitin-proteosome pathway in auxin response. Trends. Biochem. Sci. 25: 133–138.

Guzman, P. and Ecker, J.R. 1990. Exploiting the triple response of *Arabidopsis* to identify ethylene related mutants. Plant Cell 2: 513–523.

Hagen, G. and Guilfoyle, T. 2002. Auxin-responsive gene expression: genes, promoters and regulatory factors. Plant Mol. Biol. 49: 373–385.

Hare, P.D. and van Staden, J. 1994. Cytokinin oxidase: biochemical features and physiological significance. Physiol. Plant. 91: 128–136.

Harper, R.M., Stowe-Evans, E.L., Luesse, D.R., Muto, H., Tatematsu, K., Watahiki, M.K., Yamamoto, K. and Liscum, E. 2000. The NPH4 locus encodes the auxin response factor ARF7, a conditional regulator of differential growth in aerial *Arabidopsis* tissue. Plant Cell 12: 757–770.

Hemerly, A.S., Ferreira, P., de Almeida Engler, J., Van Montagu, M., Engler, G. and Inze D. 1993. *cdc2a* expression in *Arabidopsis* is linked with competence for cell division. Plant Cell 5: 1711–1723.

Hobbie, H., Candace, T. Estelle, M. 1994. Molecular genetics of auxin and cytokinin. Plant Mol. Biol. 26: 1499–1519.

Hobbie, L. and Estelle, M. 1995. The *axr4* auxin resistant mutants of *Arabidopsis thaliana* define a gene important for root gravitropism and lateral root initiation. Plant J. 7: 211–220.

Hsieh, H.L., Okamoto, H., Wang, M.L., Ang, L.H., Matsui, M., Goodman, H. and Deng, X.W. 2000. FIN219, an auxin-regulated gene, defines a link between phytochrome A and the downstream

424

regulator COP1 in light control of *Arabidopsis* development. Genes Dev. 14: 1958–1970.

Iino, M. and Carr, D.J. 1982. Sources of free IAA in the mesocotyl of etiolated maize seedlings. Plant Physiol. 69: 1109–1112.

Ikeda, T., Yakushiji, H., Oda, M., Taji, A. and Imada, S. 1999. Growth dependence of ovaries of facultatively parthenocarpic eggplant *in vitro* on indole-3-acetic acid content. Sci. Hort. 79: 143–150.

Inoue, T., Higuchi, M., Hashimoto, Y., Seki, M., Kobayashi, M., Kato, T., Tabata, S., Shinozaki, K. and Kakimoto, T. 2001. Identification of CRE1 as a cytokinin receptor from *Arabidopsis*. Nature 409: 1060–1063.

Irving, H.R., Dyson, G., McConchie, R., Parish, R.W. and Gehring, C.A. 1999. Effects of exogenously applied jasmontes on growth and intracellular pH in maize coleoptile segments. J. Plant Growth Regul. 18: 93–100.

Ishitani, M., Xiong, L.M., Stevenson, B. and Zhu, J.K. 1997. Genetic analysis of osmotic and cold stress signal transduction in *Arabidopsis*: interactions and convergence of abscisic acid-dependent and abscisic acid-independent pathways. Plant Cell 9: 1935–1949.

Jensen, P.J., Hangarter, R.P. and Estelle, M. 1998. Auxin transport is required for hypocotyl elongation in light-grown but not dark-grown *Arabidopsis*. Plant Physiol. 116: 455–462.

John, P.C.L., Zhang, K., Dong, C., Diederich, L. and Wightman, F. 1993. p34^{cdc2} related proteins in control of cell cycle progression, the switch between division and differentiation in tissue development, and stimulation of division by auxin and cytokinin. Aust. J. Plant. Physiol. 20: 503–526.

Jones, A.M., Cochran, D.S., Lamerson, P.M., Evans, M.L. and Cohen, J.D. 1991. Red light-regulated growth.1. Changes in the abundance of indoleacetic-acid and a 22-kilodalton auxin-binding protein in the maize mesocotyl. Plant Physiol. 97: 352–358.

Kakimoto, T. 1996. CKI1, a histidine kinase homolog implicated in cytokinin signal transduction. Science 274: 982–985.

Kende, H. and Zeevaart, J.A.D. 1997. The five 'classical' plant hormones. Plant Cell 9: 1197–1210.

Kieber, J.J. 1997. The ethylene response pathway in *Arabidopsis*. Annu. Rev. Plant Physiol. Plant Mol. Biol. 48: 277–296.

Kieber, J.J., Haberer, G., D'Agostino, I., Hutchison, C., Deruere, J. and Carson, S. 2001. Cytokinin signalling in *Arabidopsis*: role of the A type response regulators. Proceedings the 17th International Conference on Plant Growth Substances, Brno, p. 46.

Kieber, J.J., Rothenberg, M., Roman, G., Feldmann, K.A. and Ecker, J.R. 1993. *Ctr1*, a negative regulator of the ethylene response pathway in *Arabidopsis*, encodes a member of the raf family of protein-kinases. Cell 72: 427–441.

Kim, S.K., Chang, S.C., Lee, E.J., Chung, W.S., Kim, Y.S., Hwang, S. and Lee, J.S. 2000. Involvement of brassinosteroids in the gravitropic response of primary root of maize. Plant Physiol. 123: 997–1004.

Klee, H.J., Horsch, R.B., Hinchee, M.A., Hein, M.B. and Hoffmann, N.L. 1987. The effects of overproduction of 2 *Agrobacterium tumefaciens* T-DNA auxin biosynthetic gene products in transgenic petunia plants. Genes. Dev. 1: 86–96.

Kotov, A.A. and Kotova, L.M. 2000. The contents of auxins and cytokinins in pea internodes as related to the growth of lateral buds. J. Plant Physiol. 156: 438–448.

Kraepiel, Y., Marrec, K., Sotta, B., Caboche, M. and Miginiac, E. 1995. *In-vitro* morphogenic characteristics of phytochrome mutants in *Nicotiana plumbaginifolia* are modified and correlated to high indole-3-acetic-acid levels. Planta 197: 142–146.

Kuhn, E. 2001. From library screening to microarray technology: strategies to determine gene expression profiles and to identify differentially regulated genes in plants. Ann. Bot. 87: 139–155.

Law, D.M. and Davies, P.J. 1990. Comparative indole-3-acetic-acid levels in the slender pea and other pea phenotypes. Plant Physiol. 93: 1539–1543.

Lehman, A., Black, R. and Ecker, J.R. 1996. HOOKLESS1, an ethylene response gene, is required for differential cell elongation in the *Arabidopsis* hypocotyl. Cell 85: 183–194.

Lester, D.R., Ross, J.J., Davies, P.J. and Reid, J.B. 1997. Mendel's stem length gene (*Le*) encodes a gibberellin 3-β-hydroxylase. Plant Cell 9: 1435–1443.

Letham, D.S, Goodwin, P.B. and Higgins, T.J.V. 1978. Phytohormones and Related Compounds: A Comprehensive Treatise. Elsevier/North-Holland Biomedical Press, Amsterdam, Netherlands.

Leyser, H.M.O., Lincoln, C.A., Timpte, C., Lammer, D., Turner, J. and Estelle, M. 1993. *Arabidopsis* auxin-resistance gene *Axr1* encodes a protein related to ubiquitin-activating enzyme-E1. Nature 364: 161–164.

Leyser, H.M.O., Pickett, F.B., Dharmasiri, S. and Estelle, M. 1996. Mutations in the AXR3 gene of *Arabidopsis* result in altered auxin response including ectopic expression from the SAUR-AC1 promoter. Plant J. 10: 403–413.

Liang, X., Abel, S., Keller, J.A., Shen, N.F. and Theologis, A. 1992. The 1-aminocyclopropane-1-carboxylate synthase gene family of *Arabidopsis thaliana*. Proc. Natl. Acad. Sci. USA 89: 11046–11050.

Lincoln, C., Britton, J.H. and Estelle, M. 1990. Growth and development of the *axr1* mutants of *Arabidopsis*. Plant Cell 2: 1071–1080.

Liscum, E. and Briggs, W.R. 1996. Mutations of *Arabidopsis* in potential transduction and response components of the phototropic signaling pathway. Plant Physiol. 112: 291–296.

Liscum, E. and Reed, J.W. 2002. Genetics of Aux/IAA and ARF action in plant growth and development. Plant Mol. Biol. 49: 387–400.

Liu, Z.B., Ulmasov, T., Shi., X.Y., Hagen, G. and Guilfoyle, T.J. 1994. Soybean GH3 promoter contains multiple auxin-inducible elements. Plant Cell 6: 645–657.

Luschnig, C., Gaxiola, R.A., Grisafi, P. and Fink, G.R. 1998. EIR1: a root specific protein involved in auxin transport, is required for gravitropism in *Arabidopsis thaliana*. Genes Dev. 12: 2175–2187.

Maher, E.P. and Martindale, S.J.B. 1980. Mutants of *Arabidopsis thaliana* with altered responses to auxins and gravity. Biochem. Genet. 18: 1041–1053.

Marchant, A., Kargul, J., May, S.T., Muller, P., Delbarre, A., Perrot-Rechenmann, C. and Bennett, M.J. 1999. AUX1 regulates root gravitropism in *Arabidopsis* by facilitating auxin uptake within root apical tissues. EMBO J. 18: 2066–2073.

Masucci, J.D. and Schiefelbein, J. 1994. The *rhd6* mutation of *Arabidopsis thaliana* alters root epidermal-cell polarity and is rescued by auxin. Dev. Biol. 163: 554–554.

McCallum, C.M., Comai, L., Green, E.A. and Henikoff, S. 2000a. Targeted screening for induced mutations. Nature Biotechnol. 18: 455–457.

McCallum, C.M., Comai, L., Green, E.A. and Henikoff, S. 2000b. Targeting Induced Local Lesions IN Genomes (TILLING) for plant functional genomics. Plant Physiol.123: 439–442.

McCourt, P. 1999. Genetic analysis of hormone signalling. Annu. Rev. Plant Physiol. Plant Mol. Biol. 50: 219–243.

McNellis, T.W., von Arnim, A.G., Araki, T., Komeda, Y., Misera, S. and Deng, X.-W. 1994. Genetic and molecular analysis of an allelic series of *cop1* mutants suggests functional roles for the multiple protein domains. Plant Cell 6: 487–500.

Medford, J.I., Horgan, R., Elsawi, Z. and Klee, H.J. 1989. Alterations of endogenous cytokinins in transgenic plants using a chimeric isopentenyl transferase gene. Plant Cell 1: 403–413.

Miller, C.O., Scoog, F., Okumura, F.S., von Saltza, M.H. and Strong, F.M. 1956. J. Am. Chem. Soc. 78:1375–1380.

Morelli, G. and Ruperti, I. 2000. Shade avoidance responses. Driving auxin along lateral routes. Plant Physiol. 122: 621–626.

Müller, A., Guan, C., Gälweiler, L., Tänzler, P., Huijser, P., Marchant, A., Parry, G., Bennett, M.J., Wisman, E. and Palme, K. 1998. *AtPIN2* defines a locus of *Arabidopsis* for root gravitropism control. EMBO J. 17: 6903–6911.

Nagpal, P., Walker, L., Young, J., Sonawala, A., Timpte, C., Estelle, M. and Reed, J. 2000. AXR2 encodes a member of the Aux/IAA p[rotein family. Plant Physiol. 123: 563–573.

Nakazawa, M., Yabe, N., Ichikawa, T., Yamamoto, Y.Y., Yoshizumi, T., Hasunuma, K. and Matsui, M. 2001. DFL1, an auxin-responsive GH3 gene homologue, negatively regulates shoot cell elongation and lateral root formation, and positively regulates the light response of hypocotyl length. Plant J. 25: 213–221.

Nogue, F., Grandjean, O., Craig, S., Dennis, S. and Chaudhury, M. 2000. Higher levels of cell proliferation rate and cyclin CycD3 expression in the *Arabidopsis amp1* mutant. Plant Growth Regul. 32: 275–283.

Osterlund, M.T., Hardtke, C.S., Wei, N. and Deng, X.W. 2000. Targeted destabilization of HY5 during light-regulated development of *Arabidopsis*. Nature 405: 462–466.

Oyama,T., Shimura, Y. and Okada, K. 1997. The *Arabidopsis* HY5 gene encodes a bZIP protein that regulates stimulus-induced development of root and hypocotyl. Gene. Dev. 11: 2983–2995.

Palni, L.M.S., Burch, L. and Horgan, R. 1988. The effect of auxin concentration on cytokinin stability and metabolism. Planta 194: 439–442.

Pickett, F.B., Wilson, A.K. and Estelle, M. 1990. The *aux1* mutation of *Arabidopsis* confers both auxin and ethylene resistance. Plant Physiol. 94: 1462–1466.

Pitts, J.P., Cernac, A. and Estelle, M. 1998. Auxin and ethylene promote root hair elongation in *Arabidopsis*. Plant J. 16: 553–560.

Rahman, A., Amakawa, T., Goto, N. and Tsurumi, S. 2001. Auxin is a positive regulator for ethylene-mediated response in the growth of *arabidopsis* roots. Plant Cell Physiol. 42: 301–307.

Raz, V. and Ecker, J.R. 1999. Regulation of differential growth in the apical hook of *Arabidopsis*. Development 126: 3661–3668.

Richards, D.E., King, K.E., Ait-ali, T. and Harberd, N.P. 2001. How gibberellin regulates plant growth and development: A molecular genetic analysis of gibberellin signaling. Annu. Rev. Plant Physiol. Plant Mol. Biol. 52: 67–88.

Riou-Khamlichi, C., Huntley, R., Jacqmard, A. and Murray, J.A.H. 1999. Cytokinin activation of *Arabidopsis* cell division through a D-type cyclin. Science 283: 1541–1544.

Rogg, L.E., Lasswell, J. and Bartel, B. 2001. A gain of function mutation in IAA28 suppresses lateral root development. Plant Cell 13: 465–480.

Roman, G., Lubarsky, B., Kieber, J.J., Rothenberg, M. and Ecker, J.R. 1995. Genetic analysis of ethylene signal transduction in *Arabidopsis thaliana*: five novel mutant loci integrated into a stress response pathway. Genetics 139: 1393–1409.

Romano, C.P., Hein, M.B. and Klee, H.J. 1991. Inactivation of auxin in tobacco transformed with the indoleacetic acid lysine synthetase gene of *Pseudomonas savastanoi*. Genes Dev. 5: 438–446.

Romano, C., Cooper, M. and Klee, H. 1993. Uncoupling auxin and ethylene effects in transgenic tobacco and *Arabidopsis* plants. Plant Cell 5: 181–189.

Romano, C.P., Robson, P.R.H., Smith, H., Estelle, M. and Klee, H. 1995. Transgene-mediated auxin overproduction in *Arabidopsis*: hypocotyl elongation phenotype and interactions with the *hy6-1* hypocotyl elongation and *axr1* auxin-resistant mutants. Plant Mol. Biol. 27: 1071–1083.

Ross, J.J. 1998. Effects of auxin transport inhibitors on gibberellins in pea. J. Plant Growth Regul. 17: 141–146.

Ross, J.J. and O'Neill, D. 2001. New interactions between classical plant hormones. Trends Plant Sci. 6: 2–4.

Ross, J.J., O'Neill, D.P., Smith, J.J., Kerckhoffs, L.H.J. and Elliott, R.C. 2000. Evidence that auxin promotes gibberellin A(1) biosynthesis in pea. Plant J. 21: 547–552.

Rouse, D., Mackay, P., Stirnberg, P., Estelle, M. and Leyser, O. 1998. Changes in auxin response from mutations in an AUX/IAA gene. Science 279: 1371–1373.

Ruegger, M., Dewey, E., Hobbie, L., Brown, D., Bernasconi, P., Turner, J., Muday, G. and Estelle, M. 1997. Reduced naphthylphthalamic acid binding in the *tir3* mutant of *Arabidopsis* is associated with a reduction in polar auxin transport and diverse morphological defects. Plant Cell 9: 745–757.

Ruegger, M., Dewey, E., Gray, W., Hobbie, L., Turner, J. and Estelle, M. 1988. The TIR1 protein of *Arabidopsis* functions in auxin response and is related human SKP2 and yeast Grr1p. Genes Dev. 12: 198–207.

Sastry, T. and Muir, R. 1963. Gibberellins: affect on diffusible auxin in fruit development. Science 140: 494–495.

Schmulling, T. 2001. CREam of cytokinin signalling: receptor identified. Trends Plant Sci. 6: 281–284.

Schwechheimer, C., Serino, G., Callis, J., Crosby, W.L., Lyapina, S., Deshaies, R.J., Gray, W.M., Estelle, M. and Deng, X.W. 2001. Interactions of the COP9 signalosome with the E3 ubiquitin ligase SCFTIR1 in mediating auxin response. Science 292: 1379–1382.

Skoog, F. and Miller, C.O. 1957. Chemical regulation of growth and organ formation in plant tissues cultured *in vitro*. Symp. Soc. Exp. Biol. 11: 118–131.

Skoog, F. and Tsui, C. 1948. Formation of adventitious shoots and roots in tobacco. Am. J. Bot. 35: 782–787.

Smalle, J., Haegman, M., Kurepa, J., Van Montagu, M. and Vanderstraeten, D. 1997. Ethylene can stimulate *Arabidopsis* hypocotyl elongation in the light. Proc. Natl. Acad. Sci. USA 94: 2756–2761.

Smith, H. and Whitelam, G.C. 1997. The shade avoidance syndrome: multiple responses mediated by multiple phytochromes. Plant Cell Envir. 20: 840–844.

Soni, R., Carmichael, J.P., Shah, Z.H. and Murray, J.A.H. 1995. A family of cyclin D homologues from plants differentially controlled by growth regulators and containing the conserved retinoblastoma protein interaction motif. Plant Cell 7: 85–103.

Sponsel, V.M., Schmidt, F.M., Porter, S.G., Nakayama, M., Kohlstruk, S. and Estelle, M. 1997. Characterization of new gibberellin-responsive semidwarf mutants of *Arabidopsis*. Plant Physiol. 115: 1009–1020.

Steindler, C., Carabelli, M., Borello, U., Morelli, G. and Ruberti, I. 1997. Phytochrome A, phytochrome B and other phytochrome(s) regulate ATHB-2 gene expression in etiolated and green *Arabidopsis* plants. Plant Cell Envir. 20: 759–763.

Steindler, C., Matteucci, A., Sessa, G., Weimar, T., Ohgishi, M., Aoyama, T., Morelli, G. and Ruberti, I. 1999. Shade avoidance responses are mediated by the ATHB-2 HD-Zip protein, a negative regulator of gene expression. Development 126: 4235–4245.

Stirnberg, P., Chatfield, S.P. and Leyser, H.M.O. 1999. AXR1 acts after lateral bud formation to inhibit lateral bud growth in *Arabidopsis*. Plant Physiol. 121: 839–847.

Tamimi, S. and Firn, R.D. 1985. The basipetal auxin transport-system and the control of cell elongation in hypocotyls. J. Exp. Bot. 36: 955–962.

Tang, Z.J., Sadka, A., Morishige, D.T., and Mullet, J.E. 2001. Homeodomain leucine zipper proteins bind to the phosphate response domamin of the soybean VspB tripartite promoter. Plant Physiol. 125: 797–809.

Tian, Q. and Reed, J.W. 1999. Control pf auxin regulated root development by the *Arabidopsis* thaliana SHY2/IAA3 gene. Development 126: 711–721.

Timpte, C., Wilson, A. and Estelle, M. 1994. The *axr2-1* mutation of *Arabidopsis thaliana* is a gain of function mutation that disrupts an early step in auxin response. Genetics 138: 1239–1249.

Torii, K.U., McNellis, T.W. and Deng, X.W. 1998. Functional dissection of *Arabidopsis* COP1 reveals specific roles of its three structural modules in light control of seedling development. EMBO J. 17: 5577–5587.

Ueda, J., Miyamoto, K. and Aoki, M. 1994. Jasmonic acid inhibits the IAA-induced elongation of oat coleoptile segments – a possible mechanism involving the metabolism of cell-wall polysaccharides. Plant Cell Physiol. 35: 1065–1070.

Vivian-Smith, A. and Koltunow, A.M. 1999. Genetic analysis of growth-regulator-induced parthenocarpy in *Arabidopsis*. Plant Physiol. 121: 437–451

von Arnim, A.G., Osterlund, M.T., Kwok, S.F. and Deng, X.W. 1997. Genetic and developmental control of nuclear accumulation of COP1, a repressor of photomorphogenesis in *Arabidopsis*. Plant Physiol. 114: 779–788.

Walton, J.D and Ray, P.M. 1981. Evidence for receptor function of auxin binding sites in maize. Red light inhibition of mesophyill elongation and auxin binding. Plant Physiol. 68: 1334–1338.

Wang, C.X., Jarlfords, U. and Hildebrand, D.F. 1999. Regulation and subcellular localisation of auxin-induced lipoxygenases. Plant Sci. 148: 147–153.

Watahiki, M.K. and Yamamoto, K.T. 1997. The *massugu1* mutation of *Arabidopsis* identified with failure of auxin-induced growth curvature of hypocotyl confers auxin insensitivity to hypocotyl and leaf. Plant Physiol. 115: 419–426.

Wei, N. and Deng, X.W. 1999. Making sense of the COP9 signalosome: a regulatory protein complex conserved from *Arabidopsis* to human. Trends Genet. 15: 98–103.

Went, F.W. and Thimann, K.V. 1937. Phytohormones. MacMillan, New York.

Wickson, M. and Thimann, K.V. 1958. The antagonism of auxin and kinetin in apical dominance. Physiol. Plant. 11: 62–74.

Williams, D.M. and Cole, P.A. 2001. Kinase chips hit the proteomics era. Trends Biochem. Sci. 26: 271–273.

Wilson, A.K., Pickett, F.B., Turner, J.C. and Estelle, M. 1990. A dominant mutation in *Arabidopsis* confers resistance to auxin, ethylene and abscisic acid. Mol. Gen. Genet. 222: 377–383.

Woeste, K.E., Vogel, J.P. and Kieber, J.J. 1999. Factors regulating ethylene biosynthesis in etiolated *Arabidopsis thaliana* seedlings. Physiol. Plant. 105: 478–484.

Worley, C.K., Zenser, N., Ramos, J., Rouse, D., Leyser, O., Theologis, A. and Callis, J. 2000. Degradation of Aux/IAA proteins is essential for normal auxin signalling. Plant J. 21: 553–562.

Xie, D.X. Feys, B.F., James, S., Nieto-Rostro. M. and Turner, J.G. 1998. COI1: an *Arabidopsis* gene required for jasmonate-regulated defence and fertility. Science 280: 1091–1094.

Yamamuro, C., Ihara, Y., Wu, X., Noguchi, T., Fujioka, S., Takatsuto, S., Ashikari, M., Kitano, H. and Matsuoka, M. 2000. Loss of function of a rice brassinosteroid insensitive1 homolog prevents internode elongation and bending of the lamina joint. Plant Cell 12: 1591–1605.

Yang, S.F. and Hoffman, N.E. 1984. Ethylene biosynthesis and its regulation in higher plants. Annu. Rev. Plant Physiol. Plant Mol. Biol. 35: 155–189.

Yip, W. and Yang, S.F.B. 1986. Effect of thidiazuron, a cytokinin-active urea derivative, in cytokinin-dependent ethylene production systems. Plant Physiol. 80: 515–519.

Yi, H.C., Joo, S., Nam, K.H., Lee, J.S., Kang, B.G. and Kim, W.T. 1999. Auxin and brassinosteroid differentially regulate the expression of three members of the 1-aminocyclopropane-1-carboxylate synthase gene family in mung bean (*Vigna radiata* L.). Plant Mol. Biol. 41: 443–454.

Yu, S.J., Kim, S., Lee, J.S. and Lee, D.H. 1998. Differential accumulation of transcripts for ACC synthase and ACC oxidase homologs in etiolated mung bean hypocotyls in response to various stimuli. Mol. Cells 8: 350–358.

Zhang, R. Zhang, X., Wang, J., Letham, D.S., McKinney, S.A. and Higgins, T.J.V.1995. The effect of auxin on cytokinin levels and metabolism in transgenic tobacco tissue expressing an *ipt* gene. Planta 196: 84–94.

Zobel, R.W. 1974. Control of morphogenesis in the ethylene requiring tomato mutant, *diageotropica*. Can. J. Bot. 52: 735–741.

Index, Vol. 49 Nos. 3–4 (2002)